Universitext

Universitext

Universitext is a series of textbooks that presents material from a wide variety of mathematical disciplines at master's level and beyond. The books, often well class-tested by their author, may have an informal, personal, even experimental approach to their subject matter. Some of the most successful and established books in the series have evolved through several editions, always following the evolution of teaching curricula, into very polished texts.

Thus as research topics trickle down into graduate-level teaching, first textbooks written for new, cutting-edge courses may make their way into *Universitext*.

For further volumes:
www.springer.com/series/223

Dirk van Dalen

Logic and Structure

Fifth Edition

 Springer

Dirk van Dalen
Department of Philosophy
Utrecht University
Utrecht, The Netherlands

Based on a previous edition of the Work:

Logic and Structure, 4th edition by Dirk van Dalen
Copyright © Springer-Verlag Berlin Heidelberg 2004, 1994, 1983, 1980

ISSN 0172-5939   ISSN 2191-6675 (electronic)
Universitext
ISBN 978-1-4471-4557-8   ISBN 978-1-4471-4558-5 (eBook)
DOI 10.1007/978-1-4471-4558-5
Springer London Heidelberg New York Dordrecht

Library of Congress Control Number: 2012953020

Mathematics Subject Classification: 03-01, 03B05, 03B10, 03B15, 03B20, 03C07, 03C20

Printed on acid-free paper

Springer is part of Springer Science+Business Media (www.springer.com)

Preface

Logic appears in a "sacred" and in a "profane" form; the sacred form is dominant in proof theory, the profane form in model theory. The phenomenon is not unfamiliar, one also observes this dichotomy in other areas, e.g. set theory and recursion theory. Some early catastrophes, such as the discovery of the set theoretical paradoxes (Cantor, Russell), or the definability paradoxes (Richard, Berry), make us treat a subject for some time with the utmost awe and diffidence. Sooner or later, however, people start to treat the matter in a more free and easy way. Being raised in the "sacred" tradition, my first encounter with the profane tradition was something like a culture shock. Hartley Rogers introduced me to a more relaxed world of logic by his example of teaching recursion theory to mathematicians as if it were just an ordinary course in, say, linear algebra or algebraic topology. In the course of time I have come to accept this viewpoint as the didactically sound one: before going into esoteric niceties one should develop a certain feeling for the subject and obtain a reasonable amount of plain working knowledge. For this reason this introductory text sets out in the profane vein and tends towards the sacred only at the end.

The present book has developed from courses given at the Mathematics Department of Utrecht University. The experience drawn from these courses and the reaction of the participants suggested strongly that one should not practice and teach logic in isolation. As soon as possible examples from everyday mathematics should be introduced; indeed, first-order logic finds a rich field of applications in the study of groups, rings, partially ordered sets, etc.

The role of logic in mathematics and computer science is twofold—a tool for applications in both areas, and a technique for laying the foundations. The latter role will be neglected here, we will concentrate on the daily matters of formalized (or formalizable) science. Indeed, I have opted for a practical approach—I will cover the basics of proof techniques and semantics, and then go on to topics that are less abstract. Experience has taught us that the natural deduction technique of Gentzen lends itself best to an introduction; it is close enough to actual informal reasoning to enable students to devise proofs by themselves. Hardly any artificial tricks are involved, and at the end there is the pleasing discovery that the system has striking structural properties; in particular it perfectly suits the constructive interpretation of logic and it allows normal forms. The latter topic has been added to this edition in view of its importance in theoretical computer science. In Chap. 4 we already have enough technical power to obtain some of the traditional and (even today) surprising model theoretic results.

The book is written for beginners without knowledge of more advanced topics; no esoteric set theory or recursion theory is required. The basic ingredients are natural deduction and semantics, and the latter is presented in constructive and classical form.

In Chap. 6 intuitionistic logic is treated on the basis of natural deduction without the rule of reductio ad absurdum, and of Kripke semantics. Intuitionistic logic has gradually freed itself from the image of eccentricity and now it is recognized for its usefulness in e.g. topos theory and type theory; hence its inclusion in an introductory text is fully justified. The chapter on normalization has been added for the same reasons; normalization plays an important role in certain parts of computer science. Traditionally normalization (and cut elimination) belong to proof theory, but gradually applications in other areas have been introduced. In Chap. 7 we consider only weak normalization, and a number of easy applications are given.

Various people have contributed to the shaping of the text at one time or another; Dana Scott, Jane Bridge, Henk Barendregt and Jeff Zucker have been most helpful for the preparation of the first edition. Since then many colleagues and students have spotted mistakes and suggested improvements; this edition benefited from the remarks of Eleanor McDonnell, A. Scedrov and Karst Koymans. To all of these critics and advisers I am grateful. Progress has dictated that the traditional typewriter should be replaced by more modern devices; this book has been redone in LATEX by Addie Dekker and my wife, Doke. Addie led the way with the first three sections of Chap. 2 and Doke finished the rest of the manuscript; I am indebted to both of them, especially to Doke who found time and courage to master the secrets of the LATEX trade. Thanks go to Leen Kievit for putting together the derivations and for adding the finer touches required for a LATEX manuscript. Paul Taylor's macro for proof trees has been used for the natural deduction derivations.

June 1994   Dirk van Dalen

The conversion to TEX has introduced a number of typos that are corrected in the present new printing. Many readers have been kind enough to send me their collection of misprints, and I am grateful to them for their help. In particular I want to thank Jan Smith, Vincenzo Scianna, A. Ursini, Mohammad Ardeshir and Norihiro Kamide. Here in Utrecht my logic classes have been very helpful; in particular Marko Hollenberg, who taught part of a course, has provided me with useful comments. Thanks go to them too.

I have used the occasion to incorporate a few improvements. The definition of "subformula" has been streamlined—together with the notion of positive and negative occurrence. There is also a small addendum on "induction on the rank of a formula".

January 1997   Dirk van Dalen

At the request of users I have added a chapter on the incompleteness of arithmetic. It makes the book more self-contained, and adds useful information on basic recursion theory and arithmetic. The coding of formal arithmetic makes use of the exponential; this is not the most efficient coding, but for the heart of the argument that is not of the utmost importance. In order to avoid extra work the formal system of arithmetic contains the exponential. As the proof technique of the book is that of natural deduction, the coding of the notion of derivability is also based on it. There are of course many other approaches. The reader is encouraged to consult the literature.

The material of this chapter is by and large that of a course given in Utrecht in 1993. Students have been most helpful in commenting on the presentation, and in preparing T$_E$X versions. W. Dean has kindly pointed out some more corrections in the old text.

The final text has benefited from the comments and criticism of a number of colleagues and students. I am grateful for the advice of Lev Beklemishev, John Kuiper, Craig Smoryński and Albert Visser. Thanks are due to Xander Schrijen, whose valuable assistance helped to overcome the T$_E$X problems.

May 2003   Dirk van Dalen

A number of corrections have been provided by Tony Hurkens; furthermore, I am indebted to him and Harold Hodes for pointing out that the definition of "free for" was in need of improvement. Sjoerd Zwart found a nasty typo that had escaped me and all (or most) readers.

April 2008   Dirk van Dalen

To the fifth edition a new section on ultraproducts has been added. The topic has a long history and it presents an elegant and instructive approach to the role of models in logic.

Again I have received comments and suggestions from readers. It is a pleasure to thank Diego Barreiro, Victor Krivtsov, Einam Livnat, Thomas Opfer, Masahiko Rokuyama, Katsuhiko Sano, Patrick Skevik and Iskender Tasdelen.

2012   Dirk van Dalen

Contents

Chapter 1
Introduction

Without adopting one of the various views advocated in the foundations of mathematics, we may agree that mathematicians need and use a language, if only for the communication of their results and their problems. While mathematicians have been claiming the greatest possible exactness for their methods, they have been less sensitive as to their means of communication. It is well known that Leibniz proposed to put the practice of mathematical communication and mathematical reasoning on a firm base; it was, however, not before the nineteenth century that those enterprises were (more) successfully undertaken by G. Frege and G. Peano. No matter how ingeniously and rigorously Frege, Russell, Hilbert, Bernays and others developed mathematical logic, it was only in the second half of this century that logic and its language showed any features of interest to the general mathematician. The sophisticated results of Gödel were of course immediately appreciated, but for a long time they remained technical highlights without practical use. Even Tarski's result on the decidability of elementary algebra and geometry had to bide its time before any applications turned up.

Nowadays the applications of logic to algebra, analysis, topology, etc. are numerous and well recognized. It seems strange that quite a number of simple facts, within the grasp of any student, were overlooked for such a long time. It is not possible to give proper credit to all those who opened up this new territory. Any list would inevitably show the preferences of the author, and neglect some fields and persons.

Let us note that mathematics has a fairly regular, canonical way of formulating its material, partly by its nature, partly under the influence of strong schools, like the one of Bourbaki. Furthermore the crisis at the beginning of this century has forced mathematicians to pay attention to the finer details of their language and to their assumptions concerning the nature and the extent of the mathematical universe. This attention started to pay off when it was discovered that there was in some cases a close connection between classes of mathematical structures and their syntactical description. Here is an example:

It is well known that a subset of a group G which is closed under multiplication and inverse, is a group; however, a subset of an algebraically closed field F which

D. van Dalen, *Logic and Structure*, Universitext, DOI 10.1007/978-1-4471-4558-5_1,
© Springer-Verlag London 2013

is closed under sum, product, minus and inverse, is in general not an algebraically closed field. This phenomenon is an instance of something quite general: an axiomatizable class of structures is axiomatized by a set of universal sentences (of the form $\forall x_1, \ldots, x_n \varphi$, with φ quantifier free) iff it is closed under substructures. If we check the axioms of group theory we see that indeed all axioms are universal, while not all the axioms of the theory of algebraically closed fields are universal. The latter fact could of course be accidental, it could be the case that we were not clever enough to discover a universal axiomatization of the class of algebraically closed fields. The above theorem of Tarski and Los tells us, however, that it is impossible to find such an axiomatization!

The point of interest is that for some properties of a class of structures we have simple syntactic criteria. We can, so to speak, read the behavior of the real mathematical world (in some simple cases) from its syntactic description.

There are numerous examples of the same kind, e.g. *Lyndon's theorem*: an axiomatizable class of structures is closed under homomorphisms iff it can be axiomatized by a set of positive sentences (i.e. sentences which, in prenex normal form with the open part in disjunctive normal form, do not contain negations).

The most basic and at the same time monumental example of such a connection between syntactical notions and the mathematical universe is of course *Gödel's completeness theorem*, which tells us that provability in the familiar formal systems is extensionally identical with *truth* in all structures. That is to say, although provability and truth are totally different notions (the first is combinatorial in nature, the latter set theoretical), they determine the same class of sentences: φ is provable iff φ is true in all structures.

Given the fact that the study of logic involves a great deal of syntactical toil, we will set out by presenting an efficient machinery for dealing with syntax. We use the technique of *inductive definitions* and as a consequence we are rather inclined to see trees wherever possible; in particular we prefer natural deduction in the tree form to the linear versions that are here and there in use.

One of the amazing phenomena in the development of the foundations of mathematics is the discovery that the language of mathematics itself can be studied by mathematical means. This is far from a futile play: Gödel's incompleteness theorems, for instance, lean heavily on a mathematical analysis of the language of arithmetic, and the work of Gödel and Cohen in the field of the independence proofs in set theory requires a thorough knowledge of the mathematics of mathematical language. Set theory remains beyond the scope of this book, but a simple approach to the incompleteness of arithmetic has been included. We will aim at a thorough treatment, in the hope that the reader will realize that all these things which he suspects to be trivial, but cannot see why, are perfectly amenable to proof. It may help the reader to think of himself as a computer with great mechanical capabilities, but with no creative insight, in those cases where he is puzzled because "why should we prove something so utterly evident"! On the other hand the reader should keep in mind that he is not a computer and that, certainly when he gets beyond Chap. 3, certain details should be recognized as trivial.

For the actual practice of mathematics predicate logic is doubtlessly the perfect tool, since it allows us to handle individuals. All the same we start this book with an exposition of propositional logic. There are various reasons for this choice.

In the first place propositional logic offers in miniature the problems that we meet in predicate logic, but there the additional difficulties obscure some of the relevant features; e.g. the completeness theorem for propositional logic already uses the concept of "maximal consistent set", but without the complications of the Henkin axioms.

In the second place there are a number of truly propositional matters that would be difficult to treat in a chapter on predicate logic without creating a impression of discontinuity that borders on chaos. Finally it seems a matter of sound pedagogy to let propositional logic precede predicate logic. The beginner can, in a simple context, get used to the proof theoretical, algebraic and model theoretic skills that would be overbearing in a first encounter with predicate logic.

All that has been said about the role of logic in mathematics can be repeated for computer science; the importance of syntactical aspects is even more pronounced than in mathematics, but it does not stop there. The literature of theoretical computer science abounds with logical systems, completeness proofs and the like. In the context of type theory (typed lambda calculus) intuitionistic logic has gained an important role, whereas the technique of normalization has become a staple diet for computer scientists.

Chapter 2
Propositional Logic

2.1 Propositions and Connectives

Traditionally, logic is said to be the art (or study) of reasoning; so in order to describe logic in this tradition, we have to know what "reasoning" is. According to some traditional views reasoning consists of the building of chains of linguistic entities by means of a certain relation "... follows from ...", a view which is good enough for our present purpose. The linguistic entities occurring in this kind of reasoning are taken to be *sentences*, i.e. entities that express a complete thought, or state of affairs. We call those sentences *declarative*. This means that, from the point of view of natural language, our class of acceptable linguistic objects is rather restricted.

Fortunately this class is wide enough when viewed from the mathematician's point of view. So far logic has been able to get along pretty well under this restriction. True, one cannot deal with questions, or imperative statements, but the role of these entities is negligible in pure mathematics. I must make an exception for performative statements, which play an important role in programming; think of instructions as "goto, if ... then, else ...", etc. For reasons given below, we will, however, leave them out of consideration.

The sentences we have in mind are of the kind "27 is a square number", "every positive integer is the sum of four squares", "there is only one empty set". A common feature of all those declarative sentences is the possibility of assigning them a truth value, *true* or *false*. We do not require the actual determination of the truth value in concrete cases, such as for instance Goldbach's conjecture or Riemann's hypothesis. It suffices that we can "in principle" assign a truth value.

Our so-called *two-valued* logic is based on the assumption that every sentence is either true or false; it is the cornerstone of the practice of truth tables.

Some sentences are minimal in the sense that there is no proper part which is also a sentence, e.g. $5 \in \{0, 1, 2, 5, 7\}$, or $2 + 2 = 5$; others can be taken apart into smaller parts, e.g. "c is rational or c is irrational" (where c is some constant). Conversely, we can build larger sentences from smaller ones by using *connectives*. We know many connectives in natural language; the following list is by no means meant to be exhaustive: *and, or, not, if ... then ..., but, since, as, for, although, neither ... nor*

D. van Dalen, *Logic and Structure*, Universitext, DOI 10.1007/978-1-4471-4558-5_2,
© Springer-Verlag London 2013

... . In ordinary discourse, and also in informal mathematics, one uses these connectives incessantly; however, in formal mathematics we will economize somewhat on the connectives we admit. This is mainly for reason of exactness. Compare, for example, the following two sentences: "π is irrational, but it is not algebraic", "Max is a Marxist, but he is not humorless". In the second statement we may discover a suggestion of some contrast, as if we should be surprised that Max is not humorless. In the first case such a surprise cannot be so easily imagined (unless, e.g. one has just read that almost all irrationals are algebraic); without changing the meaning one can transform this statement into "π is irrational and π is not algebraic". So why use (in a formal text) a formulation that carries vague, emotional undertones? For these and other reasons (e.g. of economy) we stick in logic to a limited number of connectives, in particular those that have shown themselves to be useful in the daily routine of formulating and proving.

Note, however, that even here ambiguities loom. Each of the connectives already has one or more meanings in natural language. We will give some examples:

1. John drove on and hit a pedestrian.
2. John hit a pedestrian and drove on.
3. If I open the window then we'll have fresh air.
4. If I open the window then $1 + 3 = 4$.
5. If $1 + 2 = 4$, then we'll have fresh air.
6. John is working or he is at home.
7. Euclid was a Greek or a mathematician.

From 1 and 2 we conclude that "and" may have an ordering function in time. Not so in mathematics; "π is irrational and 5 is positive" simply means that both parts are the case. Time just does not play a role in formal mathematics. We could not very well say "π was neither algebraic nor transcendent before 1882". What we would want to say is "before 1882 it was unknown whether π was algebraic or transcendent".

In examples 3–5 we consider the implication. Example 3 will be generally accepted, it displays a feature that we have come to accept as inherent to implication: there is a relation between the premise and conclusion. This feature is lacking in examples 4 and 5. Nonetheless we will allow cases such as 4 and 5 in mathematics. There are various reasons to do so. One is the consideration that meaning should be left out of syntactical considerations. Otherwise syntax would become unwieldy and we would run into an esoteric practice of exceptional cases. This general implication, in use in mathematics, is called *material implication*. Some other implications have been studied under the names of *strict implication, relevant implication*, etc.

Finally 6 and 7 demonstrate the use of "or". We tend to accept 6 and to reject 7. One mostly thinks of "or" as something exclusive. In 6 we more or less expect John not to work at home, while 7 is unusual in the sense that we as a rule do not use "or" when we could actually use "and". Also, we normally hesitate to use a disjunction if we already know which of the two parts is the case, e.g. "32 is a prime or 32 is not a prime" will be considered artificial (to say the least) by most of us, since we already know that 32 is not a prime. Yet mathematics freely uses such superfluous disjunctions, for example "$2 \geq 2$" (which stands for "$2 > 2$ or $2 = 2$").

In order to provide mathematics with a precise language we will create an artificial, formal language, which will lend itself to mathematical treatment. First we will define a language for propositional logic, i.e. the logic which deals only with *propositions* (sentences, statements). Later we will extend our treatment to a logic which also takes properties of individuals into account.

The process of *formalization* of propositional logic consists of two stages: (1) present a formal language, (2) specify a procedure for obtaining *valid* or *true* propositions.

We will first describe the language, using the technique of *inductive definitions*. The procedure is quite simple: *First* give the smallest propositions, which are not decomposable into smaller propositions; *next* describe how composite propositions are constructed out of already given propositions.

Definition 2.1.1 The language of propositional logic has an alphabet consisting of

  (i) *proposition symbols*: $p_0, p_1, p_2, \ldots$,
 (ii) *connectives*: $\wedge, \vee, \rightarrow, \neg, \leftrightarrow, \bot$,
(iii) *auxiliary symbols*: (,).

The connectives carry traditional names:

$\wedge$ – *and*	– *conjunction*
$\vee$ – *or*	– *disjunction*
$\rightarrow$ – *if* ..., *then* ...	– *implication*
$\neg$ – *not*	– *negation*
$\leftrightarrow$ – *iff*	– *equivalence, bi-implication*
$\bot$ – *falsity*	– *falsum, absurdum*

The proposition symbols and $\bot$ stand for the indecomposable propositions, which we call *atoms*, or *atomic propositions*.

Definition 2.1.2 The set *PROP* of propositions is the smallest set X with the properties

  (i) $p_i \in X (i \in N), \bot \in X$,
 (ii) $\varphi, \psi \in X \Rightarrow (\varphi \wedge \psi), (\varphi \vee \psi), (\varphi \rightarrow \psi), (\varphi \leftrightarrow \psi) \in X$,
(iii) $\varphi \in X \Rightarrow (\neg \varphi) \in X$.

The clauses describe exactly the possible ways of building propositions. In order to simplify clause (ii) we write $\varphi, \psi \in X \Rightarrow (\varphi \square \psi) \in X$, where $\square$ is one of the connectives $\wedge, \vee, \rightarrow, \leftrightarrow$.

A warning to the reader is in order here. We have used Greek letters φ, ψ in the definition; are they propositions? Clearly we did not intend them to be so, as we want only those strings of symbols obtained by combining symbols of the alphabet in a correct way. Evidently no Greek letters come in at all! The explanation is that φ and ψ are used as variables for propositions. Since we want to study logic, we must use a language in which to discuss it. As a rule this language is plain, everyday English.

We call the language used to discuss logic our *meta-language* and φ and ψ are *meta-variables* for propositions. We could do without meta-variables by handling (ii) and (iii) verbally: if two propositions are given, then a new proposition is obtained by placing the connective $\wedge$ between them and by adding brackets in front and at the end, etc. This verbal version should suffice to convince the reader of the advantage of the mathematical machinery.

Note that we have added a rather unusual connective, $\bot$. It is unusual in the sense that it does not connect anything. *Logical constant* would be a better name. For uniformity we stick to our present usage. $\bot$ is added for convenience; one could very well do without it, but it has certain advantages. One may note that there is something lacking, namely a symbol for the true proposition; we will indeed add another symbol, $\top$, as an abbreviation for the "true" proposition.

Examples

$$(p_7 \to p_0), \; ((\bot \vee p_{32}) \wedge (\neg p_2)) \in PROP,$$

$$p_1 \leftrightarrow p_7, \; \neg\neg\bot, \; ((\to \wedge \notin PROP.$$

It is easy to show that something belongs to *PROP* (just carry out the construction according to Definition 2.1.2); it is somewhat harder to show that something does not belong to *PROP*. We will do one example:

$$\neg\neg \bot \notin PROP.$$

Suppose $\neg\neg \bot \in X$ and X satisfies (i), (ii), (iii) of Definition 2.1.2. We claim that $Y = X - \{\neg\neg \bot\}$ also satisfies (i), (ii) and (iii). Since $\bot, p_i \in X$, also $\bot, p_i \in Y$. If $\varphi, \psi \in Y$, then $\varphi, \psi \in X$. Since X satisfies (ii) $(\varphi \Box \psi) \in X$. From the form of the expressions it is clear that $(\varphi \Box \psi) \neq \neg\neg \bot$ (look at the brackets), so $(\varphi \Box \psi) \in X - \{\neg\neg \bot\} = Y$. Likewise one shows that Y satisfies (iii). Hence X is not the smallest set satisfying (i), (ii) and (iii), so $\neg\neg \bot$ cannot belong to *PROP*.

Properties of propositions are established by an inductive procedure analogous to Definition 2.1.2: first deal with the atoms, and then go from the parts to the composite propositions. This is made precise in the following theorem.

Theorem 2.1.3 (Induction Principle) *Let A be a property, then $A(\varphi)$ holds for all $\varphi \in PROP$ if*

  (i) $A(p_i)$, *for all i, and $A(\bot)$,*
 (ii) $A(\varphi), A(\psi) \Rightarrow A((\varphi \Box \psi))$,
(iii) $A(\varphi) \Rightarrow A((\neg\varphi))$.

Proof Let $X = \{\varphi \in PROP \mid A(\varphi)\}$, then X satisfies (i), (ii) and (iii) of Definition 2.1.2. So $PROP \subseteq X$, i.e. for all $\varphi \in PROP$ $A(\varphi)$ holds. $\qquad\qquad\square$

We call an application of Theorem 2.1.3 a *proof by induction on φ*. The reader will note an obvious similarity between the above theorem and the principle of complete induction in arithmetic.

The above procedure for obtaining all propositions, and for proving properties of propositions is elegant and perspicuous; there is another approach, however, which has its own advantages (in particular for coding): Consider propositions as the result of a linear step-by-step construction. For example $((\neg p_0) \to \bot)$ is constructed by assembling it from its basic parts by using previously constructed parts: $p_0 \ldots \bot \ldots (\neg p_0) \ldots ((\neg p_0) \to \bot)$. This is formalized as follows.

Definition 2.1.4 A sequence $\varphi_0, \ldots, \varphi_n$ is called a *formation sequence* of φ if $\varphi_n = \varphi$ and for all $i \leq n$ φ_i is atomic, or

$$\varphi_i = (\varphi_j \Box \varphi_k) \quad \text{for certain } j, k < i, \quad \text{or}$$

$$\varphi_i = (\neg \varphi_j) \quad \text{for certain } j < i.$$

Observe that in this definition we are considering strings φ of symbols from the given alphabet; this mildly abuses our notational convention.

Examples $\bot, p_2, p_3, (\bot \vee p_2), (\neg(\bot \vee p_2)), (\neg p_3)$ and $p_3, (\neg p_3)$ are both formation sequences of $(\neg p_3)$. Note that formation sequences may contain "garbage".

We now give some trivial examples of proof by induction. In practice we actually only verify the clauses of the proof by induction and leave the conclusion to the reader.

1. *Each proposition has an even number of brackets.*

Proof

   (i) Each atom has 0 brackets and 0 is even.
   (ii) Suppose φ and ψ have $2n$, resp. $2m$ brackets, then $(\varphi \Box \psi)$ has $2(n + m + 1)$ brackets.
   (iii) Suppose φ has $2n$ brackets, then $(\neg \varphi)$ has $2(n + 1)$ brackets.   □

2. *Each proposition has a formation sequence.*

Proof

   (i) If φ is an atom, then the sequence consisting of just φ is a formation sequence of φ.
   (ii) Let $\varphi_0, \ldots, \varphi_n$ and $\psi_0, \ldots, \psi_m$ be formation sequences of φ and ψ, then one easily sees that $\varphi_0, \ldots, \varphi_n, \psi_0, \ldots, \psi_m, (\varphi_n \Box \psi_m)$ is a formation sequence of $(\varphi \Box \psi)$.
   (iii) This is left to the reader.   □

We can improve on 2.

Theorem 2.1.5 *PROP is the set of all expressions having formation sequences.*

Proof Let F be the set of all expressions (i.e. strings of symbols) having formation sequences. We have shown above that $PROP \subseteq F$.

Let φ have a formation sequence $\varphi_0, \ldots, \varphi_n$, we show $\varphi \in PROP$ by induction on n.

$$n = 0 : \varphi = \varphi_0 \text{ and by definition } \varphi \text{ is atomic, so } \varphi \in PROP.$$

Suppose that all expressions with formation sequences of length $m < n$ are in *PROP*. By definition $\varphi_n = (\varphi_i \square \varphi_j)$ for $i, j < n$, or $\varphi_n = (\neg \varphi_i)$ for $i < n$, or φ_n is atomic. In the first case φ_i and φ_j have formation sequences of length $i, j < n$, so by the induction hypothesis $\varphi_i, \varphi_j \in PROP$. As *PROP* satisfies the clauses of Definition 2.1.2, also $(\varphi_i \square \varphi_j) \in PROP$. Treat negation likewise. The atomic case is trivial. Conclusion $F \subseteq PROP$. □

Theorem 2.1.5 is in a sense a justification of the definition of formation sequence. It also enables us to establish properties of propositions by ordinary induction on the length of formation sequences.

In arithmetic one often defines functions by recursion, e.g. exponentiation is defined by $x^0 = 1$ and $x^{y+1} = x^y \cdot x$, or the factorial function by $0! = 1$ and $(x + 1)! = x! \cdot (x + 1)$.

The justification is rather immediate: each value is obtained by using the preceding values (for positive arguments). There is an analogous principle in our syntax.

Example The number $b(\varphi)$ of brackets of φ, can be defined as follows:

$$\begin{cases} b(\varphi) = 0 & \text{for } \varphi \text{ atomic,} \\ b((\varphi \square \psi)) = b(\varphi) + b(\psi) + 2, \\ b((\neg \varphi)) = b(\varphi) + 2. \end{cases}$$

The value of $b(\varphi)$ can be computed by successively computing $b(\psi)$ for its subformulas ψ.

We can give this kind of definition for all sets that are defined by induction. The principle of "definition by recursion" takes the form of "there is a unique function such that ...". The reader should keep in mind that the basic idea is that one can "compute" the function value for a composition in a prescribed way from the function values of the composing parts.

The general principle behind this practice is laid down in the following theorem.

Theorem 2.1.6 (Definition by Recursion) *Let mappings* $H_\square : A^2 \rightarrow A$ *and* $H_\neg : A \rightarrow A$ *be given and let* H_{at} *be a mapping from the set of atoms into* A, *then there exists exactly one mapping* $F : PROP \rightarrow A$ *such that*

$$\begin{cases} F(\varphi) = H_{at}(\varphi) & \text{for } \varphi \text{ atomic,} \\ F((\varphi \square \psi)) = H_\square(F(\varphi), F(\psi)), \\ F((\neg \varphi)) = H_\neg(F(\varphi)). \end{cases}$$

In concrete applications it is usually rather easily seen to be a correct principle. However, in general one has to prove the existence of a unique function satisfying the above equations. The proof is left as an exercise, cf. Exercise 11.

Here are some examples of definition by recursion.

1. The (parsing) *tree* of a proposition φ is defined by

$$T(\varphi) = \quad \bullet \varphi \quad \text{for atomic } \varphi$$

$$T((\varphi \square \psi)) = \quad \underset{T(\varphi) \quad T(\psi)}{\overset{\bullet \ (\varphi \square \psi)}{\bigwedge}}$$

$$T((\neg \varphi)) = \quad \underset{T(\varphi)}{\overset{\bullet \ (\neg \varphi)}{\uparrow}}$$

Examples

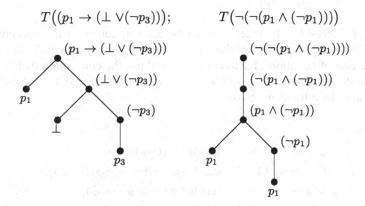

$$T\big((p_1 \to (\bot \vee (\neg p_3)))\big); \qquad T\big(\neg(\neg(p_1 \wedge (\neg p_1))))\big)$$

A simpler way to exhibit the trees consists of listing the atoms at the bottom, and indicating the connectives at the nodes.

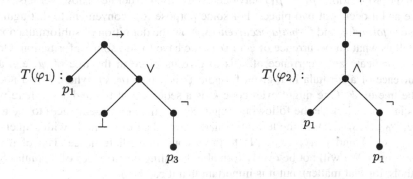

2. The *rank* $r(\varphi)$ of a proposition φ is defined by

$$\begin{cases} r(\varphi) = 0 & \text{for atomic } \varphi, \\ r((\varphi \Box \psi)) = \max(r(\varphi), r(\psi)) + 1, \\ r((\neg \varphi)) = r(\varphi) + 1. \end{cases}$$

We now use the technique of definition by recursion to define the notion of subformula.

Definition 2.1.7 The set of subformulas $Sub(\varphi)$ is given by

$$Sub(\varphi) = \{\varphi\} \quad \text{for atomic } \varphi$$
$$Sub(\varphi_1 \Box \varphi_2) = Sub(\varphi_1) \cup Sub(\varphi_2) \cup \{\varphi_1 \Box \varphi_2\}$$
$$Sub(\neg \varphi) = Sub(\varphi) \cup \{\neg \varphi\}.$$

We say that ψ *is a subformula of* φ if $\psi \in Sub(\varphi)$.

Examples p_2 is a subformula of $((p_7 \vee (\neg p_2)) \to p_1)$; $(p_1 \to \bot)$ is a subformula of $(((p_2 \vee (p_1 \wedge p_0)) \leftrightarrow (p_1 \to \bot))$.

Notational Convention In order to simplify our notation we will economize on brackets. We will always discard the outermost brackets and we will discard brackets in the case of negations. Furthermore we will use the convention that $\wedge$ and $\vee$ bind more strongly than $\to$ and $\leftrightarrow$ (cf. $\cdot$ and $+$ in arithmetic), and that $\neg$ binds more strongly than the other connectives.

Examples

$\neg \varphi \vee \varphi$	stands for $((\neg \varphi) \vee \varphi)$,
$\neg(\neg\neg\neg\varphi \wedge \bot)$	stands for $(\neg((\neg(\neg(\neg \varphi))) \wedge \bot))$,
$\varphi \vee \psi \to \varphi$	stands for $((\varphi \vee \psi) \to \varphi)$,
$\varphi \to \varphi \vee (\psi \to \chi)$	stands for $(\varphi \to (\varphi \vee (\psi \to \chi)))$.

Warning Note that those abbreviations are, properly speaking, not propositions.

In the proposition $(p_1 \to p_1)$ only one atom is used to define it; however it is used twice and it occurs at two places. For some purpose it is convenient to distinguish between *formulas* and *formula occurrences*. Now the definition of subformula does not tell us what an occurrence of φ in ψ is, we have to add some information. One way to indicate an occurrence of φ is to give its place in the tree of ψ, e.g. an occurrence of a formula in a given formula ψ is a pair (φ, k), where k is a node in the tree of ψ. One might even code k as a sequence of 0's and 1's, where we associate to each node the following sequence: $\langle \rangle$ (the empty sequence) to the top node, $\langle s_0, \ldots, s_{n-1}, 0 \rangle$ to the left immediate descendant of the node with sequence $\langle s_0, \ldots, s_{n-1} \rangle$ and $\langle s_0, \ldots, s_{n-1}, 1 \rangle$ to the second immediate descendant of it (if there is one). We will not be overly formal in handling occurrences of formulas (or symbols, for that matter), but it is important that it can be done.

The introduction of the rank function above is not a mere illustration of the "definition by recursion", it also allows us to prove facts about propositions by means of plain *complete induction* (or *mathematical induction*). We have, so to speak, reduced the tree structure to that of the straight line of natural numbers. Note that other "measures" will do just as well, e.g. the number of symbols. For completeness we will spell out the *Rank-Induction Principle*:

Theorem 2.1.8 (Rank-Induction Principle) *If for all φ $[A(\psi)$ for all ψ with rank less than $r(\varphi)] \Rightarrow A(\varphi)$, then $A(\varphi)$ holds for all $\varphi \in PROP$.*

Let us show that induction on φ and induction on the rank of φ are equivalent.[1]

First we introduce a convenient notation for the rank induction: write $\varphi \prec \psi$ $(\varphi \preceq \psi)$ for $r(\varphi) < r(\psi)$ $(r(\varphi) \leq r(\psi))$. So $\forall \psi \preceq \varphi A(\psi)$ stands for "$A(\psi)$ holds for all ψ with rank at most $r(\varphi)$".

The *Rank-Induction Principle* now reads

$$\forall \varphi (\forall \psi \prec \varphi A(\psi) \Rightarrow A(\varphi)) \Rightarrow \forall \varphi A(\varphi)$$

We will now show that the rank-induction principle follows from the induction principle. Let

$$\forall \varphi (\forall \psi \prec \varphi A(\psi) \Rightarrow A(\varphi)) \tag{2.1}$$

be given. In order to show $\forall \varphi A(\varphi)$ we will indulge in a bit of induction loading. Put $B(\varphi) := \forall \psi \preceq \varphi A(\psi)$. Now show $\forall \varphi B(\varphi)$ by induction on φ.

1. For atomic φ $\forall \psi \prec \varphi A(\psi)$ is vacuously true, hence by (2.1) $A(\varphi)$ holds. Therefore $A(\psi)$ holds for all ψ with rank ≤ 0. So $B(\varphi)$.
2. $\varphi = \varphi_1 \square \varphi_2$. Induction hypothesis: $B(\varphi_1)$, $B(\varphi_2)$. Let ρ be any proposition with $r(\rho) = r(\varphi) = n + 1$ (for a suitable n). We have to show that ρ and all propositions with rank less than $n + 1$ have the property A. Since $r(\varphi) = \max(r(\varphi_1), r(\varphi_2)) + 1$, one of φ_1 and φ_2 has rank n—say φ_1. Now pick an arbitrary ψ with $r(\psi) \leq n$, then $\psi \preceq \varphi_1$. Therefore, by $B(\varphi_1)$, $A(\psi)$. This shows that $\forall \psi \prec \rho A(\psi)$, so by (2.1) $A(\rho)$ holds. This shows $B(\varphi)$.
3. $\varphi = \neg \varphi_1$. Similar argument.

An application of the induction principle yields $\forall \varphi B(\varphi)$, and as a consequence $\forall \varphi A(\varphi)$.

Conversely, the rank-induction principle implies the induction principle. We assume the premises of the induction principle. In order to apply the rank-induction principle we have to show (2.1). Now pick an arbitrary φ; there are three cases:

1. φ atomic. Then (2.1) holds trivially.
2. $\varphi = \varphi_1 \square \varphi_2$. Then $\varphi_1, \varphi_2 \prec \varphi$ (see Exercise 6). Our assumption is $\forall \psi \prec \varphi A(\psi)$, so $A(\varphi_1)$ and $A(\varphi_2)$. Therefore $A(\varphi)$.
3. $\varphi = \neg \varphi_1$. Similar argument.

This establishes (2.1). So by rank induction we get $\forall \varphi A(\varphi)$.

---

[1] The reader may skip this proof at first reading. He will do well to apply induction on rank naively.

Exercises

1. Give formation sequences of

$$(\neg p_2 \to (p_3 \lor (p_1 \leftrightarrow p_2))) \land \neg p_3,$$
$$(p_7 \to \neg \perp) \leftrightarrow ((p_4 \land \neg p_2) \to p_1),$$
$$(((p_1 \to p_2) \to p_1) \to p_2) \to p_1.$$

2. Show that $((\to \notin PROP$.
3. Show that the relation "is a subformula of" is transitive.
4. Let φ be a subformula of ψ. Show that φ occurs in each formation sequence of ψ.
5. If φ occurs in a shortest formation sequence of ψ then φ is a subformula of ψ.
6. Let r be the rank function.
    (a) Show that $r(\varphi) \leq$ number of occurrences of connectives of φ.
    (b) Give examples of φ such that $<$ or $=$ holds in (a).
    (c) Find the rank of the propositions in Exercise 1.
    (d) Show that $r(\varphi) < r(\psi)$ if φ is a proper subformula of ψ.
7. (a) Determine the trees of the propositions in Exercise 1.
    (b) Determine the propositions with the following trees.

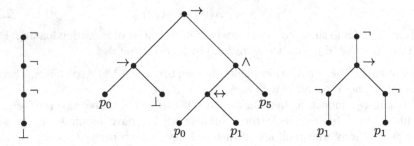

8. Let $\#(T(\varphi))$ be the number of nodes of $T(\varphi)$. By the "number of connectives in φ" we mean the number of occurrences of connectives in φ. (In general $\#(A)$ stands for the number of elements of a (finite) set A.)
    (a) If φ does not contain $\perp$, show: number of connectives of φ + number of atoms of $\varphi \leq \#(T(\varphi))$.
    (b) $\#(\mathrm{sub}(\varphi)) \leq \#(T(\varphi))$.
    (c) A branch of a tree is a maximal linearly ordered set.
        The length of a branch is the number of its nodes minus one. Show that $r(\varphi)$ is the length of a longest branch in $T(\varphi)$.
    (d) Let φ not contain $\perp$. Show: the number of connectives in φ + the number of atoms of $\varphi \leq 2^{r(\varphi)+1} - 1$.
9. Show that a proposition with n connectives has at most $2n + 1$ subformulas.
10. Show that for PROP we have a unique decomposition theorem: for each non-atomic proposition σ either there are two propositions φ and ψ such that $\sigma = \varphi \Box \psi$, or there is a proposition φ such that $\sigma = \neg \varphi$.
11. (a) Give an inductive definition of the function F, defined by recursion on *PROP* from the functions H_{at}, $H_\Box$, $H_\neg$, as a set F^* of pairs.

(b) Formulate and prove for F^* the induction principle.
(c) Prove that F^* is indeed a function on *PROP*.
(d) Prove that it is the unique function on *PROP* satisfying the recursion equations.

2.2 Semantics

The task of interpreting propositional logic is simplified by the fact that the entities considered have a simple structure. The propositions are built up from rough blocks by adding connectives.

The simplest parts (atoms) are of the form "grass is green", "Mary likes Goethe", "$6 - 3 = 2$", which are simply *true* or *false*. We extend this assignment of *truth values* to composite propositions, by reflection on the meaning of the logical connectives.

Let us agree to use 1 and 0 instead of "true" and "false". The problem we are faced with is how to interpret $\varphi \square \psi$, $\neg \varphi$, given the truth values of φ and ψ.

We will illustrate the solution by considering the in-out table for Messrs. Smith and Jones.

Conjunction A visitor who wants to see both Smith and Jones wants the table to be in the position shown here, i.e.

	in	out
Smith	×	
Jones	×	

"Smith is in" ∧ "Jones is in" is true iff
"Smith is in" is true and "Jones is in" is true.

We write $v(\varphi) = 1$ (resp. 0) for "φ is true" (resp. false). Then the above consideration can be stated as $v(\varphi \wedge \psi) = 1$ iff $v(\varphi) = v(\psi) = 1$, or $v(\varphi \wedge \psi) = \min(v(\varphi), v(\psi))$.

One can also write it in the form of a *truth table*:

∧	0	1
0	0	0
1	0	1

One reads the truth table as follows: the first argument is taken from the leftmost column and the second argument is taken from the top row.

Disjunction If a visitor wants to see one of the partners, no matter which one, he wants the table to be in one of the positions

	in	out
Smith	×	
Jones		×

	in	out
Smith		×
Jones	×	

	in	out
Smith	×	
Jones	×	

In the last case he can make a choice, but that is no problem; he wants to see at least one of the gentlemen, no matter which one.

In our notation, the interpretation of $\vee$ is given by

$$v(\varphi \vee \psi) = 1 \quad \text{iff} \quad v(\varphi) = 1 \quad \text{or} \quad v(\psi) = 1.$$

Shorter: $v(\varphi \vee \psi) = \max(v(\varphi), v(\psi))$.

In truth table form:

$\vee$	0	1
0	0	1
1	1	1

Negation The visitor who is solely interested in our Smith will state that "Smith is not in" if the table is in the position:

	in	out
Smith		$\times$

So "Smith is not in" is true if "Smith is in" is false. We write this as $v(\neg\varphi) = 1$ iff $v(\varphi) = 0$, or $v(\neg\varphi) = 1 - v(\varphi)$.

In truth table form:

$\neg$	
0	1
1	0

Implication Our legendary visitor has been informed that "Jones is in if Smith is in". Now he can at least predict the following positions of the table:

	in	out
Smith	$\times$	
Jones	$\times$	

	in	out
Smith		$\times$
Jones		$\times$

If the table is in the position

	in	out
Smith	$\times$	
Jones		$\times$

then he knows that the information was false.

The remaining case,

	in	out
Smith		$\times$
Jones	$\times$	

cannot be dealt with in such a simple way. There evidently is no reason to consider the information false, rather "not very helpful", or "irrelevant". However, we have committed ourselves to the position that each statement is true or false, so we decide to call "If Smith is in, then Jones is in" true also in this particular case. The reader should realize that we have made a deliberate choice here; a choice that will prove a happy one in view of the elegance of the system that results. There is no compelling reason, however, to stick to the notion of implication that we just introduced. Various other notions have been studied in the literature; for mathematical purposes our notion (also called "material implication") is, however, perfectly suitable.

Note that there is just one case in which an implication is false (see the truth table below), and one should keep this observation in mind for future application — it helps to cut down calculations.

In our notation the interpretation of implication is given by $v(\varphi \to \psi) = 0$ iff $v(\varphi) = 1$ and $v(\psi) = 0$.

Its truth table is:

$\to$	0	1
0	1	1
1	0	1

Equivalence If our visitor knows that "Smith is in if and only if Jones is in", then he knows that they are either both in, or both out. Hence $v(\varphi \leftrightarrow \psi) = 1$ iff $v(\varphi) = v(\psi)$.

The truth table of $\leftrightarrow$ is:

$\leftrightarrow$	0	1
0	1	0
1	0	1

Falsum An absurdity, such as "$0 \neq 0$", "some odd numbers are even", "I am not myself", cannot be true. So we put $v(\bot) = 0$.

Strictly speaking we should add one more truth table, i.e. the table for $\top$, the opposite of *falsum*.

Verum This symbol stands for a manifestly true proposition such as $1 = 1$; we put $v(\top) = 1$ for all v.

We collect the foregoing in the following definition.

Definition 2.2.1 A mapping $v : PROP \to \{0, 1\}$ is a *valuation* if

$$v(\varphi \wedge \psi) = \min(v(\varphi), v(\psi)),$$
$$v(\varphi \vee \psi) = \max(v(\varphi), v(\psi)),$$
$$v(\varphi \to \psi) = 0 \quad \Leftrightarrow \quad v(\varphi) = 1 \text{ and } v(\psi) = 0,$$
$$v(\varphi \leftrightarrow \psi) = 1 \quad \Leftrightarrow \quad v(\varphi) = v(\psi),$$
$$v(\neg\varphi) = 1 - v(\varphi)$$
$$v(\bot) = 0.$$

If a valuation is only given for atoms then it is, by virtue of the definition by recursion, possible to extend it to all propositions. Hence we get the following.

Theorem 2.2.2 *If v is a mapping from the atoms into $\{0, 1\}$, satisfying $v(\bot) = 0$, then there exists a unique valuation $[\![\cdot]\!]_v$, such that $[\![\varphi]\!]_v = v(\varphi)$ for atomic φ.*

It has become common practice to denote valuations as defined above by $[\![\varphi]\!]$, so we will adopt this notation. Since $[\![\cdot]\!]$ is completely determined by its values on the atoms, $[\![\varphi]\!]$ is often denoted by $[\![\varphi]\!]_v$. Whenever there is no confusion we will delete the index v.

Theorem 2.2.2 tells us that each of the mappings v and $[\![\cdot]\!]_v$ determines the other one uniquely, therefore we also call v a valuation (or an *atomic valuation*, if necessary). From this theorem it appears that there are many valuations (cf. Exercise 4).

It is also obvious that the *value* $[\![\varphi]\!]_v$ of φ under v only depends on the values of v on its atomic subformulas.

Lemma 2.2.3 *If $v(p_i) = v'(p_i)$ for all p_i occurring in φ, then $[\![\varphi]\!]_v = [\![\varphi]\!]_{v'}$.*

Proof An easy induction on φ.   □

An important subset of *PROP* is that of all propositions φ which are *always true*, i.e. true under all valuations.

Definition 2.2.4

(i) φ is a *tautology* if $[\![\varphi]\!]_v = 1$ for all valuations v.
(ii) $\models \varphi$ stands for "φ is a tautology".
(iii) Let Γ be a set of propositions, then $\Gamma \models \varphi$ iff for all v: $([\![\psi]\!]_v = 1$ for all $\psi \in \Gamma) \Rightarrow [\![\varphi]\!]_v = 1$.

In words: $\Gamma \models \varphi$ holds iff φ is true under all valuations that make all ψ in Γ true. We say that φ is a semantical consequence of Γ. We write $\Gamma \not\models \varphi$ if $\Gamma \models \varphi$ is not the case.

Convention $\varphi_1, \ldots, \varphi_n \models \psi$ stands for $\{\varphi_1, \ldots, \varphi_n\} \models \psi$.

Note that "$[\![\varphi]\!]_v = 1$ for all v" is another way of saying "$[\![\varphi]\!] = 1$ for all valuations".

Examples

(i) $\models \varphi \to \varphi$; $\models \neg\neg\varphi \to \varphi$; $\models \varphi \vee \psi \leftrightarrow \psi \vee \varphi$,
(ii) $\varphi, \psi \models \varphi \wedge \psi$; $\varphi, \varphi \to \psi \models \psi$; $\varphi \to \psi, \neg\psi \models \neg\varphi$.

One often has to substitute propositions for subformulas; it turns out to be sufficient to define substitution for atoms only.

We write $\varphi[\psi/p_i]$ for the proposition obtained by replacing all occurrences of p_i in φ by ψ. As a matter of fact, substitution of ψ for p_i defines a mapping of *PROP* into *PROP*, which can be given by recursion (on φ).

Definition 2.2.5

$$\varphi[\psi/p_i] = \begin{cases} \varphi & \text{if } \varphi \text{ atomic and } \varphi \neq p_i \\ \psi & \text{if } \varphi = p_i \end{cases}$$

$$(\varphi_1 \square \varphi_2)[\psi/p_i] = \varphi_1[\psi/p_i] \square \varphi_2[\psi/p_i]$$
$$(\neg\varphi)[\psi/p_i] = \neg\varphi[\psi/p_i].$$

The following theorem spells out the basic property of the substitution of equivalent propositions.

Theorem 2.2.6 (Substitution Theorem) *If* $\models \varphi_1 \leftrightarrow \varphi_2$, *then* $\models \psi[\varphi_1/p] \leftrightarrow \psi[\varphi_2/p]$, *where p is an atom.*

The substitution theorem is actually a consequence of a slightly stronger one.

Lemma 2.2.7 $[\![\varphi_1 \leftrightarrow \varphi_2]\!]_v \leq [\![\psi[\varphi_1/p] \leftrightarrow \psi[\varphi_2/p]]\!]_v$ *and* $\models (\varphi_1 \leftrightarrow \varphi_2) \rightarrow (\psi[\varphi_1/p] \leftrightarrow \psi[\varphi_2/p])$.

Proof Induction on ψ. We only have to consider $[\![\varphi_1 \leftrightarrow \varphi_2]\!]_v = 1$ (why?).

- ψ atomic. If $\psi = p$, then $\psi[\varphi_i/p] = \varphi_i$ and the result follows immediately. If $\psi \neq p$, then $\psi[\varphi_i/p] = \psi$, and $[\![\psi[\varphi_1/p] \leftrightarrow \psi[\varphi_2/p]]\!]_v = [\![\psi \leftrightarrow \psi]\!]_v = 1$.
- $\psi = \psi_1 \square \psi_2$. Induction hypothesis: $[\![\psi_i[\varphi_1/p]]\!]_v = [\![\psi_i[\varphi_2/p]]\!]_v$. Now the value of $[\![(\psi_1 \square \psi_2)[\varphi_i/p]]\!]_v = [\![\psi_1[\varphi_i/p] \square \psi_2[\varphi_i/p]]\!]_v$ is uniquely determined by its parts $[\![\psi_j[\varphi_i/p]]\!]_v$, hence $[\![(\psi_1 \square \psi_2)[\varphi_1/p]]\!]_v = [\![(\psi_1 \square \psi_2)[\varphi_2/p]]\!]_v$.
- $\psi = \neg\psi_1$. Left to the reader.

The proof of the second part essentially uses the fact that $\models \varphi \rightarrow \psi$ iff $[\![\varphi]\!]_v \leq [\![\psi]\!]_v$ for all v (cf. Exercise 6). $\qquad\square$

The proof of the substitution theorem now immediately follows. $\qquad\square$

The substitution theorem says in plain English that *parts may be replaced by equivalent parts.*

There are various techniques for testing tautologies. One such (rather slow) technique uses truth tables. We give one example:

$$(\varphi \rightarrow \psi) \leftrightarrow (\neg\psi \rightarrow \neg\varphi)$$

φ	ψ	$\neg\varphi$	$\neg\psi$	$\varphi \rightarrow \psi$	$\neg\psi \rightarrow \neg\varphi$	$(\varphi \rightarrow \psi) \leftrightarrow (\neg\psi \rightarrow \neg\varphi)$
0	0	1	1	1	1	1
0	1	1	0	1	1	1
1	0	0	1	0	0	1
1	1	0	0	1	1	1

The last column consists of 1's only. Since, by Lemma 2.2.3 only the values of φ and ψ are relevant, we had to check 2^2 cases. If there are n (atomic) parts we need 2^n lines.

One can compress the above table a bit, by writing it in the following form:

$(\varphi$	$\rightarrow$	$\psi)$	$\leftrightarrow$	$(\neg\psi$	$\rightarrow$	$\neg\varphi)$
0	1	0	1	1	1	1
0	1	1	1	0	1	1
1	0	0	1	1	0	0
1	1	1	1	0	1	0

Let us make one more remark about the role of the two 0-ary connectives, $\bot$ and $\top$. Clearly, $\models \top \leftrightarrow (\bot \rightarrow \bot)$, so we can define $\top$ from $\bot$. On the other hand, we cannot define $\bot$ from $\top$ and $\rightarrow$; we note that from $\top$ we can never get anything but a proposition equivalent to $\top$ by using $\wedge, \vee, \rightarrow$, but from $\bot$ we can generate $\bot$ and $\top$ by applying $\wedge, \vee, \rightarrow$.

Exercises

1. Check by the truth table method which of the following propositions are tautologies:
   (a) $(\neg\varphi \vee \psi) \leftrightarrow (\psi \rightarrow \varphi)$,
   (b) $\varphi \rightarrow ((\psi \rightarrow \sigma) \rightarrow ((\varphi \rightarrow \psi) \rightarrow (\varphi \rightarrow \sigma)))$,
   (c) $(\varphi \rightarrow \neg\varphi) \leftrightarrow \neg\varphi$,
   (d) $\neg(\varphi \rightarrow \neg\varphi)$,
   (e) $(\varphi \rightarrow (\psi \rightarrow \sigma)) \leftrightarrow ((\varphi \wedge \psi) \rightarrow \sigma)$,
   (f) $\varphi \vee \neg\varphi$ (*principle of the excluded third*),
   (g) $\bot \leftrightarrow (\varphi \wedge \neg\varphi)$,
   (h) $\bot \rightarrow \varphi$ (*ex falso sequitur quodlibet*).
2. Show
   (a) $\varphi \models \varphi$,
   (b) $\varphi \models \psi$ and $\psi \models \sigma \Rightarrow \varphi \models \sigma$,
   (c) $\models \varphi \rightarrow \psi \Leftrightarrow \varphi \models \psi$.
3. Determine $\varphi[\neg p_0 \rightarrow p_3/p_0]$ for $\varphi = p_1 \wedge p_0 \rightarrow (p_0 \rightarrow p_3)$; $\varphi = (p_3 \leftrightarrow p_0) \vee (p_2 \rightarrow \neg p_0)$.
4. Show that there are $2^{\aleph_0}$ valuations.
5. Show

$$[\![\varphi \wedge \psi]\!]_v = [\![\varphi]\!]_v \cdot [\![\psi]\!]_v,$$

$$[\![\varphi \vee \psi]\!]_v = [\![\varphi]\!]_v + [\![\psi]\!]_v - [\![\varphi]\!]_v \cdot [\![\psi]\!]_v,$$

$$[\![\varphi \rightarrow \psi]\!]_v = 1 - [\![\varphi]\!]_v + [\![\varphi]\!]_v \cdot [\![\psi]\!]_v,$$

$$[\![\varphi \leftrightarrow \psi]\!]_v = 1 - |[\![\varphi]\!]_v - [\![\psi]\!]_v|.$$

6. Show $[\![\varphi \rightarrow \psi]\!]_v = 1 \Leftrightarrow [\![\varphi]\!]_v \leq [\![\psi]\!]_v$.

2.3 Some Properties of Propositional Logic

On the basis of the previous sections we can already prove a lot of theorems about propositional logic. One of the earliest discoveries in modern propositional logic was its similarity with algebras.

Following Boole, an extensive study of the algebraic properties was made by a number of logicians. The purely algebraic aspects have since then been studied in *Boolean algebra*.

We will just mention a few of those algebraic laws.

Theorem 2.3.1 *The following propositions are tautologies*:

$$(\varphi \vee \psi) \vee \sigma \leftrightarrow \varphi \vee (\psi \vee \sigma) \qquad (\varphi \wedge \psi) \wedge \sigma \leftrightarrow \varphi \wedge (\psi \wedge \sigma)$$

associativity

$$\varphi \vee \psi \leftrightarrow \psi \vee \varphi \qquad \varphi \wedge \psi \leftrightarrow \psi \wedge \varphi$$

commutativity

$$\varphi \vee (\psi \wedge \sigma) \leftrightarrow (\varphi \vee \psi) \wedge (\varphi \vee \sigma) \qquad \varphi \wedge (\psi \vee \sigma) \leftrightarrow (\varphi \wedge \psi) \vee (\varphi \wedge \sigma)$$

distributivity

$$\neg(\varphi \vee \psi) \leftrightarrow \neg\varphi \wedge \neg\psi \qquad \neg(\varphi \wedge \psi) \leftrightarrow \neg\varphi \vee \neg\psi$$

De Morgan's laws

$$\varphi \vee \varphi \leftrightarrow \varphi \qquad \varphi \wedge \varphi \leftrightarrow \varphi$$

idempotency

$$\neg\neg\varphi \leftrightarrow \varphi$$

double negation law

Proof Check the truth tables or do a little computation. For example, De Morgan's law: $[\![\neg(\varphi \vee \psi)]\!] = 1 \Leftrightarrow [\![\varphi \vee \psi]\!] = 0 \Leftrightarrow [\![\varphi]\!] = [\![\psi]\!] = 0 \Leftrightarrow [\![\neg\varphi]\!] = [\![\neg\psi]\!] = 1 \Leftrightarrow [\![\neg\varphi \wedge \neg\psi]\!] = 1$.

So $[\![\neg(\varphi \vee \psi)]\!] = [\![\neg\varphi \wedge \neg\psi]\!]$ for all valuations, i.e. $\models \neg(\varphi \vee \psi) \leftrightarrow \neg\varphi \wedge \neg\psi$.

The remaining tautologies are left to the reader. $\square$

In order to apply the previous theorem in "logical calculations" we need a few more equivalences. This is demonstrated in the simple equivalence $\models \varphi \wedge (\varphi \vee \psi) \leftrightarrow \varphi$ (an exercise for the reader). For, by the distributive law $\models \varphi \wedge (\varphi \vee \psi) \leftrightarrow (\varphi \wedge \varphi) \vee (\varphi \wedge \psi)$ and $\models (\varphi \wedge \varphi) \vee (\varphi \wedge \psi) \leftrightarrow \varphi \vee (\varphi \wedge \psi)$, by idempotency and the substitution theorem. So $\models \varphi \wedge (\varphi \vee \psi) \leftrightarrow \varphi \vee (\varphi \wedge \psi)$. Another application of the distributive law will bring us back to start, so just applying the above laws will not eliminate ψ!

Therefore, we list a few more convenient properties.

Lemma 2.3.2 *If* $\models \varphi \rightarrow \psi$, *then*

$$\models \varphi \wedge \psi \leftrightarrow \varphi \quad and$$
$$\models \varphi \vee \psi \leftrightarrow \psi$$

Proof By Exercise 6 of Sect. 2.2 $\models \varphi \rightarrow \psi$ implies $[\![\varphi]\!]_v \leq [\![\psi]\!]_v$ for all v. So $[\![\varphi \wedge \psi]\!]_v = \min([\![\varphi]\!]_v, [\![\psi]\!]_v) = [\![\varphi]\!]_v$ and $[\![\varphi \vee \psi]\!]_v = \max([\![\varphi]\!]_v, [\![\psi]\!]_v) = [\![\psi]\!]_v$ for all v. $\square$

Lemma 2.3.3

(a) $\models \varphi \Rightarrow \models \varphi \wedge \psi \leftrightarrow \psi$,

(b) $\models \varphi \Rightarrow \models \neg\varphi \vee \psi \leftrightarrow \psi$,

(c) $\models \perp \vee \psi \leftrightarrow \psi$,

(d) $\models \top \wedge \psi \leftrightarrow \psi$.

Proof Left to the reader.   □

The following theorem establishes some equivalences involving various connectives. It tells us that we can "define" up to logical equivalence all connectives in terms of $\{\vee, \neg\}$, or $\{\rightarrow, \neg\}$, or $\{\wedge, \neg\}$, or $\{\rightarrow, \perp\}$.

That is, we can find e.g. a proposition involving only $\vee$ and $\neg$, which is equivalent to $\varphi \leftrightarrow \psi$, etc.

Theorem 2.3.4

(a) $\models (\varphi \leftrightarrow \psi) \leftrightarrow (\varphi \rightarrow \psi) \wedge (\psi \rightarrow \varphi)$,

(b) $\models (\varphi \rightarrow \psi) \leftrightarrow (\neg\varphi \vee \psi)$,

(c) $\models \varphi \vee \psi \leftrightarrow (\neg\varphi \rightarrow \psi)$,

(d) $\models \varphi \vee \psi \leftrightarrow \neg(\neg\varphi \wedge \neg\psi)$,

(e) $\models \varphi \wedge \psi \leftrightarrow \neg(\neg\varphi \vee \neg\psi)$,

(f) $\models \neg\varphi \leftrightarrow (\varphi \rightarrow \perp)$,

(g) $\models \perp \leftrightarrow \varphi \wedge \neg\varphi$.

Proof Compute the truth values of the left-hand and right-hand sides.   □

We now have enough material to handle logic as if it were algebra. For convenience we write $\varphi \approx \psi$ for $\models \varphi \leftrightarrow \psi$.

Lemma 2.3.5 $\approx$ *is an equivalence relation on PROP, i.e.*

$$\varphi \approx \varphi \ (reflexivity),$$
$$\varphi \approx \psi \quad \Rightarrow \quad \psi \approx \varphi \ (symmetry),$$
$$\varphi \approx \psi \ and \ \psi \approx \sigma \quad \Rightarrow \quad \varphi \approx \sigma \ (transitivity).$$

Proof Use $\models \varphi \leftrightarrow \psi$ iff $[\![\varphi]\!]_v = [\![\psi]\!]_v$ for all v.   □

We give some examples of algebraic computations, which establish a chain of equivalences.

1. $\models [\varphi \to (\psi \to \sigma)] \leftrightarrow [\varphi \wedge \psi \to \sigma]$,

$$\varphi \to (\psi \to \sigma) \approx \neg\varphi \vee (\psi \to \sigma), \quad \text{(Theorem 2.3.4(b))}$$
$$\neg\varphi \vee (\psi \to \sigma) \approx \neg\varphi \vee (\neg\psi \vee \sigma), \quad \text{(Theorem 2.3.4(b) and Subst. Thm.)}$$
$$\neg\varphi \vee (\neg\psi \vee \sigma) \approx (\neg\varphi \vee \neg\psi) \vee \sigma, \quad \text{(ass.)}$$
$$(\neg\varphi \vee \neg\psi) \vee \sigma \approx \neg(\varphi \wedge \psi) \vee \sigma, \quad \text{(\textit{De Morgan} and Subst. Thm.)}$$
$$\neg(\varphi \wedge \psi) \vee \sigma \approx (\varphi \wedge \psi) \to \sigma, \quad \text{(Theorem 2.3.4(b))}$$

So $\varphi \to (\psi \to \sigma) \approx (\varphi \wedge \psi) \to \sigma$.

We now leave out the references to the facts used, and make one long string. We just calculate until we reach a tautology.

2. $\models (\varphi \to \psi) \leftrightarrow (\neg\psi \to \neg\varphi)$,

$$\neg\psi \to \neg\varphi \approx \neg\neg\psi \vee \neg\varphi \approx \psi \vee \neg\varphi \approx \neg\varphi \vee \psi \approx \varphi \to \psi$$

3. $\models \varphi \to (\psi \to \varphi)$,

$$\varphi \to (\psi \to \varphi) \approx \neg\varphi \vee (\neg\psi \vee \varphi) \approx (\neg\varphi \vee \varphi) \vee \neg\psi.$$

We have seen that $\vee$ and $\wedge$ are associative, therefore we adopt the convention, also used in algebra, to delete brackets in iterated disjunctions and conjunctions; i.e. we write $\varphi_1 \vee \varphi_2 \vee \varphi_3 \vee \varphi_4$, etc. This is all right, since no matter how we restore (syntactically correctly) the brackets, the resulting formula is determined uniquely up to equivalence.

Have we introduced *all* connectives so far? Obviously not. We can easily invent new ones. Here is a famous one, introduced by Sheffer: $\varphi | \psi$ stands for "not both φ and ψ". More precise: $\varphi | \psi$ is given by the following truth table:

Sheffer stroke

$\mid$	0	1
0	1	1
1	1	0

Let us say that an n-ary logical connective \$ is *defined* by its truth table, or by its valuation function, if $[\![\$(p_1, \ldots, p_n)]\!] = f([\![p_1]\!], \ldots, [\![p_n]\!])$ for some function f.

Although we can apparently introduce many new connectives in this way, there are no surprises in stock for us, as all of those connectives are definable in terms of $\vee$ and $\neg$.

Theorem 2.3.6 *For each n-ary connective \$ defined by its valuation function, there is a proposition τ, containing only $p_1, \ldots, p_n$, $\vee$ and $\neg$, such that $\models \tau \leftrightarrow$ $\$(p_1, \ldots, p_n)$.*

Proof Induction on n. For $n = 1$ there are 4 possible connectives with truth tables

$\$_1$	
0	0
1	0

$\$_2$	
0	1
1	1

$\$_3$	
0	0
1	1

$\$_4$	
0	1
1	0

One easily checks that the propositions $\neg(p \vee \neg p)$, $p \vee \neg p$, p and $\neg p$ will meet the requirements.

Suppose that for all n-ary connectives propositions have been found.
Consider $\$(p_1, \ldots, p_n, p_{n+1})$ with truth table:

p_1	$p_2 \cdots p_n$		p_{n+1}	$\$(p_1, \ldots, p_n, p_{n+1})$
0	0	0	0	i_1
.	.	0	1	i_2
.	0	1	.	.
.	1	1	.	.
0	.	.	.	.
.	1	.	.	.
. . . .
1	0	.	.	.
.	.	.	.	.
.	.	.	.	.
.	0	.	.	.
.	1	0	.	.
.	.	0	.	.
1	.	1	0	.
.	.	1	1	$i_{2^{n+1}}$

where $i_k \leq 1$.

We consider two auxiliary connectives $\$_1$ and $\$_2$ defined by

$$\$_1(p_2, \ldots, p_{n+1}) = \$(\bot, p_2, \ldots, p_{n+1}) \quad \text{and}$$
$$\$_2(p_2, \ldots, p_{n+1}) = \$(\top, p_2, \ldots, p_{n+1}), \quad \text{where } \top = \neg \bot$$

(as given by the upper and lower halves of the above table).

By the induction hypothesis there are propositions σ_1 and σ_2, containing only
$p_2, \ldots, p_{n+1}, \vee$ and $\neg$ so that $\models \$_i(p_2, \ldots, p_{n+1}) \leftrightarrow \sigma_i$.

From those two propositions we can construct the proposition τ:

$$[\tau := (p_1 \to \sigma_2) \wedge (\neg p_1 \to \sigma_1).$$

Claim $\models \$(p_1, \ldots, p_{n+1}) \leftrightarrow \tau$.

If $[\![p_1]\!]_v = 0$, then $[\![p_1 \to \sigma_2]\!]_v = 1$, so $[\![\tau]\!]_v = [\![\neg p_1 \to \sigma_1]\!]_v = [\![\sigma_1]\!]_v = [\![\$_1(p_2, \ldots, p_{n+1})]\!]_v = [\![\$(p_1, p_2, \ldots, p_{n+1})]\!]_v$, using $[\![p_1]\!]_v = 0 = [\![\bot]\!]_v$.
The case $[\![p_1]\!]_v = 1$ is similar.

Now expressing $\to$ and $\wedge$ in terms of $\vee$ and $\neg$ (2.3.4), we have $[\![\tau']\!] = [\![\$(p_1, \ldots, p_{n+1})]\!]$ for all valuations (another use of Lemma 2.3.5), where $\tau' \approx \tau$ and τ' contains only the connectives $\vee$ and $\neg$. $\square$

For another solution see Exercise 7.
The above theorem and Theorem 2.3.4 are pragmatic justifications for our choice
of the truth table for $\to$: we get an extremely elegant and useful theory. Theorem 2.3.6 is usually expressed by saying that $\vee$ and $\neg$ form a *functionally complete*

set of connectives. Likewise $\wedge, \neg$ and $\rightarrow, \neg$ and $\perp, \rightarrow$ form functionally complete sets.

In analogy to the $\sum$ and $\prod$ from algebra we introduce finite disjunctions and conjunctions.

Definition 2.3.7

$$\begin{cases} \bigwedge_{i \leq 0} \varphi_i = \varphi_0 \\ \bigwedge_{i \leq n+1} \varphi_i = \bigwedge_{i \leq n} \varphi_i \wedge \varphi_{n+1} \end{cases} \qquad \begin{cases} \bigvee_{i \leq 0} \varphi_i = \varphi_0 \\ \bigvee_{i \leq n+1} \varphi_i = \bigvee_{i \leq n} \varphi_i \vee \varphi_{n+1} \end{cases}$$

Definition 2.3.8 If $\varphi = \bigwedge_{i \leq n} \bigvee_{j \leq m_i} \varphi_{ij}$, where φ_{ij} is atomic or the negation of an atom, then φ is a *conjunctive normal form*. If $\varphi = \bigvee_{i \leq n} \bigwedge_{j \leq m_i} \varphi_{ij}$, where φ_{ij} is atomic or the negation of an atom, then φ is a *disjunctive normal form*.

The normal forms are analogous to the well-known normal forms in algebra: $ax^2 + byx$ is "normal", whereas $x(ax + by)$ is not. One can obtain normal forms by simply "multiplying", i.e. repeated application of distributive laws. In algebra there is only one "normal form"; in logic there is a certain duality between $\wedge$ and $\vee$, so that we have two normal form theorems.

Theorem 2.3.9 *For each φ there are conjunctive normal forms $\varphi^\wedge$ and disjunctive normal forms $\varphi^\vee$, such that $\models \varphi \leftrightarrow \varphi^\wedge$ and $\models \varphi \leftrightarrow \varphi^\vee$.*

Proof First eliminate all connectives other than $\perp, \wedge, \vee$ and $\neg$. Then prove the theorem by induction on the resulting proposition in the restricted language of $\perp, \wedge, \vee$ and $\neg$. In fact, $\perp$ plays no role in this setting; it could just as well be ignored.

(a) φ is atomic. Then $\varphi^\wedge = \varphi^\vee = \varphi$.
(b) $\varphi = \psi \wedge \sigma$. Then $\varphi^\wedge = \psi^\wedge \wedge \sigma^\wedge$. In order to obtain a disjunctive normal form we consider $\psi^\vee = \bigvee \psi_i, \sigma^\vee = \bigvee \sigma_j$, where the ψ_i's and σ_j's are conjunctions of atoms and negations of atoms.

Now $\varphi = \psi \wedge \sigma \approx \psi^\vee \wedge \sigma^\vee \approx \bigvee_{i,j} (\psi_i \wedge \sigma_j)$.
The last proposition is in normal form, so we equate $\varphi^\vee$ to it.

(c) $\varphi = \psi \vee \sigma$. Similar to (b).
(d) $\varphi = \neg \psi$. By the induction hypothesis ψ has normal forms $\psi^\vee$ and $\psi^\wedge$. $\neg \psi \approx \neg \psi^\wedge \approx \neg \bigvee \bigwedge \psi_{ij} \approx \bigwedge \bigvee \neg \psi_{ij} \approx \bigwedge \bigvee \psi'_{ij}$, where $\psi'_{ij} = \neg \psi_{ij}$ if ψ_{ij} is atomic, and $\psi_{ij} = \neg \psi'_{ij}$ if ψ_{ij} is the negation of an atom. (Observe $\neg \neg \psi_{ij} \approx \psi_{ij}$.) Clearly $\bigwedge \bigvee \psi'_{ij}$ is a conjunctive normal form for φ. The disjunctive normal form is left to the reader.

For another proof of the normal form theorems see Exercise 7. $\square$

When looking at the algebra of logic in Theorem 2.3.1, we saw that $\vee$ and $\wedge$ behaved in a very similar way, to the extent that the same laws hold for both. We will make this "duality" precise. For this purpose we consider a language with only the connectives $\vee, \wedge$ and $\neg$.

Definition 2.3.10 Define an auxiliary mapping $*: PROP \to PROP$ recursively by

$$\varphi^* = \neg\varphi \quad \text{if } \varphi \text{ is atomic,}$$
$$(\varphi \wedge \psi)^* = \varphi^* \vee \psi^*,$$
$$(\varphi \vee \psi)^* = \varphi^* \wedge \psi^*,$$
$$(\neg\varphi)^* = \neg\varphi^*.$$

Example $((p_0 \wedge \neg p_1) \vee p_2)^* = (p_0 \wedge \neg p_1)^* \wedge p_2^* = (p_0^* \vee (\neg p_1)^*) \wedge \neg p_2 = (\neg p_0 \vee \neg p_1^*) \wedge \neg p_2 = (\neg p_0 \vee \neg\neg p_1) \wedge \neg p_2 \approx (\neg p_0 \vee p_1) \wedge \neg p_2.$

Note that the effect of the *-translation boils down to taking the negation and applying De Morgan's laws.

Lemma 2.3.11 $[\![\varphi^*]\!] = [\![\neg\varphi]\!]$.

Proof Induction on φ. For atomic φ $[\![\varphi^*]\!] = [\![\neg\varphi]\!]$. $[\![(\varphi \wedge \psi)^*]\!] = [\![\varphi^* \vee \psi^*]\!] = [\![\neg\varphi \vee \neg\psi]\!]) = [\![\neg(\varphi \wedge \psi)]\!])$. $[\![(\varphi \vee \psi)^*]\!]$ and $[\![(\neg\varphi)^*]\!]$ are left to the reader. □

Corollary 2.3.12 $\models \varphi^* \leftrightarrow \neg\varphi$.

Proof The proof is immediate from Lemma 2.3.11. □

So far this is not the proper duality we have been looking for. We really just want to interchange $\wedge$ and $\vee$. So we introduce a new translation.

Definition 2.3.13 The duality mapping $^d: PROP \to PROP$ is recursively defined by

$$\varphi^d = \varphi \quad \text{for } \varphi \text{ atomic,}$$
$$(\varphi \wedge \psi)^d = \varphi^d \vee \psi^d,$$
$$(\varphi \vee \psi)^d = \varphi^d \wedge \psi^d,$$
$$(\neg\varphi)^d = \neg\varphi^d.$$

Theorem 2.3.14 (Duality Theorem) $\models \varphi \leftrightarrow \psi \Leftrightarrow \models \varphi^d \leftrightarrow \psi^d$.

Proof We use the *-translation as an intermediate step. Let us introduce the notion of simultaneous substitution to simplify the proof.

$\sigma[\tau_0, \ldots, \tau_n / p_0, \ldots, p_n]$ is obtained by substituting τ_i for p_i for all $i \leq n$ simultaneously (see Exercise 15). Observe that $\varphi^* = \varphi^d[\neg p_0, \ldots, \neg p_n / p_0, \ldots, p_n]$, so $\varphi^*[\neg p_0, \ldots, \neg p_n / p_0, \ldots, p_n] = \varphi^d[\neg\neg p_0, \ldots, \neg\neg p_n / p_0, \ldots, p_n]$, where the atoms of φ occur among the $p_0, \ldots, p_n$.

By the Substitution Theorem $\models \varphi^d \leftrightarrow \varphi^*[\neg p_0, \ldots, \neg p_n / p_0, \ldots, p_n]$. The same equivalence holds for ψ.

By Corollary 2.3.12 $\models \varphi^* \leftrightarrow \neg\varphi$, $\models \psi^* \leftrightarrow \neg\psi$. Since $\models \varphi \leftrightarrow \psi$, also $\models$ $\neg\varphi \leftrightarrow \neg\psi$. Hence $\models \varphi^* \leftrightarrow \psi^*$, and therefore $\models \varphi^*[\neg p_0, \ldots, \neg p_n/p_0, \ldots, p_n] \leftrightarrow$ $\psi^*[\neg p_0, \ldots, \neg p_n/p_0, \ldots, p_n]$.

Using the above relation between φ^d and φ^* we now obtain $\models \varphi^d \leftrightarrow \psi^d$. The converse follows immediately, as $\varphi^{dd} = \varphi$.   □

The Duality Theorem gives us one identity for free for each identity we establish.

Exercises

1. Show by "algebraic" means:

$$\models \quad (\varphi \to \psi) \leftrightarrow (\neg\psi \to \neg\varphi), \quad \text{contraposition,}$$

$$\models \quad (\varphi \to \psi) \wedge (\psi \to \sigma) \to (\varphi \to \sigma), \quad \text{transitivity of } \to,$$

$$\models \quad (\varphi \to (\psi \wedge \neg\psi)) \to \neg\varphi,$$

$$\models \quad (\varphi \to \neg\varphi) \to \neg\varphi,$$

$$\models \quad \neg(\varphi \wedge \neg\varphi),$$

$$\models \quad \varphi \to (\psi \to \varphi \wedge \psi),$$

$$\models \quad ((\varphi \to \psi) \to \varphi) \to \varphi, \quad \text{Peirce's law.}$$

2. Simplify the following propositions (i.e. find a simpler equivalent proposition):

   (a) $(\varphi \to \psi) \wedge \varphi$,   (b) $(\varphi \to \psi) \vee \neg\varphi$,   (c) $(\varphi \to \psi) \to \psi$,

   (d) $\varphi \to (\varphi \wedge \psi)$,   (e) $(\varphi \wedge \psi) \vee \varphi$,   (f) $(\varphi \to \psi) \to \varphi$.

3. Show that $\{\neg\}$ is not a functionally complete set of connectives. Idem for $\{\to, \vee\}$ (hint: show that for each formula φ with only $\to$ and $\vee$ there is a valuation v such that $[\![\varphi]\!]_v = 1$).

4. Show that the Sheffer stroke, $|$, forms a functionally complete set (hint: $\models \neg\varphi \leftrightarrow \varphi \mid \varphi$).

5. Show that the connective $\downarrow$ (φ *nor* ψ), with valuation function $[\![\varphi\downarrow\psi]\!] = 1$ iff $[\![\varphi]\!] = [\![\psi]\!] = 0$, forms a functionally complete set.

6. Show that $|$ and $\downarrow$ are the only binary connectives $ such that $\{$\}$ is functionally complete.

7. The functional completeness of $\{\vee, \neg\}$ can be shown in an alternative way. Let $ be an n-ary connective with valuation function $[\![\$(p_1, \ldots, p_n)]\!] = f([\![p_1]\!], \ldots, [\![p_n]\!])$. We want a proposition τ (in $\vee, \neg$) such that $[\![\tau]\!] = f([\![p_1]\!], \ldots, [\![p_n]\!])$.

   Suppose $f([\![p_1]\!], \ldots, [\![p_n]\!]) = 1$ at least once. Consider all tuples $([\![p_1]\!], \ldots, [\![p_n]\!])$ with $f([\![p_1]\!], \ldots, [\![p_n]\!]) = 1$ and form corresponding conjunctions $\bar{p}_1 \wedge \bar{p}_2 \wedge \cdots \wedge \bar{p}_n$ such that $\bar{p}_i = p_i$ if $[\![p_i]\!] = 1$, $\bar{p}_i = \neg p_i$ if $[\![p_i]\!] = 0$. Then show $\models (\bar{p}_1^1 \wedge \bar{p}_2^1 \wedge \cdots \wedge \bar{p}_n^1) \vee \cdots \vee (\bar{p}_1^k \wedge \bar{p}_2^k \wedge \cdots \wedge \bar{p}_n^k) \leftrightarrow \$(p_1, \ldots, p_n)$, where the disjunction is taken over all n-tuples such that $f([\![p_1]\!], \ldots, [\![p_n]\!]) = 1$.

   Alternatively, we can consider the tuples for which $f([\![p_1]\!], \ldots, [\![p_n]\!]) = 0$. Carry out the details. Note that this proof of the functional completeness at the same time proves the normal form theorems.

8. Let the ternary connective $ be defined by $[\![\$(\varphi_1, \varphi_2, \varphi_3)]\!] = 1 \Leftrightarrow [\![\varphi_1]\!] + [\![\varphi_2]\!] + [\![\varphi_3]\!] \geq 2$ (the majority connective). Express $ in terms of $\vee$ and $\neg$.
9. Let the binary connective # be defined by

#	0	1
0	0	1
1	1	0

   Express # in terms of $\vee$ and $\neg$.
10. Determine conjunctive and disjunctive normal forms for $\neg(\varphi \leftrightarrow \psi)$, $((\varphi \to \psi) \to \psi) \to \psi$, $(\varphi \to (\varphi \wedge \neg\psi)) \wedge (\psi \to (\psi \wedge \neg\varphi))$.
11. Give a criterion for a conjunctive normal form to be a tautology.
12. Prove

$$\bigwedge_{i \leq n} \varphi_i \vee \bigwedge_{j \leq m} \psi_j \approx \bigwedge_{\substack{i \leq n \\ j \leq m}} (\varphi_i \vee \psi_j)$$

and

$$\bigvee_{i \leq n} \varphi_i \wedge \bigvee_{j \leq m} \psi_j \approx \bigvee_{\substack{i \leq n \\ j \leq m}} (\varphi_i \wedge \psi_j).$$

13. The set of all valuations, thought of as the set of all 0–1-sequences, forms a topological space, called the Cantor space $\mathcal{C}$. The basic open sets are finite unions of sets of the form $\{v \mid [\![p_{i_1}]\!]_v = \cdots = [\![p_{i_n}]\!]_v = 1$ and $[\![p_{j_1}]\!]_v = \cdots = [\![p_{j_m}]\!]_v = 0\}$, $i_k \neq j_p$ for $k \leq n$; $p \leq m$.

    Define a function $[\![\ ]\!] : PROP \to \mathcal{P}(\mathcal{C})$ (subsets of the Cantor space) by: $[\![\varphi]\!] = \{v \mid [\![\varphi]\!]_v = 1\}$.
    (a) Show that $[\![\varphi]\!]$ is a basic open set (which is also closed),
    (b) $[\![\varphi \vee \psi]\!] = [\![\varphi]\!] \cup [\![\psi]\!]$; $[\![\varphi \wedge \psi]\!] = [\![\varphi]\!] \cap [\![\psi]\!]$; $[\![\neg\varphi]\!] = [\![\varphi]\!]^c$,
    (c) $\models \varphi \Leftrightarrow [\![\varphi]\!] = C$; $[\![\bot]\!] = \emptyset$; $\models \varphi \to \psi \Leftrightarrow [\![\varphi]\!] \subseteq [\![\psi]\!]$.
        Extend the mapping to sets of propositions Γ by $[\![\Gamma]\!] = \{v \mid [\![\varphi]\!]_v = 1$ for all $\varphi \in \Gamma\}$. Note that $[\![\Gamma]\!]$ is closed.
    (d) $\Gamma \models \varphi \Leftrightarrow [\![\Gamma]\!] \subseteq [\![\varphi]\!]$.
14. We can view the relation $\models \varphi \to \psi$ as a kind of ordering. Put $\varphi \sqsubset \psi :=$ $\models \varphi \to \psi$ and $\not\models \psi \to \varphi$.
    (i) For each φ, ψ such that $\varphi \sqsubset \psi$, find σ with $\varphi \sqsubset \sigma \sqsubset \psi$.
    (ii) Find $\varphi_1, \varphi_2, \varphi_3, \ldots$ such that $\varphi_1 \sqsubset \varphi_2 \sqsubset \varphi_3 \sqsubset \varphi_4 \sqsubset \cdots$,
    (iii) and show that for each φ, ψ with φ and ψ incomparable, there is a least σ with $\varphi, \psi \sqsubset \sigma$.
15. Give a recursive definition of the simultaneous substitution $\varphi[\psi, \ldots, \psi_n/p_1, \ldots, p_n]$ and formulate and prove the appropriate analogue of the Substitution Theorem (Theorem 2.2.6).

2.4 Natural Deduction

In the preceding sections we have adopted the view that propositional logic is based on truth tables; i.e. we have looked at logic from a semantical point of view. This, however, is not the only possible point of view. If one thinks of logic as a codification of (exact) reasoning, then it should stay close to the practice of inference making, instead of basing itself on the notion of truth. We will now explore the non-semantic approach, by setting up a system for deriving conclusions from premises. Although this approach is of a formal nature, i.e. it abstains from interpreting the statements and rules, it is advisable to keep some interpretation in mind. We are going to introduce a number of derivation rules, which are, in a way, the atomic steps in a derivation. These derivation rules are designed (by Gentzen), to render the intuitive meaning of the connectives as faithfully as possible.

There is one minor problem, which at the same time is a major advantage, namely: our rules express the constructive meaning of the connectives. This advantage will not be exploited now, but it is good to keep it in mind when dealing with logic (it is exploited in intuitionistic logic).

One small example: the principle of the excluded third tells us that $\models \varphi \vee \neg\varphi$, i.e., assuming that φ is a definite mathematical statement, either it or its negation must be true. Now consider some unsolved problem, e.g. Riemann's hypothesis, call it R. Then either R is true, or $\neg R$ is true. However, we do not know which of the two is true, so the constructive content of $R \vee \neg R$ is nil. Constructively, one would require a method to find out which of the alternatives holds.

The propositional connective which has a strikingly different meaning in a constructive and in a non-constructive approach is the disjunction. Therefore we restrict our language for the moment to the connectives $\wedge$, $\rightarrow$ and $\perp$. This is no real restriction as $\{\rightarrow, \perp\}$ is a functionally complete set.

Our derivations consist of very simple steps, such as "from φ and $\varphi \rightarrow \psi$ conclude ψ", written as:

$$\frac{\varphi \quad \varphi \rightarrow \psi}{\psi}$$

The propositions above the line are *premises*, and the one below the line is the *conclusion*. The above example *eliminated* the connective $\rightarrow$. We can also *introduce* connectives. The derivation rules for $\wedge$ and $\rightarrow$ are separated into

Introduction Rules   Elimination Rules

$$(\wedge I) \quad \frac{\varphi \quad \psi}{\varphi \wedge \psi} \wedge I \qquad (\wedge E) \quad \frac{\varphi \wedge \psi}{\varphi} \wedge E \quad \frac{\varphi \wedge \psi}{\psi} \wedge E$$

$$[\varphi]$$

$$(\rightarrow I) \quad \begin{array}{c} \vdots \\ \dfrac{\psi}{\varphi \rightarrow \psi} \rightarrow I \end{array} \qquad (\rightarrow E) \quad \frac{\varphi \quad \varphi \rightarrow \psi}{\psi} \rightarrow E$$

We have two rules for $\bot$, both of which eliminate $\bot$, but introduce a formula.

$$[\neg\varphi]$$

$$(\bot) \, \frac{\bot}{\varphi} \, \bot \qquad (RAA) \quad \begin{array}{c} \vdots \\ \bot \\ \hline \varphi \end{array} \, RAA$$

As usual "$\neg\varphi$" is used here as an abbreviation for "$\varphi \to \bot$".

The rules for $\wedge$ are evident: if we have φ and ψ we may conclude $\varphi \wedge \psi$, and if we have $\varphi \wedge \psi$ we may conclude φ (or ψ). The introduction rule for implication has a different form. It states that, if we can derive ψ from φ (as a hypothesis), then we may conclude $\varphi \to \psi$ (without the hypothesis φ). This agrees with the intuitive meaning of implication: $\varphi \to \psi$ means "ψ follows from φ". We have written the rule ($\to$ I) in the above form to suggest a derivation. The notation will become clearer after we have defined derivations. For the time being we will write the premises of a rule in the order that suits us best, later we will become more fastidious.

The rule ($\to E$) is also evident on the meaning of implication. If φ is given and we know that ψ follows from φ, then we also have ψ. The *falsum rule*, ($\bot$), expresses that from an absurdity we can derive everything (ex falso sequitur quodlibet), and the *reductio ad absurdum rule*, (RAA), is a formulation of the *principle of proof by contradiction*: if one derives a contradiction from the hypothesis $\neg\varphi$, then one has a derivation of φ (without the hypothesis $\neg\varphi$, of course). In both ($\to I$) and (RAA) hypotheses disappear, which is indicated by the striking out of the hypothesis. We say that such a hypothesis is *canceled*. Let us digress for a moment on the cancellation of hypotheses. We first consider implication introduction. There is a well-known theorem in plane geometry which states that "if a triangle is isosceles, then the angles opposite the equal sides are equal to one another" (Euclid's Elements, Book I, Proposition 5). This is shown as follows: we suppose that we have an isosceles triangle and then, in a number of steps, we deduce that the angles at the base are equal. Thence we conclude that *the angles at the base are equal if the triangle is isosceles*.

Query 1: do we still need the hypothesis that the triangle is isosceles? Of course not! We have, so to speak, incorporated this condition in the statement itself. It is precisely the role of conditional statements, such as "if it rains I will use my umbrella", to get rid of the obligation to require (or verify) the condition. In abstracto: if we can deduce ψ using the hypothesis φ, then $\varphi \to \psi$ is the case *without the hypothesis φ* (there may be other hypotheses, of course).

Query 2: is it forbidden to maintain the hypothesis? Answer: no, but it clearly is superfluous. As a matter of fact we usually experience superfluous conditions as confusing or even misleading, but that is rather a matter of the psychology of problem solving than of formal logic. Usually we want the best possible result, and it is intuitively clear that the more hypotheses we state for a theorem, the weaker our result is. Therefore we will as a rule cancel as many hypotheses as possible.

In the case of (RAA) we also deal with cancellation of hypotheses. Again, let us consider an example.

In analysis we introduce the notion of a *convergent sequence* (a_n) and subsequently the notion "a is a limit of (a_n)". The next step is to prove that for each convergent sequence there is a unique limit; we are interested in the part of the proof that shows that there is at most one limit. Such a proof may run as follows: we suppose that there are two distinct limits a and a', and from this hypothesis, $a \neq a'$, we derive a contradiction. Conclusion: $a = a'$. In this case we of course drop the hypothesis $a \neq a'$; this time it is not a case of being superfluous, but of being in conflict! So, both in the case ($\rightarrow I$) and in (RAA), it is sound practice to cancel all occurrences of the hypothesis concerned.

In order to master the technique of natural deduction, and to become familiar with the technique of cancellation, one cannot do better than to look at a few concrete cases. So before we go on to the notion of *derivation* we consider a few examples.

$$
\mathbf{I} \qquad
\dfrac{
  \dfrac{\dfrac{[\varphi \wedge \psi]^1}{\psi} \wedge E \quad \dfrac{[\varphi \wedge \psi]^1}{\varphi} \wedge E}{\psi \wedge \varphi} \wedge I
}{\varphi \wedge \psi \rightarrow \psi \wedge \varphi} \rightarrow I_1
\qquad\qquad
\mathbf{II} \qquad
\dfrac{
  \dfrac{\dfrac{[\varphi]^2 \quad [\varphi \rightarrow \bot]^1}{\bot} \rightarrow E}{(\varphi \rightarrow \bot) \rightarrow \bot} \rightarrow I_1
}{\varphi \rightarrow ((\varphi \rightarrow \bot) \rightarrow \bot)} \rightarrow I_2
$$

$$
\mathbf{III} \qquad
\dfrac{
  \dfrac{
    \dfrac{\dfrac{[\varphi \wedge \psi]^1}{\psi} \wedge E \quad \dfrac{\dfrac{[\varphi \wedge \psi]^1}{\varphi} \wedge E \quad [\varphi \rightarrow (\psi \rightarrow \sigma)]^2}{\psi \rightarrow \sigma} \rightarrow E}{\sigma} \rightarrow E}{\varphi \wedge \psi \rightarrow \sigma} \rightarrow I_1
}{(\varphi \rightarrow (\psi \rightarrow \sigma)) \rightarrow (\varphi \wedge \psi \rightarrow \sigma)} \rightarrow I_2
$$

If we use the customary abbreviation "$\neg\varphi$" for "$\varphi \rightarrow \bot$", we can bring some derivations into a more convenient form. (Recall that $\neg\varphi$ and $\varphi \rightarrow \bot$, as given in 2.2, are semantically equivalent.) We rewrite derivation II using the abbreviation:

$$
\mathbf{II}' \qquad
\dfrac{
  \dfrac{\dfrac{[\varphi]^2 \quad [\neg\varphi]^1}{\bot} \rightarrow E}{\neg\neg\varphi} \rightarrow I_1
}{\varphi \rightarrow \neg\neg\varphi} \rightarrow I_2
$$

In the following example we use the negation sign and also the bi-implication; $\varphi \leftrightarrow \psi$ for $(\varphi \rightarrow \psi) \wedge (\psi \rightarrow \varphi)$.

$$\textbf{IV}\qquad \frac{\dfrac{[\varphi]^1 \quad \dfrac{[\varphi \leftrightarrow \neg\varphi]^3}{\varphi \to \neg\varphi}\wedge E}{\dfrac{\neg\varphi}{\dfrac{\bot}{\neg\varphi}\to I_1}\to E}\quad \dfrac{\dfrac{[\varphi \leftrightarrow \neg\varphi]^3}{\neg\varphi \to \varphi}\wedge E}{\varphi}\to E \quad \dfrac{[\varphi]^2 \quad \dfrac{\dfrac{[\varphi \leftrightarrow \neg\varphi]^3}{\varphi \to \neg\varphi}\wedge E \quad [\varphi]^2}{\dfrac{\neg\varphi}{\dfrac{\bot}{\neg\varphi}\to I_2}\to E}}{}\to E}{\dfrac{\bot}{\neg(\varphi \leftrightarrow \neg\varphi)}\to I_3}$$

The examples show us that derivations have the form of trees. We show the trees below:

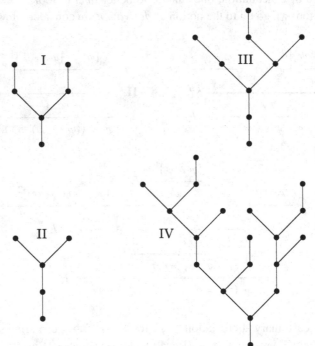

One can just as well present derivations as (linear) strings of propositions. We will stick, however, to the tree form, the idea being that what comes naturally in tree form should not be put in a linear straightjacket.

We now strive to define the notion of *derivation* in general. We will use an inductive definition to produce trees.

Notation If $\dfrac{\mathcal{D}}{\varphi}, \dfrac{\mathcal{D'}}{\varphi'}$ are derivations with conclusions φ, φ', then $\dfrac{\dfrac{\mathcal{D}}{\varphi}}{\psi}, \dfrac{\dfrac{\mathcal{D}}{\varphi}\ \dfrac{\mathcal{D'}}{\varphi'}}{\psi}$ are derivations obtained by applying a derivation rule to φ (and φ and φ'). The cancellation of a hypothesis is indicated as follows: if $\overset{\psi}{\underset{\varphi}{\mathcal{D}}}$ is a derivation with hypothesis ψ, then

$$\dfrac{\begin{array}{c}[\psi]\\ \mathcal{D}\\ \varphi\end{array}}{\sigma}$$ is a derivation with ψ canceled.

With respect to the cancellation of hypotheses, we note that one does not necessarily cancel *all* occurrences of such a proposition ψ. This clearly is justified, as one feels that adding hypotheses does not make a proposition underivable (irrelevant information may always be added). It is a matter of prudence, however, to cancel as much as possible. Why carry more hypotheses than necessary?

Furthermore one may apply ($\to$ I) if there is no hypothesis available for cancellation; e.g. $\dfrac{\varphi}{\psi\to\varphi}\to I$ is a correct derivation, using just ($\to$ I). To sum up: given a derivation tree of ψ (or $\bot$), we obtain a derivation tree of $\varphi\to\psi$ (or φ) at the bottom of the tree and strike out some (or all) occurrences, if any, of φ (or $\neg\varphi$) on top of a tree.

A few words on the practical use of natural deduction: if you want to give a derivation for a proposition it is advisable to devise some kind of strategy, just as in a game. Suppose that you want to show $[\varphi\wedge\psi\to\sigma]\to[\varphi\to(\psi\to\sigma)]$ (Example III), then (since the proposition is an implicational formula) the rule ($\to$ I) suggests itself. So try to derive $\varphi\to(\psi\to\sigma)$ from $\varphi\wedge\psi\to\sigma$.

Now we know where to start and where to go to. To make use of $\varphi\wedge\psi\to\sigma$ we want $\varphi\wedge\psi$ (for ($\to$ E)), and to get $\varphi\to(\psi\to\sigma)$ we want to derive $\psi\to\sigma$ from φ. So we may add φ as a hypothesis and look for a derivation of $\psi\to\sigma$. Again, this asks for a derivation of σ from ψ, so add ψ as a hypothesis and look for a derivation of σ. By now we have the following hypotheses available: $\varphi\wedge\psi\to\sigma$, φ and ψ. Keeping in mind that we want to eliminate $\varphi\wedge\psi$ it is evident what we should do. The derivation III shows in detail how to carry out the derivation. After making a number of derivations one gets the practical conviction that one should first take propositions apart from the bottom upwards, and then construct the required propositions by putting together the parts in a suitable way. This practical conviction is confirmed by the *Normalization Theorem*, to which we will return later. There is a particular point which tends to confuse novices:

$$\dfrac{\begin{array}{c}[\varphi]\\ \vdots\\ \bot\end{array}}{\neg\varphi}\to I \qquad\text{and}\qquad \dfrac{\begin{array}{c}[\neg\varphi]\\ \vdots\\ \bot\end{array}}{\varphi}\ RAA$$

look very much alike. Are they not both cases of reductio ad absurdum? As a matter of fact the leftmost derivation tells us (informally) that the assumption of φ leads to a contradiction, so φ *cannot be the case*. This is in our terminology the meaning of "not φ". The rightmost derivation tells us that the assumption of $\neg\varphi$ leads to a contradiction, hence (by the same reasoning) $\neg\varphi$ cannot be the case. So, on account of the meaning of negation, we only would get $\neg\neg\varphi$. It is by no means clear that $\neg\neg\varphi$ is equivalent to φ (indeed, this is denied by the intuitionists), so it is an extra property of our logic. (This is confirmed in a technical sense: $\neg\neg\varphi \to \varphi$ is not derivable in the system without RAA.)

We now return to our theoretical notions.

Definition 2.4.1 The set of derivations is the smallest set X such that

(1) The one-element tree φ belongs to X for all $\varphi \in PROP$.

$$(2\wedge) \quad \text{If } \begin{matrix} \mathcal{D} \\ \varphi \end{matrix}, \begin{matrix} \mathcal{D}' \\ \varphi' \end{matrix} \in X, \text{ then } \frac{\begin{matrix} \mathcal{D} & \mathcal{D}' \\ \varphi & \varphi' \end{matrix}}{\varphi \wedge \varphi'} \in X.$$

$$\text{If } \begin{matrix} \mathcal{D} \\ \varphi \wedge \psi \end{matrix} \in X, \text{ then } \frac{\begin{matrix} \mathcal{D} \\ \varphi \wedge \psi \end{matrix}}{\varphi}, \frac{\begin{matrix} \mathcal{D} \\ \varphi \wedge \psi \end{matrix}}{\psi} \in X.$$

$$(2\to) \quad \text{If } \begin{matrix} \varphi \\ \mathcal{D} \\ \psi \end{matrix} \in X, \text{ then } \frac{\begin{matrix} [\varphi] \\ \mathcal{D} \\ \psi \end{matrix}}{\varphi \to \psi} \in X.$$

$$\text{If } \begin{matrix} \mathcal{D} \\ \varphi \end{matrix}, \begin{matrix} \mathcal{D}' \\ \varphi \to \psi \end{matrix} \in X, \text{ then } \frac{\begin{matrix} \mathcal{D} & \mathcal{D}' \\ \varphi & \varphi \to \psi \end{matrix}}{\psi} \in X.$$

$$(2\bot) \quad \text{If } \begin{matrix} \mathcal{D} \\ \bot \end{matrix} \in X, \text{ then } \frac{\bot}{\varphi} \in X.$$

$$\text{If } \begin{matrix} \neg\varphi \\ \mathcal{D} \\ \bot \end{matrix} \in X, \text{ then } \frac{\begin{matrix} [\neg\varphi] \\ \mathcal{D} \\ \bot \end{matrix}}{\varphi} \in X.$$

The bottom formula of a derivation is called its *conclusion*. Since the class of derivations is inductively defined, we can mimic the results of Sect. 2.1.

For example, we have a *principle of induction on* $\mathcal{D}$: let A be a property. If $A(\mathcal{D})$ holds for one-element derivations and A is preserved under the clauses $(2\wedge)$, $(2\to)$ and $(2\bot)$, then $A(\mathcal{D})$ holds for all derivations. Likewise we can define mappings on the set of derivations by recursion (cf. Exercises 6, 7, 9).

Definition 2.4.2 The relation $\Gamma \vdash \varphi$ between sets of propositions and propositions is defined as follows: there is a derivation with conclusion φ and with all (uncanceled) hypotheses in Γ. (See also Exercise 6.)

We say that φ is *derivable* from Γ. Note that by definition Γ may contain many superfluous "hypotheses". The symbol $\vdash$ is called the *turnstile*.

If $\Gamma = \emptyset$, we write $\vdash \varphi$, and we say that φ is a *theorem*.

We could have avoided the notion of "derivation" and taken instead the notion of "derivability" as fundamental, see Exercise 10. The two notions, however, are closely related.

Lemma 2.4.3

(a) $\Gamma \vdash \varphi$ if $\varphi \in \Gamma$,
(b) $\Gamma \vdash \varphi, \Gamma' \vdash \psi \Rightarrow \Gamma \cup \Gamma' \vdash \varphi \wedge \psi$,
(c) $\Gamma \vdash \varphi \wedge \psi \Rightarrow \Gamma \vdash \varphi$ and $\Gamma \vdash \psi$,
(d) $\Gamma \cup \{\varphi\} \vdash \psi \Rightarrow \Gamma \vdash \varphi \rightarrow \psi$,
(e) $\Gamma \vdash \varphi, \Gamma' \vdash \varphi \rightarrow \psi \Rightarrow \Gamma \cup \Gamma' \vdash \psi$,
(f) $\Gamma \vdash \bot \Rightarrow \Gamma \vdash \varphi$,
(g) $\Gamma \cup \{\neg \varphi\} \vdash \bot \Rightarrow \Gamma \vdash \varphi$.

Proof Immediate from the definition of derivation.   □

We now list some theorems. $\neg$ and $\leftrightarrow$ are used as abbreviations.

Theorem 2.4.4

(1) $\vdash \varphi \rightarrow (\psi \rightarrow \varphi)$,
(2) $\vdash \varphi \rightarrow (\neg \varphi \rightarrow \psi)$,
(3) $\vdash (\varphi \rightarrow \psi) \rightarrow [(\psi \rightarrow \sigma) \rightarrow (\varphi \rightarrow \sigma)]$,
(4) $\vdash (\varphi \rightarrow \psi) \leftrightarrow (\neg \psi \rightarrow \neg \varphi)$,
(5) $\vdash \neg \neg \varphi \leftrightarrow \varphi$,
(6) $\vdash [\varphi \rightarrow (\psi \rightarrow \sigma)] \leftrightarrow [\varphi \wedge \psi \rightarrow \sigma]$,
(7) $\vdash \bot \leftrightarrow (\varphi \wedge \neg \varphi)$.

Proof

$$
1. \quad \cfrac{\cfrac{[\varphi]^1}{\psi \rightarrow \varphi} \rightarrow I}{\varphi \rightarrow (\psi \rightarrow \varphi)} \rightarrow I_1
\qquad
2. \quad \cfrac{\cfrac{\cfrac{\cfrac{[\varphi]^2 \quad [\neg \varphi]^1}{\bot} \rightarrow E}{\cfrac{}{\bot} \bot}{\psi}}{\neg \varphi \rightarrow \psi} \rightarrow I_1}{\varphi \rightarrow (\neg \varphi \rightarrow \psi)} \rightarrow I_2
$$

$$
3. \quad \cfrac{\cfrac{\cfrac{\cfrac{\cfrac{[\varphi]^1 \quad [\varphi \rightarrow \psi]^3}{\psi} \rightarrow E \qquad [\psi \rightarrow \sigma]^2}{\sigma} \rightarrow E}{\varphi \rightarrow \sigma} \rightarrow I_1}{(\psi \rightarrow \sigma) \rightarrow (\varphi \rightarrow \sigma)} \rightarrow I_2}{(\varphi \rightarrow \psi) \rightarrow ((\psi \rightarrow \sigma) \rightarrow (\varphi \rightarrow \sigma))} \rightarrow I_3
$$

4. For one direction, substitute $\bot$ for σ in 3, then $\vdash (\varphi \to \psi) \to (\neg\psi \to \neg\varphi)$. Conversely:

$$
\cfrac{
  \cfrac{
    \cfrac{[\neg\psi]^1 \quad [\neg\psi \to \neg\varphi]^3}{\neg\varphi} \to E \qquad [\varphi]^2
  }{
    \cfrac{\cfrac{\bot}{\psi} \text{RAA}_1}{\cfrac{\varphi \to \psi}{\;}} \to I_2
  } \to E
}{(\neg\psi \to \neg\varphi) \to (\varphi \to \psi)} \to I_3
$$

So now we have

$$
\cfrac{
  \overset{\textstyle \mathcal{D}}{(\varphi \to \psi) \to (\neg\psi \to \neg\varphi)} \qquad \overset{\textstyle \mathcal{D}'}{(\neg\psi \to \neg\varphi) \to (\varphi \to \psi)}
}{(\varphi \to \psi) \leftrightarrow (\neg\psi \to \neg\varphi)} \land I
$$

5. We already proved $\varphi \to \neg\neg\varphi$ as an example. Conversely:

$$
\cfrac{
  \cfrac{
    \cfrac{[\neg\varphi]^1 \quad [\neg\neg\varphi]^2}{\bot} \to E
  }{\cfrac{\varphi}{\;} \text{RAA}_1}
}{\neg\neg\varphi \to \varphi} \to I_2
$$

The result now follows. Numbers 6 and 7 are left to the reader.   □

The system outlined in this section is called the "calculus of natural deduction" for a good reason: its manner of making inferences corresponds to the reasoning we intuitively use. The rules present means to take formulas apart, or to put them together. A derivation then consists of a skillful manipulation of the rules, the use of which is usually suggested by the form of the formula we want to prove.

We will discuss one example in order to illustrate the general strategy of building derivations. Let us consider the converse of our previous example **III**.

To prove $(\varphi \land \psi \to \sigma) \to [\varphi \to (\psi \to \sigma)]$ there is just one initial step: assume $\varphi \land \psi \to \sigma$ and try to derive $\varphi \to (\psi \to \sigma)$. Now we can either look at the assumption or at the desired result. Let us consider the latter one first: to show $\varphi \to (\psi \to \sigma)$, we should assume φ and derive $\psi \to \sigma$, but for the latter we should assume ψ and derive σ.

So, altogether we may assume $\varphi \land \psi \to \sigma$ and φ and ψ. Now the procedure suggests itself: derive $\varphi \land \psi$ from φ and ψ, and σ from $\varphi \land \psi$ and $\varphi \land \psi \to \sigma$.

Put together, we get the following derivation:

$$
\frac{\dfrac{[\varphi]^2 \quad [\psi]^1}{\varphi \wedge \psi} \wedge I \qquad [\varphi \wedge \psi \to \sigma]^3}{\dfrac{\dfrac{\dfrac{\sigma}{\psi \to \sigma} \to I_1}{\varphi \to (\psi \to \sigma)} \to I_2}{(\varphi \wedge \psi \to \sigma) \to (\varphi \to (\psi \to \sigma))} \to I_3} \to E
$$

Had we considered $\varphi \wedge \psi \to \sigma$ first, then the only way to proceed would be to add $\varphi \wedge \psi$ and apply $\to E$. Now $\varphi \wedge \psi$ either remains an assumption, or it is obtained from something else. It immediately occurs to the reader to derive $\varphi \wedge \psi$ from φ and ψ. But now he will build up the derivation we obtained above.

Simple as this example seems, there are complications. In particular the rule of RAA is not nearly as natural as the other ones. Its use must be learned by practice; also a sense for the distinction between *constructive* and *non-constructive* will be helpful when trying to decide on when to use it.

Finally, we recall that $\top$ is an abbreviation for $\neg\bot$ (i.e. $\bot \to \bot$).

Exercises

1. Show that the following propositions are derivable:

   (a) $\varphi \to \varphi$,   (d) $(\varphi \to \psi) \leftrightarrow \neg(\varphi \wedge \neg\psi)$,
   (b) $\bot \to \varphi$,   (e) $(\varphi \wedge \psi) \leftrightarrow \neg(\varphi \to \neg\psi)$,
   (c) $\neg(\varphi \wedge \neg\varphi)$,   (f) $\varphi \to (\psi \to (\varphi \wedge \psi))$.

2. Do the same for

   (a) $(\varphi \to \neg\varphi) \to \neg\varphi$,
   (b) $[\varphi \to (\psi \to \sigma)] \leftrightarrow [\psi \to (\varphi \to \sigma)]$,
   (c) $(\varphi \to \psi) \wedge (\varphi \to \neg\psi) \to \neg\varphi$,
   (d) $(\varphi \to \psi) \to [(\varphi \to (\psi \to \sigma)) \to (\varphi \to \sigma)]$.

3. Show

   (a) $\varphi \vdash \neg(\neg\varphi \wedge \psi)$,   (d) $\vdash \varphi \Rightarrow \vdash \psi \to \varphi$,
   (b) $\neg(\varphi \wedge \neg\psi), \varphi \vdash \psi$,   (e) $\neg\varphi \vdash \varphi \to \psi$.
   (c) $\neg\varphi \vdash (\varphi \to \psi) \leftrightarrow \neg\varphi$,

4. Show

   $$\vdash [(\varphi \to \psi) \to (\varphi \to \sigma)] \to [(\varphi \to (\psi \to \sigma))],$$
   $$\vdash ((\varphi \to \psi) \to \varphi) \to \varphi.$$

5. Show

   $$\Gamma \vdash \varphi \Rightarrow \Gamma \cup \Delta \vdash \varphi,$$
   $$\Gamma \vdash \varphi; \ \Delta, \varphi \vdash \psi \Rightarrow \Gamma \cup \Delta \vdash \psi.$$

6. Give a recursive definition of the function Hyp which assigns to each derivation $\mathcal{D}$ its set of hypotheses $Hyp(\mathcal{D})$ (this is a bit stricter than the notion in Definition 2.4.2, since it is the smallest set of hypotheses, i.e. hypotheses without "garbage").

7. Analogous to the substitution operator for propositions we define a substitution operator for derivations. $\mathcal{D}[\varphi/p]$ is obtained by replacing each occurrence of p in each proposition in $\mathcal{D}$ by φ. Give a recursive definition of $\mathcal{D}[\varphi/p]$. Show that $\mathcal{D}[\varphi/p]$ is a derivation if $\mathcal{D}$ is one, and that $\Gamma \vdash \sigma \Rightarrow \Gamma[\varphi/p] \vdash \sigma[\varphi/p]$. Remark: for several purposes finer notions of substitution are required, but this one will do for us.

8. (*Substitution Theorem*) $\vdash (\varphi_1 \leftrightarrow \varphi_2) \rightarrow (\psi[\varphi_1/p] \leftrightarrow \psi[\varphi_2/p])$.
   Hint: use induction on ψ; the theorem will also follow from the Substitution Theorem for $\models$, once we have established the Completeness Theorem.

9. The *size*, $s(\mathcal{D})$, of a derivation is the number of proposition occurrences in $\mathcal{D}$. Give an inductive definition of $s(\mathcal{D})$. Show that one can prove properties of derivations by *induction on size*.

10. Give an inductive definition of the relation $\vdash$ (use the list of Lemma 2.4.3), and show that this relation coincides with the derived relation of Definition 2.4.2. Conclude that each Γ with $\Gamma \vdash \varphi$ contains a finite Δ, such that also $\Delta \vdash \varphi$.

11. Show

$$(a) \quad \vdash \top,$$
$$(b) \quad \vdash \varphi \Leftrightarrow \vdash \varphi \leftrightarrow \top,$$
$$(c) \quad \vdash \neg\varphi \Leftrightarrow \vdash \varphi \leftrightarrow \bot.$$

2.5 Completeness

In the present section we will show that "truth" and "derivability" coincide; to be precise: the relations "$\models$" and "$\vdash$" coincide. The easy part of the claim is: "derivability" implies "truth"; for derivability is established by the existence of a derivation. The latter motion is inductively defined, so we can prove the implication by induction on the derivation.

Lemma 2.5.1 (Soundness) $\Gamma \vdash \varphi \Rightarrow \Gamma \models \varphi$.

Proof Since, by Definition 2.4.2, $\Gamma \vdash \varphi$ iff there is a derivation $\mathcal{D}$ with all its hypotheses in Γ, it suffices to show: for each derivation $\mathcal{D}$ with conclusion φ and hypotheses in Γ we have $\Gamma \models \varphi$. We now use induction on $\mathcal{D}$.

(*basis*) If $\mathcal{D}$ has one element, then evidently $\varphi \in \Gamma$. The reader easily sees that $\Gamma \models \varphi$.

($\wedge$ I) Induction hypothesis: $\dfrac{\mathcal{D}}{\varphi}$ and $\dfrac{\mathcal{D}'}{\varphi'}$ are derivations and for each Γ, Γ' containing the hypotheses of $\mathcal{D}, \mathcal{D}', \Gamma \models \varphi, \Gamma' \models \varphi'$.

Now let Γ'' contain the hypotheses of $\dfrac{\begin{array}{cc} \mathcal{D} & \mathcal{D}' \\ \varphi & \varphi' \end{array}}{\varphi \wedge \varphi'}$.

Choosing Γ and Γ' to be precisely the set of hypotheses of $\mathcal{D}, \mathcal{D}'$, we see that $\Gamma'' \supseteq \Gamma \cup \Gamma'$.

So $\Gamma'' \models \varphi$ and $\Gamma'' \models \varphi'$. Let $\llbracket \psi \rrbracket_v = 1$ for all $\psi \in \Gamma''$, then $\llbracket \varphi \rrbracket_v = \llbracket \varphi' \rrbracket_v = 1$, hence $\llbracket \varphi \wedge \varphi' \rrbracket_v = 1$. This shows $\Gamma'' \models \varphi \wedge \varphi'$.

($\wedge$ E) Induction hypothesis: for any Γ containing the hypotheses of $\begin{smallmatrix}\mathcal{D}\\\varphi\wedge\psi\end{smallmatrix}$ we have $\Gamma \models \varphi \wedge \psi$. Consider a Γ containing all hypotheses of $\dfrac{\mathcal{D}}{\varphi\wedge\psi}$ and $\dfrac{\mathcal{D}}{\varphi\wedge\psi}$. It is left to

the reader to show $\Gamma \models \varphi$ and $\Gamma \models \psi$.

($\rightarrow$ I) Induction hypothesis: for any Γ containing all hypotheses of $\begin{smallmatrix}\varphi\\\mathcal{D}\\\psi\end{smallmatrix}$, $\Gamma \models \psi$.

Let Γ' contain all hypotheses of $\dfrac{\begin{smallmatrix}[\varphi]\\\mathcal{D}\\\psi\end{smallmatrix}}{\varphi\rightarrow\psi}$. Now $\Gamma' \cup \{\varphi\}$ contains all hypotheses of

$\begin{smallmatrix}\varphi\\\mathcal{D}\\\psi\end{smallmatrix}$, so if $\llbracket \varphi \rrbracket = 1$ and $\llbracket \chi \rrbracket = 1$ for all χ in Γ', then $\llbracket \psi \rrbracket = 1$. Therefore the truth table of $\rightarrow$ tells us that $\llbracket \varphi \rightarrow \psi \rrbracket = 1$ if all propositions in Γ' have value 1. Hence $\Gamma' \models \varphi \rightarrow \psi$.

($\rightarrow$ E) An exercise for the reader.

($\perp$) Induction hypothesis: for each Γ containing all hypotheses of $\begin{smallmatrix}\mathcal{D}\\\perp\end{smallmatrix}$, $\Gamma \models \perp$. Since $\llbracket \perp \rrbracket = 0$ for all valuations, there is no valuation such that $\llbracket \psi \rrbracket = 1$ for all $\psi \in \Gamma$. Let Γ' contain all hypotheses of $\dfrac{\begin{smallmatrix}\mathcal{D}\\\perp\end{smallmatrix}}{\varphi}$ and suppose that $\Gamma' \not\models \varphi$, then $\llbracket \psi \rrbracket = 1$

for all $\psi \in \Gamma'$ and $\llbracket \varphi \rrbracket = 0$ for some valuation. Since Γ' contains all hypotheses of the first derivation we have a contradiction.

(RAA) Induction hypothesis: for each Γ containing all hypotheses of $\begin{smallmatrix}\neg\varphi\\\mathcal{D}\\\perp\end{smallmatrix}$, we have $\Gamma \models \perp$. Let Γ' contain all hypotheses of $\dfrac{\begin{smallmatrix}[\neg\varphi]\\\mathcal{D}\\\perp\end{smallmatrix}}{\varphi}$ and suppose $\Gamma' \not\models \varphi$, then there

exists a valuation such that $\llbracket \psi \rrbracket = 1$ for all $\psi \in \Gamma'$ and $\llbracket \varphi \rrbracket = 0$, i.e. $\llbracket \neg\varphi \rrbracket = 1$. But $\Gamma'' = \Gamma' \cup \{\neg\varphi\}$ contains all hypotheses of the first derivation and $\llbracket \psi \rrbracket = 1$ for all $\psi \in \Gamma''$. This is impossible since $\Gamma'' \models \perp$. Hence $\Gamma' \models \varphi$. □

This lemma may not seem very impressive, but it enables us to show that some propositions are not theorems, simply by showing that they are not tautologies. Without this lemma that would have been a very awkward task. We would have to show that there is no derivation (without hypotheses) of the given proposition. In general this requires insight in the nature of derivations, something which is beyond us at the moment.

Examples $\not\vdash p_0$, $\not\vdash (\varphi \rightarrow \psi) \rightarrow \varphi \wedge \psi$.

In the first example take the constant 0 valuation. $[\![p_0]\!] = 0$, so $\not\models p_0$ and hence $\not\vdash p_0$. In the second example we are faced with a meta-proposition (a *schema*); strictly speaking it cannot be derivable (only *real* propositions can be). By $\vdash (\varphi \to \psi) \to \varphi \wedge \psi$ we mean that all propositions of that form (obtained by substituting real propositions for φ and ψ, if you like) are derivable. To refute it we need only one instance which is not derivable. Take $\varphi = \psi = p_0$.

In order to prove the converse of Lemma 2.5.1 we need a few new notions. The first one has an impressive history; it is the notion of *freedom from contradiction* or *consistency*. It was made the cornerstone of the foundations of mathematics by Hilbert.

Definition 2.5.2 A set Γ of propositions is *consistent* if $\Gamma \not\vdash \bot$.

In words: one cannot derive a contradiction from Γ. The consistency of Γ can be expressed in various other forms.

Lemma 2.5.3 *The following three conditions are equivalent*:

(i) Γ *is consistent*,
(ii) *For no* φ, $\Gamma \vdash \varphi$ *and* $\Gamma \vdash \neg\varphi$,
(iii) *There is at least one* φ *such that* $\Gamma \not\vdash \varphi$.

Proof Let us call Γ *inconsistent* if $\Gamma \vdash \bot$; then we can just as well prove the equivalence of

(iv) Γ is inconsistent,
(v)   There is a φ such that $\Gamma \vdash \varphi$ and $\Gamma \vdash \neg\varphi$,
(vi) $\Gamma \vdash \varphi$ for all φ.
(iv) $\Rightarrow$ (vi) Let $\Gamma \vdash \bot$, i.e. there is a derivation $\mathcal{D}$ with conclusion $\bot$ and hypotheses in Γ. By $(\bot)$ we can add one inference, $\bot \vdash \varphi$, to $\mathcal{D}$, so that $\Gamma \vdash \varphi$. This holds for all φ.
(vi) $\Rightarrow$ (v) Trivial.
(v) $\Rightarrow$ (iv) Let $\Gamma \vdash \varphi$ and $\Gamma \vdash \neg\varphi$. From the two associated derivations one obtains a derivation for $\Gamma \vdash \bot$ by $(\to E)$. $\square$

Clause (vi) tells us why inconsistent sets (theories) are devoid of mathematical interest. For, if everything is derivable, we cannot distinguish between "good" and "bad" propositions. Mathematics tries to find distinctions, not to blur them.

In mathematical practice one tries to establish consistency by exhibiting a model (think of the consistency of the negation of Euclid's fifth postulate and the non-euclidean geometries). In the context of propositional logic this means looking for a suitable valuation.

Lemma 2.5.4 *If there is a valuation such that* $[\![\psi]\!]_v = 1$ *for all* $\psi \in \Gamma$, *then* Γ *is consistent*.

Proof Suppose $\Gamma \vdash \bot$, then by Lemma 2.5.1 $\Gamma \models \bot$, so for any valuation v $[\![(\psi)]\!]_v = 1$ for all $\psi \in \Gamma \Rightarrow [\![\bot]\!]_v = 1$. Since $[\![\bot]\!]_v = 0$ for all valuations, there is no valuation with $[\![\psi]\!]_v = 1$ for all $\psi \in \Gamma$. Contradiction. Hence Γ is consistent. $\square$

Examples

1. $\{p_0, \neg p_1, p_1 \to p_2\}$ is consistent. A suitable valuation is one satisfying $[\![p_0]\!] = 1$, $[\![p_1]\!] = 0$.
2. $\{p_0, p_1, \ldots\}$ is consistent. Choose the constant 1 valuation.

Clause (v) of Lemma 2.5.3 tells us that $\Gamma \cup \{\varphi, \neg\varphi\}$ is inconsistent. Now, how could $\Gamma \cup \{\neg\varphi\}$ be inconsistent? It seems plausible to blame this on the derivability of φ. The following confirms this.

Lemma 2.5.5

(a) $\Gamma \cup \{\neg\varphi\}$ *is inconsistent* $\Rightarrow \Gamma \vdash \varphi$,
(b) $\Gamma \cup \{\varphi\}$ *is inconsistent* $\Rightarrow \Gamma \vdash \neg\varphi$.

Proof The assumptions of (a) and (b) yield the two derivations below: with conclusion $\bot$. By applying (RAA), and $(\to I)$, we obtain derivations with hypotheses in Γ, of φ, resp. $\neg\varphi$.

$$
\begin{array}{cc}
[\neg\varphi] & [\varphi] \\
\mathcal{D} & \mathcal{D}' \\
\dfrac{\bot}{\varphi} \ RAA & \dfrac{\bot}{\neg\varphi} \to I
\end{array}
$$
$\square$

Definition 2.5.6 A set Γ is *maximally consistent* iff

(a) Γ is consistent,
(b) $\Gamma \subseteq \Gamma'$ and Γ' consistent $\Rightarrow \Gamma = \Gamma'$.

Remark One could replace (b) by (b'): if Γ is a proper subset of Γ', then Γ' is inconsistent. That is, by just throwing in one extra proposition, the set becomes inconsistent.

Maximally consistent sets play an important role in logic. We will show that there are lots of them.

Here is one example: $\Gamma = \{\varphi | [\![\varphi]\!] = 1\}$ for a fixed valuation. By Lemma 2.5.4 Γ is consistent. Consider a consistent set Γ' such that $\Gamma \subseteq \Gamma'$. Now let $\psi \in \Gamma'$ and suppose $[\![\psi]\!] = 0$, then $[\![\neg\psi]\!] = 1$, and so $\neg\psi \in \Gamma$.

But since $\Gamma \subseteq \Gamma'$ this implies that Γ' is inconsistent. Contradiction. Therefore $[\![\psi]\!] = 1$ for all $\psi \in \Gamma'$, so by definition $\Gamma = \Gamma'$. Moreover, from the proof of Lemma 2.5.11 it follows that this basically is the only kind of maximally consistent set we may expect.

The following fundamental lemma is proved directly. The reader may recognize in it an analogue of the maximal ideal existence lemma from ring theory (or the Boolean prime ideal theorem), which is usually proved by an application of Zorn's lemma.

Lemma 2.5.7 *Each consistent set Γ is contained in a maximally consistent set Γ^*.*

Proof There are countably many propositions, so suppose we have a list $\varphi_0, \varphi_1, \varphi_2, \ldots$ of all propositions (cf. Exercise 5). We define a non-decreasing sequence of sets Γ_i such that the union is maximally consistent.

$$\Gamma_0 = \Gamma,$$
$$\Gamma_{n+1} = \begin{cases} \Gamma_n \cup \{\varphi_n\} & \text{if } \Gamma_n \cup \{\varphi_n\} \text{ is consistent,} \\ \Gamma_n & \text{else,} \end{cases}$$
$$\Gamma^* = \bigcup \{\Gamma_n \mid n \geq 0\}.$$

(a) Γ_n is consistent for all n.
   Immediate, by induction on n.
(b) Γ^* is consistent.
   Suppose $\Gamma^* \vdash \perp$ then, by the definition of $\perp$ there is derivation $\mathcal{D}$ of $\perp$ with hypotheses in Γ^*; $\mathcal{D}$ has finitely many hypotheses $\psi_0, \ldots, \psi_k$. Since $\Gamma^* = \bigcup \{\Gamma_n | n \geq 0\}$, we have for each $i \leq k$ $\psi_i \in \Gamma_{n_i}$ for some n_i. Let n be $\max\{n_i | i \leq k\}$, then $\psi_0, \ldots, \psi_k \in \Gamma_n$ and hence $\Gamma_n \vdash \perp$. But Γ_n is consistent. Contradiction.
(c) Γ^* is maximally consistent. Let $\Gamma^* \subseteq \Delta$ and Δ consistent. If $\psi \in \Delta$, then $\psi = \varphi_m$ for some m. Since $\Gamma_m \subseteq \Gamma^* \subseteq \Delta$ and Δ is consistent, $\Gamma_m \cup \{\varphi_m\}$ is consistent. Therefore $\Gamma_{m+1} = \Gamma_m \cup \{\varphi_m\}$, i.e. $\varphi_m \in \Gamma_{m+1} \subseteq \Gamma^*$. This shows $\Gamma^* = \Delta$.

Lemma 2.5.8 *If Γ is maximally consistent, then Γ is closed under derivability (i.e. $\Gamma \vdash \varphi \Rightarrow \varphi \in \Gamma$).*

Proof Let $\Gamma \vdash \varphi$ and suppose $\varphi \notin \Gamma$. Then $\Gamma \cup \{\varphi\}$ must be inconsistent. Hence $\Gamma \vdash \neg\varphi$, so Γ is inconsistent. Contradiction.   □

Lemma 2.5.9 *Let Γ be maximally consistent; then*

$$\text{for all } \varphi \quad \text{either } \varphi \in \Gamma, \text{ or } \neg\varphi \in \Gamma,$$
$$\text{for all } \varphi, \psi \quad \varphi \to \psi \in \Gamma \Leftrightarrow (\varphi \in \Gamma \Rightarrow \psi \in \Gamma).$$

Proof (a) We know that not both φ and $\neg\varphi$ can belong to Γ. Consider $\Gamma' = \Gamma \cup \{\varphi\}$. If Γ' is inconsistent, then, by Lemmas 2.5.5, 2.5.8, $\neg\varphi \in \Gamma$. If Γ' is consistent, then $\varphi \in \Gamma$ by the maximality of Γ.

(b) Let $\varphi \to \psi \in \Gamma$ and $\varphi \in \Gamma$. To show: $\psi \in \Gamma$. Since $\varphi, \varphi \to \psi \in \Gamma$ and since Γ is closed under derivability (Lemma 2.5.8), we get $\psi \in \Gamma$ by $\to E$.

Conversely: let $\varphi \in \Gamma \Rightarrow \psi \in \Gamma$. If $\varphi \in \Gamma$ then obviously $\Gamma \vdash \psi$, so $\Gamma \vdash \varphi \to \psi$. If $\varphi \notin \Gamma$, then $\neg\varphi \in \Gamma$, and hence $\Gamma \vdash \neg\varphi$. Therefore $\Gamma \vdash \varphi \to \psi$. □

Note that we automatically get the following.

Corollary 2.5.10 *If Γ is maximally consistent, then $\varphi \in \Gamma \Leftrightarrow \neg\varphi \notin \Gamma$, and $\neg\varphi \in \Gamma \Leftrightarrow \varphi \notin \Gamma$.*

Lemma 2.5.11 *If Γ is consistent, then there exists a valuation such that $[\![\psi]\!] = 1$ for all $\psi \in \Gamma$.*

Proof (a) By Lemma 2.5.7 Γ is contained in a maximally consistent Γ^*.

(b) Define $v(p_i) = \begin{cases} 1 \text{ if } p_i \in \Gamma^* \\ 0 \text{ else} \end{cases}$ and extend v to the valuation $[\![\ ]\!]_v$.

Claim: $[\![\varphi]\!] = 1 \Leftrightarrow \varphi \in \Gamma^*$. Use induction on φ.

1. For atomic φ the claim holds by definition.
2. $\varphi = \psi \wedge \sigma$. $[\![\varphi]\!]_v = 1 \Leftrightarrow [\![\psi]\!]_v = [\![\sigma]\!]_v = 1 \Leftrightarrow$ (induction hypothesis) $\psi, \sigma \in \Gamma^*$ and so $\varphi \in \Gamma^*$. Conversely $\psi \wedge \sigma \in \Gamma^* \Leftrightarrow \psi, \sigma \in \Gamma^*$ (Lemma 2.5.8). The rest follows from the induction hypothesis.
3. $\varphi = \psi \to \sigma$. $[\![\psi \to \sigma]\!]_v = 0 \Leftrightarrow [\![\psi]\!]_v = 1$ and $[\![\sigma]\!]_v = 0 \Leftrightarrow$ (induction hypothesis) $\psi \in \Gamma^*$ and $\sigma \notin \Gamma^* \Leftrightarrow \psi \to \sigma \notin \Gamma^*$ (by Lemma 2.5.9).

(c) Since $\Gamma \subseteq \Gamma^*$ we have $[\![\psi]\!]_v = 1$ for all $\psi \in \Gamma$. □

Corollary 2.5.12 *$\Gamma \nvdash \varphi \Leftrightarrow$ there is a valuation such that $[\![\psi]\!] = 1$ for all $\psi \in \Gamma$ and $[\![\varphi]\!] = 0$.*

Proof $\Gamma \nvdash \varphi \Leftrightarrow \Gamma \cup \{\neg\varphi\}$ consistent $\Leftrightarrow$ there is a valuation such that $[\![\psi]\!] = 1$ for all $\psi \in \Gamma \cup \{\neg\varphi\}$, or $[\![\psi]\!] = 1$ for all $\psi \in \Gamma$ and $[\![\varphi]\!] = 0$. □

Theorem 2.5.13 (Completeness Theorem) *$\Gamma \vdash \varphi \Leftrightarrow \Gamma \models \varphi$.*

Proof $\Gamma \nvdash \varphi \Rightarrow \Gamma \nvDash \varphi$ by Corollary 2.5.12. The converse holds by Lemma 2.5.1. □

In particular we have $\vdash \varphi \Leftrightarrow \models \varphi$, so the set of theorems is exactly the set to tautologies.

The Completeness Theorem tells us that the tedious task of making derivations can be replaced by the (equally tedious, but automatic) task of checking tautologies. This simplifies, at least in theory, the search for theorems considerably; for derivations one has to be (moderately) clever, for truth tables one has to possess perseverance.

For logical theories one sometimes considers another notion of completeness: a set Γ is called *complete* if for each φ, either $\Gamma \vdash \varphi$, or $\Gamma \vdash \neg\varphi$. This notion is closely related to "maximally consistent". From Exercise 6 it follows that

$\text{Cons}(\Gamma) = \{\sigma \mid \Gamma \vdash \sigma\}$ (*the set of consequences of* Γ) is maximally consistent if Γ is a complete set. The converse also holds (cf. Exercise 10). Propositional logic itself (i.e. the case $\Gamma = \emptyset$) is not complete in this sense, e.g. $\nvdash p_0$ and $\nvdash \neg p_0$.

There is another important notion which is traditionally considered in logic: that of *decidability*. Propositional logic is decidable in the following sense: there is an effective procedure to check the derivability of propositions φ. Stated otherwise: there is an algorithm that for each φ tests if $\vdash \varphi$.

The algorithm is simple: write down the complete truth table for φ and check if the last column contains only 1's. If so, then $\models \varphi$ and, by the Completeness Theorem, $\vdash \varphi$. If not, then $\not\models \varphi$ and hence $\nvdash \varphi$. This is certainly not the best possible algorithm, one can find more economical ones. There are also algorithms that give more information, e.g. they not only test $\vdash \varphi$, but also yield a derivation, if one exists. Such algorithms require, however, a deeper analysis of derivations, which falls outside the scope of this book.

There is one aspect of the Completeness Theorem that we want to discuss now. It does not come as a surprise that truth follows from derivability. After all we start with a combinatorial notion, defined inductively, and we end up with "being true for all valuations". A simple inductive proof does the trick.

For the converse the situation is totally different. By definition $\Gamma \models \varphi$ means that $[\![\varphi]\!]_v = 1$ for all valuations v that make all propositions of Γ true. So we know something about the behavior of *all* valuations with respect to Γ and φ. Can we hope to extract from such infinitely many set theoretical facts the finite, concrete information needed to build a derivation for $\Gamma \vdash \varphi$? Evidently the available facts do not give us much to go on. Let us therefore simplify matters a bit by cutting down the Γ; after all we use only finitely many formulas of Γ in a derivation, so let us suppose that those formulas $\psi_1, \ldots, \psi_n$ are given. Now we can hope for more success, since only finitely many atoms are involved, and hence we can consider a finite "part" of the infinitely many valuations that play a role. That is, only the restrictions of the valuations to the set of atoms occurring in $\psi_1, \ldots, \psi_n, \varphi$ are relevant. Let us simplify the problem one more step. We know that $\psi_1, \ldots, \psi_n \vdash \varphi$ ($\psi_1, \ldots, \psi_n \models \varphi$) can be replaced by $\vdash \psi_1 \wedge \cdots \wedge \psi_n \to \varphi (\models \psi_1 \wedge \cdots \wedge \psi_n \to \varphi)$, on the ground of the derivation rules (the definition of valuation). So we ask ourselves: given the truth table for a tautology σ, can we effectively find a derivation for σ? This question is not answered by the Completeness Theorem, since our proof of it is not effective (at least not prima facie so). It has been answered positively, e.g. by Post, Bernays and Kalmar (cf. Kleene 1952, IV, §29) and it is easily treated by means of Gentzen techniques, or semantic tableaux. We will just sketch a method of proof: we can effectively find a conjunctive normal form σ^* for σ such that $\vdash \sigma \leftrightarrow \sigma^*$. It is easily shown that σ^* is a tautology iff each conjunct contains an atom and its negation, or $\neg \bot$, and glue it all together to obtain a derivation of σ^*, which immediately yields a derivation of σ.

Exercises

1. Check which of the following sets are consistent:
    (a) $\{\neg p_1 \wedge p_2 \to p_0, p_1 \to (\neg p_1 \to p_2), p_0 \leftrightarrow \neg p_2\}$,

(b) $\{p_0 \to p_1, p_1 \to p_2, p_2 \to p_3, p_3 \to \neg p_0\}$,

(c) $\{p_0 \to p_1, p_0 \land p_2 \to p_1 \land p_3, p_0 \land p_2 \land p_4 \to p_1 \land p_3 \land p_5, \ldots\}$.

2. Show that the following are equivalent:

   (a) $\{\varphi_1, \ldots, \varphi_n\}$ is consistent.

   (b) $\nvdash \neg(\varphi_1 \land \varphi_2 \land \cdots \land \varphi_n)$.

   (c) $\nvdash \varphi_1 \land \varphi_2 \land \cdots \land \varphi_{n-1} \to \neg\varphi_n$.

3. φ is *independent* from Γ if $\Gamma \nvdash \varphi$ and $\Gamma \nvdash \neg\varphi$. Show that: $p_1 \to p_2$ is independent from $\{p_1 \leftrightarrow p_0 \land \neg p_2, p_2 \to p_0\}$.

4. A set Γ is *independent* if for each $\varphi \in \Gamma$ $\Gamma - \{\varphi\} \nvdash \varphi$.

   (a) Show that each finite set Γ has an independent subset Δ such that $\Delta \vdash \varphi$ for all $\varphi \in \Gamma$.

   (b) Let $\Gamma = \{\varphi_0, \varphi_1, \varphi_2, \ldots\}$. Find an equivalent set $\Gamma' = \{\psi_0, \psi_1, \ldots\}$ (i.e. $\Gamma \vdash \psi_i$ and $\Gamma' \vdash \varphi_i$ for all i) such that $\vdash \psi_{n+1} \to \psi_n$, but $\nvdash \psi_n \to \psi_{n+1}$. Note that Γ' may be finite.

   (c) Consider an infinite Γ' as in (b). Define $\sigma_0 = \psi_0, \sigma_{n+1} = \psi_n \to \psi_{n+1}$. Show that $\Delta = \{\sigma_0, \sigma_1, \sigma_2, \ldots\}$ is independent and equivalent to Γ'.

   (d) Show that each set Γ is equivalent to an independent set Δ.

   (e) Show that Δ need not be a subset of Γ (consider $\{p_0, p_0 \land p_1, p_0 \land p_1 \land p_2, \ldots\}$).

5. Find an effective way of enumerating all propositions (hint: consider sets Γ_n of all propositions of rank $\leq n$ with atoms from $p_0, \ldots, p_n$).

6. Show that a consistent set Γ is maximally consistent if either $\varphi \in \Gamma$ or $\neg\varphi \in \Gamma$ for all φ.

7. Show that $\{p_0, p_1, p_2, \ldots, p_n, \ldots\}$ is complete.

8. (*Compactness Theorem*). Show: there is a v such that $[\![\psi]\!]_v = 1$ for all $\psi \in \Gamma \Leftrightarrow$ for each finite subset $\Delta \subseteq \Gamma$ there is a v such that $[\![\sigma]\!]_v = 1$ for all $\sigma \in \Delta$.

   Formulated in terms of Exercise 13 of 2.3: $[\![\Gamma]\!] \neq \emptyset$ if $[\![\Delta]\!] \neq \emptyset$ for all finite $\Delta \subseteq \Gamma$.

9. Consider an infinite set $\{\varphi_1, \varphi_2, \varphi_3, \ldots\}$. If for each valuation there is an n such that $[\![\varphi_n]\!] = 1$, then there is an m such that $\vdash \varphi_1 \lor \cdots \lor \varphi_m$. (Hint: consider the negations $\neg\varphi_1, \neg\varphi_2 \ldots$ and apply Exercise 8.)

10. Show: $\mathrm{Cons}(\Gamma) = \{\sigma \mid \Gamma \vdash \sigma\}$ is maximally consistent $\Leftrightarrow \Gamma$ is complete.

11. Show: Γ is maximally consistent $\Leftrightarrow$ there is a unique valuation such that $[\![\psi]\!] = 1$ for all $\psi \in \Gamma$, where Γ is a theory, i.e. Γ is closed under $\vdash$ ($\Gamma \vdash \sigma \Rightarrow \sigma \in \Gamma$).

12. Let φ be a proposition containing the atom p. For convenience we write $\varphi(\sigma)$ for $\varphi[\sigma/p]$.

    As before we abbreviate $\neg\bot$ by $\top$.

    Show:

    (i) $\varphi(\top) \vdash \varphi(\top) \leftrightarrow \top$ and $\varphi(\top) \vdash \varphi(\varphi(\top))$.

    (ii) $\neg\varphi(\top) \vdash \varphi(\top) \leftrightarrow \bot$,

    $\varphi(p), \neg\varphi(\top) \vdash p \leftrightarrow \bot$,

    $\varphi(p), \neg\varphi(\top) \vdash \varphi(\varphi(\top))$.

    (iii) $\varphi(p) \vdash \varphi(\varphi(\top))$.

13. If the atoms p and q do not occur in ψ and φ respectively, then

$$\models \varphi(p) \to \psi \Rightarrow \models \varphi(\sigma) \to \psi \text{ for all } \sigma,$$
$$\models \varphi \to \psi(q) \Rightarrow \models \varphi \to \psi(\sigma) \text{ for all } \sigma.$$

14. Let $\vdash \varphi \to \psi$. We call σ an *interpolant* if $\vdash \varphi \to \sigma$ and $\vdash \sigma \to \psi$, and more-over σ contains only atoms common to φ and ψ. Consider $\varphi(p, r), \psi(r, q)$ with all atoms displayed. Show that $\varphi(\varphi(\top, r), r)$ is an interpolant (use Exercises 12, 13).

15. Prove the general *interpolation theorem* (Craig): For any φ, ψ with $\vdash \varphi \to \psi$ there exists an interpolant (iterate the procedure of Exercise 13).

2.6 The Missing Connectives

The language of Sect. 2.4 contained only the connectives $\wedge$, $\to$ and $\bot$. We already know that, from the semantical point of view, this language is sufficiently rich, i.e. the missing connectives can be defined. As a matter of fact we have already used the negation as a defined notion in the preceding sections.

It is a matter of sound mathematical practice to introduce new notions if their use simplifies our labor, and if they codify informal existing practice. This, clearly, is a reason for introducing $\neg$, $\leftrightarrow$ and $\vee$.

Now there are two ways to proceed: one can introduce the new connectives as abbreviations (of complicated propositions), or one can enrich the language by actually adding the connectives to the alphabet, and providing rules of derivation.

The first procedure was adopted above; it is completely harmless, e.g. each time one reads $\varphi \leftrightarrow \psi$, one has to replace it by $(\varphi \to \psi) \wedge (\psi \to \varphi)$. So it is nothing but a shorthand, introduced for convenience. The second procedure is of a more theoretical nature. The language is enriched and the set of derivations is enlarged. As a consequence one has to review the theoretical results (such as the Completeness Theorem) obtained for the simpler language.

We will adopt the first procedure and also outline the second approach.

Definition 2.6.1

$$\varphi \vee \psi := \neg(\neg\varphi \wedge \neg\psi),$$
$$\neg\varphi := \varphi \to \bot,$$
$$\varphi \leftrightarrow \psi := (\varphi \to \psi) \wedge (\psi \to \varphi).$$

N.B. This means that the above expressions are *not* part of the language, but abbreviations for certain propositions.

The properties of $\vee$, $\neg$ and $\leftrightarrow$ are given in the following lemma.

Lemma 2.6.2

(i) $\varphi \vdash \varphi \vee \psi, \psi \vdash \varphi \vee \psi,$

(ii) $\Gamma, \varphi \vdash \sigma$ and $\Gamma, \psi \vdash \sigma \Rightarrow \Gamma, \varphi \vee \psi \vdash \sigma,$

(iii) $\varphi, \neg\varphi \vdash \bot,$

(iv) $\Gamma, \varphi \vdash \bot \Rightarrow \Gamma \vdash \neg\varphi,$

(v) $\varphi \leftrightarrow \psi, \varphi \vdash \psi$ and $\varphi \leftrightarrow \psi, \psi \vdash \varphi,$

(vi) $\Gamma, \varphi \vdash \psi$ and $\Gamma, \psi \vdash \varphi \Rightarrow \Gamma \vdash \varphi \leftrightarrow \psi.$

Proof The only non-trivial part is (*ii*). We exhibit a derivation of σ from Γ and $\varphi \vee \psi$ (i.e. $\neg(\neg\varphi \wedge \neg\psi)$), given derivations $\mathcal{D}_1$ and $\mathcal{D}_2$ of $\Gamma, \varphi \vdash \sigma$ and $\Gamma, \psi \vdash \sigma$.

$$
\begin{array}{c}
[\varphi]^1 \qquad\qquad\qquad [\psi]^2 \\[4pt]
\mathcal{D}_1 \qquad\qquad\qquad \mathcal{D}_2 \\[4pt]
\dfrac{\sigma \quad [\neg\sigma]^3}{} \to E \qquad\quad \dfrac{\sigma \quad [\neg\sigma]^3}{} \to E \\
\dfrac{\bot}{\neg\varphi} \to I_1 \qquad\qquad \dfrac{\bot}{\neg\psi} \to I_2 \\[6pt]
\dfrac{\neg\varphi \wedge \neg\psi \qquad\qquad\qquad\qquad \neg(\neg\varphi \wedge \neg\psi)}{} \to E \\
\dfrac{\bot}{\sigma} \, RAA_3
\end{array}
$$

The remaining cases are left to the reader.   □

Note that (i) and (ii) read as introduction and elimination rules for $\vee$, (iii) and (iv) as ditto for $\neg$, (vi) and (v) as ditto for $\leftrightarrow$.

They legalize the following shortcuts in derivations:

$$
\dfrac{\varphi}{\varphi \vee \psi} \vee I \qquad \dfrac{\psi}{\varphi \vee \psi} \vee I \qquad
\begin{array}{ccc}
[\varphi] & [\psi] \\
\vdots & \vdots \\
\dfrac{\varphi \vee \psi \qquad \sigma \qquad \sigma}{\sigma} \vee E
\end{array}
$$

$$
\begin{array}{c}
[\varphi] \\
\vdots \\
\dfrac{\bot}{\neg\varphi} \neg I
\end{array}
\qquad\qquad
\dfrac{\varphi \quad \neg\varphi}{\bot} \neg E
$$

$$\frac{[\varphi] \qquad [\psi]}{}$$

$$\begin{array}{cc} \vdots & \vdots \\ \psi & \varphi \\ \hline \varphi \leftrightarrow \psi \end{array} \leftrightarrow I \qquad\qquad \frac{\varphi \quad \varphi \leftrightarrow \psi}{\psi} \qquad \frac{\psi \quad \varphi \leftrightarrow \psi}{\varphi} \leftrightarrow E$$

Consider for example an application of $\vee E$

$$\begin{array}{ccc} & [\varphi] & [\psi] \\ \mathcal{D}_0 & \mathcal{D}_1 & \mathcal{D}_2 \\ \varphi \vee \psi & \sigma & \sigma \\ \hline & \sigma & \end{array} \vee E$$

This is a mere shorthand for

$$\begin{array}{cc} [\varphi]^1 & [\psi]^2 \\ \mathcal{D}_1 & \mathcal{D}_2 \\ \sigma \quad [\neg\sigma]^3 & \sigma \quad [\neg\sigma]^3 \\ \dfrac{\bot}{\neg\varphi} 1 & \dfrac{\bot}{\neg\psi} 2 \end{array}$$

$$\mathcal{D}_0$$

$$\frac{\neg(\neg\varphi \wedge \neg\psi) \qquad\qquad \neg\varphi \wedge \neg\psi}{\dfrac{\bot}{\sigma} 3} 1$$

The reader is urged to use the above shortcuts in actual derivations, whenever convenient. As a rule, only $\vee I$ and $\vee E$ are of importance; the reader has of course recognized the rules for $\neg$ and $\leftrightarrow$ as slightly eccentric applications of familiar rules.

Examples $\vdash (\varphi \wedge \psi) \vee \sigma \leftrightarrow (\varphi \vee \sigma) \wedge (\psi \vee \sigma)$.

$$\frac{\dfrac{[\varphi \wedge \psi]^1}{\varphi}}{(\varphi \wedge \psi) \vee \sigma \quad \varphi \vee \sigma \quad [\sigma]^1 \quad \varphi \vee \sigma} 1 \qquad \frac{\dfrac{[\varphi \wedge \psi]^2}{\psi}}{(\varphi \wedge \psi) \vee \sigma \quad \psi \vee \sigma \quad [\sigma]^2 \quad \psi \vee \sigma} 2$$

$$\frac{\varphi \vee \sigma \qquad\qquad\qquad\qquad \psi \vee \sigma}{(\varphi \vee \sigma) \wedge (\psi \vee \sigma)}$$

$$(2.2)$$

Conversely

$$
\cfrac{
\cfrac{
(\varphi \vee \sigma) \wedge (\psi \vee \sigma)
\qquad
\cfrac{
(\varphi \vee \sigma) \wedge (\psi \vee \sigma)
}{
\psi \vee \sigma
}
\qquad
\cfrac{
\cfrac{[\varphi]^2 \quad [\psi]^1}{\varphi \wedge \psi}
}{
(\varphi \wedge \psi) \vee \sigma
}
\qquad
\cfrac{[\sigma]^1}{(\varphi \wedge \psi) \vee \sigma}
}{
(\varphi \wedge \psi) \vee \sigma
}\ 1
\qquad
\cfrac{[\sigma]^2}{(\varphi \wedge \psi) \vee \sigma}
}{
(\varphi \wedge \psi) \vee \sigma
}\ 2
$$

$$\varphi \vee \sigma$$

(2.3)

Combining (2.2) and (2.3) we get one derivation:

$$
\cfrac{
\cfrac{
\begin{matrix}[(\varphi \wedge \psi) \vee \sigma] \\ \mathcal{D}\end{matrix}
}{
(\varphi \vee \sigma) \wedge (\psi \vee \sigma)
}
\qquad
\cfrac{
\begin{matrix}[(\varphi \vee \sigma) \wedge (\psi \vee \sigma)] \\ \mathcal{D}'\end{matrix}
}{
(\varphi \wedge \psi) \vee \sigma
}
}{
(\varphi \wedge \psi) \vee \sigma \leftrightarrow (\varphi \vee \sigma) \wedge (\psi \vee \sigma)
}\ \leftrightarrow I
$$

$\vdash \varphi \vee \neg\varphi$

$$
\cfrac{
\cfrac{
\cfrac{
\cfrac{
\cfrac{\dfrac{[\varphi]^1}{\varphi \vee \neg\varphi}\ \vee I \qquad [\neg(\varphi \vee \neg\varphi)]^2}{\bot}\ \to E
}{\neg\varphi}\ \to I_1
}{\varphi \vee \neg\varphi}\ \vee I
\qquad [\neg(\varphi \vee \neg\varphi)]^2
}{\bot}\ \to E
}{\varphi \vee \neg\varphi}\ RAA_2
$$

$\vdash (\varphi \to \psi) \vee (\psi \to \varphi)$

$$
\cfrac{
\cfrac{
\cfrac{
\cfrac{
\cfrac{\dfrac{\dfrac{[\varphi]^1}{\psi \to \varphi}\ \to I_1}{(\varphi \to \psi) \vee (\psi \to \varphi)}\ \vee I \qquad [\neg((\varphi \to \psi) \vee (\psi \to \varphi))]^2}{\dfrac{\bot}{\psi}}\ \to E \quad \bot
}{\varphi \to \psi}\ \to I_1
}{(\varphi \to \psi) \vee (\psi \to \varphi)}\ \vee I
\qquad [\neg((\varphi \to \psi) \vee (\psi \to \varphi))]^2
}{\bot}\ \to E
}{(\varphi \to \psi) \vee (\psi \to \varphi)}\ RAA_2
$$

$\vdash \neg(\varphi \wedge \psi) \to \neg\varphi \vee \neg\psi$

$$
\cfrac{
\cfrac{
[\neg(\varphi \wedge \psi)]
\qquad
\cfrac{
\cfrac{[\neg(\neg\varphi \vee \neg\psi)] \quad \cfrac{[\neg\varphi]}{\neg\varphi \vee \neg\psi}}{\cfrac{\bot}{\varphi}}
\qquad
\cfrac{[\neg(\neg\varphi \vee \neg\psi)] \quad \cfrac{[\neg\psi]}{\neg\varphi \vee \neg\psi}}{\cfrac{\bot}{\psi}}
}{\varphi \wedge \psi}
}{\cfrac{\bot}{\neg\varphi \vee \neg\psi}}
}{\neg(\varphi \wedge \psi) \to \neg\varphi \vee \neg\psi}
$$

We now give a sketch of the second approach. We add $\vee$, $\neg$ and $\leftrightarrow$ to the language, and extend the set of propositions correspondingly. Next we add the rules for $\vee$, $\neg$ and $\leftrightarrow$ listed above to our stock of derivation rules. To be precise we should now also introduce a new derivability sign. However, we will stick to the trusted $\vdash$ in the expectation that the reader will remember that now we are making derivations in a larger system. The following holds.

Theorem 2.6.3

$$\vdash \varphi \vee \psi \leftrightarrow \neg(\neg\varphi \wedge \neg\psi).$$
$$\vdash \neg\varphi \leftrightarrow (\varphi \to \bot).$$
$$\vdash (\varphi \leftrightarrow \psi) \leftrightarrow (\varphi \to \psi) \wedge (\psi \to \varphi).$$

Proof Observe that by Lemma 2.6.2 the defined and the primitive (real) connectives obey exactly the same derivability relations (derivation rules, if you wish). This leads immediately to the desired result. Let us give one example.

$\varphi \vdash \neg(\neg\varphi \wedge \neg\psi)$ and $\psi \vdash \neg(\neg\varphi \wedge \neg\psi)$ (2.6.2 (i)), so by $\vee E$ we get

$$\varphi \vee \psi \vdash \neg(\neg\varphi \wedge \neg\psi) \dots \quad (1)$$

Conversely $\varphi \vdash \varphi \vee \psi$ and $\psi \vdash \varphi \vee \psi$ (by $\vee I$), hence by 2.6.2 (ii)

$$\neg(\neg\varphi \wedge \neg\psi) \vdash \varphi \vee \psi \dots \quad (2)$$

Apply $\leftrightarrow I$, to (1) and (2), then $\vdash \varphi \vee \psi \leftrightarrow \neg(\neg\varphi \wedge \neg\psi)$. The rest is left to the reader. $\square$

For more results the reader is directed to the exercises.

The rules for $\vee$, $\leftrightarrow$, and $\neg$ indeed capture the intuitive meaning of those connectives. Let us consider disjunction: $(\vee I)$. If we know φ then we certainly know $\varphi \vee \psi$ (we even know exactly which disjunct). The rule $(\vee E)$ captures the idea of "proof by cases": if we know $\varphi \vee \psi$ and in each of both cases we can conclude σ, then we may outright conclude σ. Disjunction intuitively calls for a decision: which of the two disjuncts is given or may be assumed? This constructive streak of $\vee$ is crudely but conveniently blotted out by the identification of $\varphi \vee \psi$ and $\neg(\neg\varphi \wedge \neg\psi)$. The latter only tells us that φ and ψ cannot both be wrong, but not which one is right. For more information on this matter of constructiveness, which plays a role in demarcating the borderline between two-valued classical logic and effective intuitionistic logic, the reader is referred to Chap. 6.

Note that with $\vee$ as a primitive connective some theorems become harder to prove. For example, $\vdash \neg(\neg\neg\varphi \wedge \neg\varphi)$ is trivial, but $\vdash \varphi \vee \neg\varphi$ is not. The following rule of thumb may be useful: going from non-effective (or no) premises to an effective conclusion calls for an application of *RAA*.

Exercises

1. Show $\vdash \varphi \vee \psi \rightarrow \psi \vee \varphi, \vdash \varphi \vee \varphi \leftrightarrow \varphi$.
2. Consider the full language $\mathcal{L}$ with the connectives $\wedge, \rightarrow, \perp, \leftrightarrow \vee$ and the restricted language $\mathcal{L}'$ with connectives $\wedge, \rightarrow, \perp$. Using the appropriate derivation rules we get the derivability notions $\vdash$ and $\vdash'$. We define an obvious translation from $\mathcal{L}$ into $\mathcal{L}'$:

$$\varphi^+ := \varphi \quad \text{for atomic } \varphi$$
$$(\varphi \square \psi)^+ := \varphi^+ \square \psi^+ \quad \text{for } \square = \wedge, \rightarrow,$$
$$(\varphi \vee \psi)^+ := \neg(\neg\varphi^+ \wedge \neg\varphi^+), \quad \text{where } \neg \text{ is an abbreviation,}$$
$$(\varphi \leftrightarrow \psi)^+ := (\varphi^+ \rightarrow \psi^+) \wedge (\psi^+ \rightarrow \varphi^+),$$
$$(\neg\varphi)^+ := \varphi^+ \rightarrow \perp.$$

Show
   (i) $\vdash \varphi \leftrightarrow \varphi^+$,
   (ii) $\vdash \varphi \Leftrightarrow \vdash' \varphi^+$,
   (iii) $\varphi^+ = \varphi$ for $\varphi \in \mathcal{L}'$.
   (iv) Show that the full logic is *conservative* over the restricted logic, i.e. for $\varphi \in \mathcal{L}' \vdash \varphi \Leftrightarrow \vdash' \varphi$.
3. Show that the Completeness Theorem holds for the full logic. Hint: use Exercise 2.
4. Show
   (a) $\vdash \top \vee \perp$.
   (b) $\vdash (\varphi \leftrightarrow \top) \vee (\varphi \leftrightarrow \perp)$.
   (c) $\vdash \varphi \leftrightarrow (\varphi \leftrightarrow \top)$.
5. Show $\vdash (\varphi \vee \psi) \leftrightarrow ((\varphi \rightarrow \psi) \rightarrow \psi)$.
6. Show
   (a) Γ is complete $\Leftrightarrow (\Gamma \vdash \varphi \vee \psi \Leftrightarrow \Gamma \vdash \varphi$ or $\Gamma \vdash \psi$, for all $\varphi, \psi)$,
   (b) Γ is maximally consistent $\Leftrightarrow \Gamma$ is a consistent theory and for all φ, ψ $(\varphi \vee \psi \in \Gamma \Leftrightarrow \varphi \in \Gamma$ or $\psi \in \Gamma)$.
7. Show in the system with $\vee$ as a primitive connective

$$\vdash (\varphi \rightarrow \psi) \leftrightarrow (\neg\varphi \vee \psi),$$
$$\vdash (\varphi \rightarrow \psi) \vee (\psi \rightarrow \varphi).$$

Gothic Alphabet

a. Aa Aa   b. Bb Bb   c. Cc Cr

d. Dd Dd   e. Ee Ee   f. Ff Ff

g. Gg Gg   h. Hh Hh   i. Ii Ii

k. Kk Kk   l. Ll Ll   m. Mm Mm

n. Nn Nn   o. Oo Oo   p. Pp Pp

q. Qq Qq   r. Rr Rr   s. Ss Ss

t. Tt Tt   u. Uu Uu   v. Vv Vv

w. Ww Ww   x. Xx Xx   y. Yy Yy

z. Zz Zz

Chapter 3
Predicate Logic

3.1 Quantifiers

In propositional logic we used large chunks of mathematical language, namely those parts that can have a truth value. Unfortunately this use of language is patently insufficient for mathematical practice. A simple argument, such as "all squares are positive, 9 is a square, therefore 9 is positive" cannot be dealt with. From the propositional point of view the above sentence is of the form $\varphi \wedge \psi \to \sigma$, and there is no reason why this sentence should be true, although we obviously accept it as true. The moral is that we have to extend the language, in such a way as to be able to discuss objects and relations. In particular we wish to introduce means to talk about *all* objects of the domain of discourse, e.g. we want to allow statements of the form "all even numbers are a sum of two odd primes". Dually, we want a means of expressing "there exists an object such that ....", e.g. in "there exists a real number whose square is 2".

Experience has taught us that the basic mathematical statements are of the form "a has the property P" or "a and b are in the relation R", etc. Examples are: "n is even", "f is differentiable", "$3 = 5$", "$7 < 12$", "B is between A and C". Therefore we build our language from symbols for *properties, relations* and *objects*. Furthermore we add *variables* to range over objects (so called individual variables), and the usual logical connectives now including the *quantifiers* $\forall$ and $\exists$ (for "for all" and "there exists").

We first give a few informal examples.

$\exists x P(x)$	there is an x with property P,
$\forall y P(y)$	for all y P holds (all y have the property P),
$\forall x \exists y (x = 2y)$	for all x there is a y such that x is two times y,
$\forall \varepsilon(\varepsilon > 0 \to \exists n(\frac{1}{n} < \varepsilon))$	for all positive ϵ there is an n such that $\frac{1}{n} < \varepsilon$,
$x < y \to \exists z(x < z \wedge z < y)$	if $x < y$, then there is a z such that $x < z$ and $z < y$,
$\forall x \exists y (x.y = 1)$	for each x there exists an inverse y.

D. van Dalen, *Logic and Structure*, Universitext, DOI 10.1007/978-1-4471-4558-5_3,
© Springer-Verlag London 2013

We know from elementary set theory that functions are a special kind of relations. It would, however, be in flagrant conflict with mathematical practice to avoid functions (or mappings). Moreover, it would be extremely cumbersome. So we will incorporate functions in our language.

Roughly speaking the language deals with two categories of syntactical entities: one for objects—the *terms*, and one for statements—the *formulas*. Examples of terms are: $17, x, (2 + 5) - 7, x^{3y+1}$.

What is the subject of predicate logic with a given language? Or, to put it differently, what are terms and formulas about? The answer is: formulas can express properties concerning a given set of relations and functions on a fixed domain of discourse. We have already met such situations in mathematics; we talked about *structures*, e.g. groups, rings, modules, ordered sets (see any algebra text). We will make structures our point of departure and we will get to the logic later.

In our logic we will speak about "all numbers" or "all elements", but not about "all ideals" or "all subsets", etc. Loosely speaking, our variables will vary over elements of a given universe (e.g. the $n \times n$ matrices over the reals), but not over properties or relations, or properties of properties, etc. For this reason the predicate logic of this book is called *first-order logic*, or also *elementary logic*. In everyday mathematics, e.g. analysis, one uses higher order logic. In a way it is a surprise that first-order logic can do so much for mathematics, as we will see. A short introduction to second-order logic will be presented in Chap. 5.

3.2 Structures

A group is a (non-empty) set equipped with two operations, a binary one and a unary one, and with a neutral element (satisfying certain laws). A partially ordered set is a set equipped with a binary relation (satisfying certain laws).

We generalize this as follows.

Definition 3.2.1 A structure is an ordered sequence $\langle A, R_1, \ldots, R_n, F_1, \ldots, F_m, \{c_i | i \in I\} \rangle$, where A is a non-empty set. $R_1, \ldots, R_n$ are relations on A, $F_1, \ldots, F_m$ are functions on A, and the c_i $(i \in I)$ are elements of A (constants).

Warning The functions F_i are *total*, i.e. defined for all arguments; sometimes this calls for tricks, as with 0^{-1} (cf. p. 82).

Examples

$\langle \mathbb{R}, +, \cdot, ^{-1}, 0, 1 \rangle$—the field of real numbers,
$\langle \mathbb{N}, < \rangle$—the ordered set of natural numbers.

We denote structures by Gothic capitals: $\mathfrak{A}, \mathfrak{B}, \mathfrak{C}, \mathfrak{D}, \ldots$. The script letters are shown on p. 52.

If we overlook for a moment the special properties of the relations and operations (e.g. commutativity of addition on the reals), then what remains is the *type* of a structure, which is given by the number of relations, functions (or operations), and their respective arguments, plus the number (cardinality) of constants.

Definition 3.2.2 The *similarity type* of a structure $\mathfrak{A} = \langle A, R_1, \ldots, R_n, F_1, \ldots, F_m, \{c_i | i \in I\}\rangle$ is a sequence, $\langle r_1, \ldots, r_n; a_1, \ldots, a_m; \kappa\rangle$, where $R_i \subseteq A^{r_i}$, $F_j : A^{a_j} \to A$, $\kappa = |\{c_i | i \in I\}|$ (cardinality of I).

The two structures in our example have (similarity) type $\langle -; 2, 2, 1; 2\rangle$ and $\langle 2; -; 0\rangle$. The absence of relations, functions is indicated by $-$. There is no objection to extending the notion of structure to contain arbitrarily many relations or functions, but the most common structures have finite types (including finitely many constants).

It would, of course, have been better to use semicolons for our structures, i.e. $\langle A; R_1, \ldots, R_n; F_1, \ldots, F_m; c_i | i \in I\rangle$, but that would be too pedantic.

If $R \subseteq A$, then we call R a property (or *unary relation*), if $R \subseteq A^2$, then we call R a *binary relation*, if $R \subseteq A^n$, then we call R an *n-ary relation*.

The set A is called the *universe* of $\mathfrak{A}$.

Notation $A = |\mathfrak{A}|$. $\mathfrak{A}$ is called (in)finite if its universe is (in)finite. We will mostly commit a slight abuse of language by writing down the constants instead of the set of constants; in the example of the field of real numbers we should have written: $\langle \mathbb{R}, +, \cdot, ^{-1}, \{0, 1\}\rangle$, but $\langle \mathbb{R}, +, \cdot, ^{-1}, 0, 1\rangle$ is more traditional. Among the relations one finds in structures, there is a very special one: the *identity* (or *equality*) *relation*.

Since mathematical structures, as a rule, are equipped with the identity relation, we do not list the relation separately. It does, therefore, not occur in the similarity type. We henceforth assume all structures to possess an identity relation and we will explicitly mention any exceptions. For purely logical investigations it makes, of course, perfect sense to consider a logic without identity, but this book caters to readers from the mathematics or computer science community.

One also considers the "limiting cases" of relations and functions, i.e. 0-ary relations and functions. An 0-ary relation is a subset of $A^\emptyset$. Since $A^\emptyset = \{\emptyset\}$ there are two such relations: $\emptyset$ and $\{\emptyset\}$ (considered as ordinals: 0 and 1). 0-ary relations can thus be seen as truth values, which makes them play the role of the interpretations of propositions. In practice 0-ary relations do not appear; e.g. they have no role to play in ordinary algebra. Most of the time the reader can joyfully forget about them, nonetheless we will allow them in our definition because they simplify certain considerations. A 0-ary function is a mapping from $A^\emptyset$ into A, i.e. a mapping from $\{\emptyset\}$ into A. Since the mapping has a singleton as domain, we can identify it with its range.

In this way 0-ary functions can play the role of constants. The advantage of the procedure is, however, negligible in the present context, so we will keep our constants.

Exercises

1. Write down the similarity type for the following structures:
   (i) $\langle \mathbb{Q}, <, 0 \rangle$
   (ii) $\langle \mathbb{N}, +, \cdot, S, 0, 1, 2, 3, 4, \ldots, n, \ldots \rangle$, where $S(x) = x + 1$,
   (iii) $\langle \mathcal{P}(\mathbb{N}), \subseteq, \cup, \cap, {}^c, \emptyset \rangle$,
   (iv) $\langle \mathbb{Z}/(5), +, \cdot, -, {}^{-1}, 0, 1, 2, 3, 4 \rangle$,
   (v) $\langle \{0, 1\}, \wedge, \vee, \rightarrow, \neg, 0, 1 \rangle$, where $\wedge, \vee, \rightarrow, \neg$ operate according to the ordinary truth tables,
   (vi) $\langle \mathbb{R}, 1 \rangle$,
   (vii) $\langle \mathbb{R} \rangle$,
   (viii) $\langle \mathbb{R}, \mathbb{N}, <, T, {}^2, | \ |, - \rangle$, where $T(a, b, c)$ is the relation "b is between a and c", ^{2} is the square function, $-$ is the subtraction function and $| \ |$ the absolute value.
2. Give structures with type $\langle 1, 1; -; 3 \rangle$, $\langle 4; -; 0 \rangle$.

3.3 The Language of a Similarity Type

The considerations of this section are generalizations of those in Sect. 2.1. Since the arguments are rather similar, we will leave a number of details to the reader. For convenience we fix the similarity type in this section: $\langle r_1, \ldots, r_n; \ a_1, \ldots, a_m; \kappa \rangle$, where we assume $r_i \geq 0, a_j > 0$.

The alphabet consists of the following symbols:

1. Predicate symbols: $P_1, \ldots, P_n, \doteq$
2. Function symbols: $f_1, \ldots, f_m$
3. Constant symbols: $\bar{c}_i$ for $i \in I$
4. Variables: $x_0, x_1, x_2, \ldots$ (countably many)
5. Connectives: $\vee, \wedge, \rightarrow, \neg, \leftrightarrow, \bot \ \forall, \exists$
6. Auxiliary symbols: (,),

$\forall$ and $\exists$ are called the *universal* and the *existential quantifier*. The curiously looking equality symbol has been chosen to avoid possible confusion. There are in fact a number of equality symbols in use: one to indicate the identity in the models, one to indicate the equality in the meta-language and the syntactic one introduced above. We will, however, practice the usual abuse of language, and use these distinctions only if it is really necessary. As a rule the reader will have no difficulty in recognizing the kind of identity involved.

Next we define the two syntactical categories.

Definition 3.3.1 TERM is the smallest set X with the properties

(i) $\bar{c}_i \in X (i \in I)$ and $x_i \in X (i \in N)$,
(ii) $t_1, \ldots, t_{a_i} \in X \Rightarrow f_i(t_1, \ldots, t_{a_i}) \in X$, for $1 \leq i \leq m$.

TERM is our *set of terms*.

Definition 3.3.2 FORM is the smallest set X with the properties:

(i) $\perp \in X$; $P_i \in X$ if $r_i = 0$; $t_1, \ldots, t_{r_i} \in TERM \Rightarrow P_i(t_1, \ldots, t_{r_i}) \in X$; $t_1, t_2 \in TERM \Rightarrow t_1 = t_2 \in X$,

(ii) $\varphi, \psi \in X \Rightarrow (\varphi \square \psi) \in X$, where $\square \in \{\wedge, \vee, \rightarrow, \leftrightarrow\}$,

(iii) $\varphi \in X \Rightarrow (\neg \varphi) \in X$,

(iv) $\varphi \in X \Rightarrow ((\forall x_i)\varphi), ((\exists x_i)\varphi) \in X$.

FORM is our *set of formulas*. We have introduced $t_1 = t_2$ separately, but we could have subsumed it under the first clause. If convenient, we will not treat equality separately. The formulas introduced in (i) are called *atoms*. We point out that (i) includes the case of 0-ary predicate symbols, conveniently called proposition symbols.

A proposition symbol is interpreted as a 0-ary relation, i.e. as 0 or 1 (cf. Definition 3.2.2). This is in accordance with the practice of propositional logic to interpret propositions as true or false. For our present purpose propositions are a luxury. In dealing with concrete mathematical situations (e.g. groups or posets) one has no reason to introduce propositions (things with a fixed truth value). However, propositions are convenient (and even important) in the context of Boolean-valued logic or Heyting-valued logic, and in syntactical considerations.

We will, however, allow a special proposition: $\perp$, the symbol for the false proposition (cf. Sect. 2.2).

The logical connectives have, what one could call "a domain of action", e.g. in $\varphi \rightarrow \psi$ the connective $\rightarrow$ yields the new formula $\varphi \rightarrow \psi$ from formulas φ and ψ, and so $\rightarrow$ bears on φ, ψ and all their parts. For propositional connectives this is not terribly interesting, but for quantifiers (and variable-binding operators in general) it is. The notion goes by the name of *scope*. So in $((\forall x)\varphi)$ and $((\exists x)\varphi)$, φ is the *scope of the quantifier*. By locating the matching brackets one can easily effectively find the scope of a quantifier. If a variable, term or formula occurs in φ, we say that it is in the scope of the quantifier in $\forall x \varphi$ or $\exists x \varphi$.

Just as in the case of PROP, we have induction principles for TERM and FORM.

Lemma 3.3.3 *Let $A(t)$ be a property of terms. If $A(t)$ holds for t a variable or a constant, and if $A(t_1), A(t_2), \ldots, A(t_n) \Rightarrow A(f(t_1, \ldots, t_n))$, for all function symbols f, then $A(t)$ holds for all $t \in TERM$.*

Proof Cf. Theorem 2.1.3. $\square$

Lemma 3.3.4 *Let $A(\varphi)$ be a property of formulas. If*

(i) $A(\varphi)$ *for atomic* φ,

(ii) $A(\varphi), A(\psi) \Rightarrow A(\varphi \square \psi)$,

(iii) $A(\varphi) \Rightarrow A(\neg \varphi)$,

(iv) $A(\varphi) \Rightarrow A((\forall x_i)\varphi), A((\exists x_i)\varphi)$ *for all i, then $A(\varphi)$ holds for all $\varphi \in FORM$.*

Proof Cf. Theorem 2.1.3. $\square$

We will straight away introduce a number of abbreviations. In the first place we adopt the bracket conventions of propositional logic. Furthermore we delete the outer brackets and the brackets round $\forall x$ and $\exists x$ whenever possible. We agree that quantifiers bind more strongly than binary connectives. Furthermore we join strings of quantifiers, e.g. $\forall x_1 x_2 \exists x_3 x_4 \varphi$ stands for $\forall x_1 \forall x_2 \exists x_3 \exists x_4 \varphi$. For better readability we will sometimes separate the quantifier and the formula by a dot: $\forall x \cdot \varphi$. We will also assume that n in $f(t_1, \ldots, t_n)$, $P(t_1, \ldots, t_n)$ always indicates the correct number of arguments. *A word of warning*: the use of $=$ might confuse a careless reader. The symbol "$=$" is used in the language L, where it is a proper syntactic object. It occurs in formulas such as $x_0 = x_7$, but it also occurs in the meta-language, e.g. in the form $x = y$, which must be read "x and y are one and the same variable". However, the identity symbol in $x = y$ can just as well be the legitimate symbol from the alphabet, i.e. $x = y$ is a meta-atom, which can be converted into a proper atom by substituting genuine variable symbols for x and y. Sometimes $\equiv$ is used for "syntactically identical", as in "x and y are the same variable". We will opt for "$=$" for the equality in structures (sets) and "$\doteq$" for the identity predicate symbol in the language. We will use $\doteq$ a few times, but we prefer to stick to a simple "$=$" trusting the alertness of the reader.

Example 3.3.5 Example of a language of type $\langle 2; 2, 1; 1 \rangle$.

predicate symbols: $L, \doteq$
function symbols: p, i
constant symbol: $\overline{e}$

Some terms: $t_1 := x_0$; $t_2 := p(x_1, x_2)$; $t_3 := p(\overline{e}, \overline{e})$; $t_4 := i(x_7)$; $t_5 := p(i(p(x_2, \overline{e})), i(x_1))$.

Some formulas:

$$\varphi_1 := x_0 \doteq x_2, \qquad \varphi_4 := (x_0 \doteq x_1 \rightarrow x_1 \doteq x_0),$$
$$\varphi_2 := t_3 \doteq t_4, \qquad \varphi_5 := (\forall x_0)(\forall x_1)(x_0 \doteq x_1 \rightarrow \neg L(v_0, x_1))$$
$$\varphi_3 := L(i(x_5), \overline{e}), \qquad \varphi_6 := (\forall x_0)(\exists x_1)(p(x_0, x_1) \doteq \overline{e}),$$
$$\varphi_7 := (\exists x_1)(\neg x_1 \doteq \overline{e} \wedge p(x_1, x_1) \doteq \overline{e}).$$

(We have chosen a suggestive notation; think of the language of ordered groups: L for "less than", p, i for "product" and "inverse".) Note that the order in which the various symbols are listed is important. In our example p has 2 arguments and i has 1.

In mathematics there are a number of *variable-binding operations*, such as summation, integration, abstraction. Consider, for example, integration, in $\int_0^1 \sin x \, dx$ the variable plays an unusual role for a variable. For x cannot "vary"; we cannot (without writing nonsense) substitute any number we like for x. In the integral the variable x is reduced to a tag. We say that the variable x is bound by the integration symbol. Analogously we distinguish in logic between *free* and *bound* variables.

A variable may occur in a formula more than once. It is quite often useful to look at a specific instance at a certain place in the string that makes up the formula. We call these *occurrences* of the variable, and we use expressions like "x occurs in the subformula ψ of φ." In general we consider occurrences of formulas, terms, quantifiers, and the like.

In defining various syntactical notions we again freely use the principle of *definition by recursion* (cf. Theorem 2.1.6). The justification is immediate: the value of a term (formula) is uniquely determined by the values of its parts. This allows us to determine the value of $H(t)$ for a mapping acting on terms, in finitely many steps.

Definition by Recursion on TERM   Let $H_0 : Var \cup Const \to A$ (i.e. H_0 is defined on variables and constants), $H_i : A^{a_i} \to A$, then there is a unique mapping $H : TERM \to A$ such that

$$\begin{cases} H(t) = H_0(t) \text{ for } t \text{ a variable or a constant,} \\ H(f_i(t_1, \ldots, t_{a_i})) = H_i(H(t_1), \ldots, H(t_{a_i})). \end{cases}$$

Definition by Recursion on FORM   Let

$$H_{at} : At \to A \quad \text{(i.e. } H_{at} \text{ is defined on atoms),}$$
$$H_\Box : A^2 \to A \quad (\Box \in \{\vee, \wedge, \to, \leftrightarrow\}),$$
$$H_\neg : A \to A,$$
$$H_\forall : A \times N \to A,$$
$$H_\exists : A \times N \to A.$$

Then there is a unique mapping $H : FORM \to A$ such that

$$\begin{cases} H(\varphi) = H_{at}(\varphi) \quad \text{for atomic } \varphi, \\ H(\varphi \Box \psi) = H_\Box(H(\varphi), H(\psi)), \\ H(\neg\varphi) = H_\neg(H(\varphi)), \\ H(\forall x_i \varphi) = H_\forall(H(\varphi), i), \\ H(\exists x_i \varphi) = H_\exists(H(\varphi), i). \end{cases}$$

Definition 3.3.6 The set $FV(t)$ of free variables of t is defined by

(i) $FV(x_i) := \{x_i\}$,
     $FV(\overline{c}_i) := \emptyset$
(ii) $FV(f(t_1, \ldots, t_n)) := FV(t_1) \cup \cdots \cup FV(t_n)$.

Remark To avoid messy notation we will usually drop the indices and tacitly assume that the number of arguments is correct. The reader can easily provide the correct details, should he wish to do so.

Definition 3.3.7 The set $FV(\varphi)$ of free variables of φ is defined by

(i) $FV(P(t_1, \ldots, t_p)) := FV(t_1) \cup \cdots \cup FV(t_p)$,
     $FV(t_1 = t_2) := FV(t_1) \cup FV(t_2)$, $FV(\bot) = FV(P) := \emptyset$ for P a proposition symbol,

(ii) $FV(\varphi \Box \psi) := FV(\varphi) \cup FV(\psi)$,
$\quad\ \ FV(\neg \varphi) := FV(\varphi)$,
(iii) $FV(\forall x_i \varphi) := FV(\exists x_i \varphi) := FV(\varphi) - \{x_i\}$.

Definition 3.3.8 t or φ is called *closed* if $FV(t) = \emptyset$, resp. $FV(\varphi) = \emptyset$. A closed formula is also called a *sentence*. A formula without quantifiers is called *open*. $TERM_c$ denotes the set of closed terms; $SENT$ denotes the set of sentences.

It is left to the reader to define the set $BV(\varphi)$ of *bound variables* of φ.

Continuation of Example 3.3.5

$FV(t_2) = \{x_1, x_2\}$; $FV(t_3) = \emptyset$; $FV(\varphi_2) = FV(t_3) \cup FV(t_4) = \{x_7\}$;
$FV(\varphi_7) = \emptyset$; $BV(\varphi_4) = \emptyset$; $BV(\varphi_6) = \{x_0, x_1\}$. $\varphi_5, \varphi_6, \varphi_7$ are sentences.

Warning $FV(\varphi) \cap BV(\varphi)$ need not be empty; in other words, the same variable may occur free *and* bound. To handle such situations one considers free (resp. bound) *occurrences* of variables. When necessary we will make informal use of occurrences of variables; see also p. 59.

Example $\forall x_1 (x_1 = x_2) \rightarrow P(x_1)$ contains x_1 both free and bound, for the occurrence of x_1 in $P(x_1)$ is not within the scope of the quantifier.

In predicate calculus we have substitution operators for terms and for formulas.

Definition 3.3.9 Let s and t be terms, then $s[t/x]$ is defined by:

(i) $\ y[t/x] := \begin{cases} y & \text{if } y \not\equiv x \\ t & \text{if } y \equiv x \end{cases}$

$\quad\ \ c[t/x] := c$

(ii) $f(t_1, \ldots, t_p)[t/x] := f(t_1[t/x], \ldots, t_p[t/x])$.

Note that in the clause (i) $y \equiv x$ means "x and y are the same variables".

Definition 3.3.10 $\varphi[t/x]$ is defined by:

(i) $\ \bot [t/x] := \bot$,

$\quad\ P[t/x] := P \quad$ for propositions P,

$\quad\ P(t_1, \ldots, t_p)[t/x] := P(t_1[t/x], \ldots, t_p[t/x])$,

$\quad\ (t_1 = t_2)[t/x] := t_1[t/x] = t_2[t/x]$,

(ii) $(\varphi \Box \psi)[t/x] := \varphi[t/x] \Box \psi[t/x]$,

$\quad\ (\neg \varphi)[t/x] := \neg \varphi[t/x]$

(iii) $(\forall y\varphi)[t/x] := \begin{cases} \forall y\varphi[t/x] & \text{if } x \not\equiv y \\ \forall y\varphi & \text{if } x \equiv y \end{cases}$

$(\exists y\varphi)[t/x] := \begin{cases} \exists y\varphi[t/x] & \text{if } x \not\equiv y \\ \exists y\varphi & \text{if } x \equiv y \end{cases}$

Substitution of formulas is defined as in the case of propositions; for convenience we use "$\$$" as a symbol for the propositional symbol (0-ary predicate symbol) which acts as a "place holder".

Definition 3.3.11 $\sigma[\varphi/\$]$ is defined by:

(i) $\sigma[\varphi/\$] := \begin{cases} \sigma & \text{if } \sigma \not\equiv \$ \\ \varphi & \text{if } \sigma \equiv \$ \end{cases}$ for atomic σ,

(ii) $(\sigma_1\Box\sigma_2)[\varphi/\$] := \sigma_1[\varphi/\$]\Box\sigma_2[\varphi/\$]$

$(\neg\sigma_1)[\sigma/\$] := \neg\sigma_1[\varphi/\$]$

$(\forall y\sigma)[\varphi/\$] := \forall y.\sigma[\varphi/\$]$

$(\exists y\sigma)[\varphi/\$] := \exists y.\sigma[\varphi/\$].$

Continuation of Example 3.3.5

$$t_4[t_2/x_1] = i(x_7); \qquad t_4[t_2/x_7] = i(p(x_1, x_2));$$
$$t_5[x_2/x_1] = p(i(p(x_2, \overline{e}), i(x_2)),$$
$$\varphi_1[t_3/x_0] = p(\overline{e}, \overline{e}) \doteq x_2; \qquad \varphi_5[t_3/x_0] = \varphi_5.$$

We will sometimes make *simultaneous substitutions*, the definition is a slight modification of Definitions 3.3.9, 3.3.10 and 3.3.11. The reader is asked to write down the formal definitions. We denote the result of a simultaneous substitution of $t_1, \ldots, t_n$ for $y_1, \ldots, y_n$ in t by $t[t_1, \ldots, t_n/y_1, \ldots, y_n]$ (similarly for φ).

Note that a simultaneous substitution is not the same as its corresponding repeated substitution.

Example $(x_0 \doteq x_1)[x_1, x_0/x_0, x_1] = (x_1 \doteq x_0)$, but $((x_0 \doteq x_1)[x_1/x_0])[x_0/x_1] = (x_1 \doteq x_1)[x_0/x_1] = (x_0 \doteq x_0)$.

The quantifier clause in Definition 3.3.10 forbids substitution for bound variables. There is, however, one more case we want to forbid: a substitution, in which some variable after the substitution becomes bound. We will give an example of such a substitution; the reason why we forbid it is that it can change the truth value in an absurd way. At this moment we do not have a truth definition, so the argument is purely heuristic.

Example $\exists x(y < x)[x/y] = \exists x (x < x)$.

Note that the right-hand side is false in an ordered structure, whereas $\exists x\,(y < x)$ may very well be true. We make our restriction precise.

Definition 3.3.12 t is *free for x in φ* if

(i) φ is atomic,
(ii) $\varphi := \varphi_1 \square \varphi_2$ (or $\varphi := \neg \varphi_1$) and t is free for x in φ_1 and φ_2 (resp. φ_1),
(iii) $\varphi := \exists y \psi$ (or $\varphi := \forall y \psi$) and if $x \in FV(\varphi)$, then $y \notin FV(t)$ and t is free for x in ψ.

Examples

1. x_2 is free for x_0 in $\exists x_3 P(x_0, x_3)$,
2. $f(x_0, x_1)$ is not free for x_0 in $\exists x_1 P(x_0, x_3)$,
3. x_5 is free for x_1 in $P(x_1, x_3) \to \exists x_1 Q(x_1, x_2)$.

Note that the use of "t is free for x in φ" comes down to the fact that the (free) variables of t are not going to be bound after substitution in φ.

Lemma 3.3.13 *t is free for x in $\varphi \Leftrightarrow$ the variables of t in $\varphi[t/x]$ are not bound by a quantifier.*

Proof Induction on φ.

- For atomic φ the lemma is evident.
- $\varphi = \varphi_1 \square \varphi_2$. t is free for x in $\varphi \overset{\text{def.}}{\Leftrightarrow} t$ is free for x in φ_1 and t is free for x in $\varphi_2 \overset{\text{i.h.}}{\Leftrightarrow}$ the variables of t in $\varphi_1[t/x]$ are not bound by a quantifier and the variables of t in $\varphi_2[t/x]$ are not bound by a quantifier $\Leftrightarrow$ the variables of t in $(\varphi_1 \square \varphi_2)[t/x]$ are not bound by a quantifier.
- $\varphi = \neg \varphi_1$, similar.
- $\varphi = \exists y \psi$. It suffices to consider the case $x \in FV(\varphi)$. t is free for x in $\varphi \overset{\text{def.}}{\Leftrightarrow} y \notin FV(t)$ and t is free for x in $\psi \overset{\text{i.h.}}{\Leftrightarrow}$ the variables of t are not in the scope of $\exists y$ and the variables of t in $\psi[t/x]$ are not bound by (another) quantifier $\Leftrightarrow$ the variables of t in $\varphi[t/x]$ are not bound by a quantifier. $\square$

There is an analogous definition and lemma for the substitution of formulas.

Definition 3.3.14 φ is free for $\$$ in σ if:

(i) σ is atomic,
(ii) $\sigma := \sigma_1 \square \sigma_2$ (or $\neg \sigma_1$) and φ is free for $\$$ in σ_1 and in σ_2 (or in σ_1),
(iii) $\sigma := \exists y \tau$ (or $\forall y \tau$) and if $\$$ occurs in σ then $y \notin FV(\varphi)$ and φ is free for $\$$ in τ.

Lemma 3.3.15 *φ is free for $\$$ in $\sigma \Leftrightarrow$ the free variables of φ are in $\sigma[\varphi/\$]$ not bound by a quantifier.*

Proof As for Lemma 3.3.13. $\square$

From now on we tacitly suppose that all our substitutions are "free for".

For convenience we introduce an informal notation that simplifies reading and writing.

Notation In order to simplify the substitution notation and to conform to an ancient suggestive tradition we will write down (meta-) expressions like $\varphi(x, y, z)$, $\psi(x, x)$, etc. This neither means that the listed variables occur free nor that no other ones occur free. It is merely a convenient way to handle substitution informally: $\varphi(t)$ is the result of replacing x by t in $\varphi(x)$; $\varphi(t)$ is called a *substitution instance* of $\varphi(x)$.

We use the languages introduced above to describe structures, or classes of structures of a given type. The predicate symbols, function symbols and constant symbols act as names for various relations, operations and constants. In describing a structure it is a great help to be able to refer to all elements of $|\mathfrak{A}|$ individually, i.e. to have *names* for all elements (if only as an auxiliary device). Therefore we introduce the following.

Definition 3.3.16 The *extended language*, $L(\mathfrak{A})$, of $\mathfrak{A}$ is obtained from the language L, of the type of $\mathfrak{A}$, by adding constant symbols for all elements of $|\mathfrak{A}|$. We denote the constant symbol, belonging to $a \in |\mathfrak{A}|$, by $\bar{a}$.

Example Consider the language L of groups; then $L(\mathfrak{A})$, for $\mathfrak{A}$ the additive group of integers, has (extra) constant symbols $\bar{0}, \bar{1}, \bar{2}, \ldots, \overline{-1}, \overline{-2}, \overline{-3}, \ldots$. Observe that in this way 0 gets two names: the old one and one of the new ones. This is no problem, why shouldn't something have more than one name?

Exercises

1. Write down an alphabet for the languages of the types given in Exercise 1 of Sect. 3.2.
2. Write down five terms of the language belonging to Exercise 1 (iii), (viii), Write down two atomic formulas of the language belonging to Exercise 1 (vii) and two closed atoms for Exercise 1 (iii), (vi).
3. Write down an alphabet for languages of types $\langle 3; 1, 1, 2; 0 \rangle$, $\langle -; 2; 0 \rangle$ and $\langle 1; -; 3 \rangle$.
4. Check which terms are free in the following cases, and carry out the substitution:

   (a) x for x in $x = x$,
   (b) y for x in $x = x$,
   (c) $x + y$ for y in $z = \bar{0}$,
   (d) $\bar{0} + y$ for y in $\exists x(y = x)$,
   (e) $x + y$ for z in $\exists w(w + x = \bar{0})$,
   (f) $x + w$ for z in $\forall w(x + z = \bar{0})$,
   (g) $x + y$ for z in $\forall w(x + z = \bar{0}) \wedge \exists y(z = x)$,
   (h) $x + y$ for z in $\forall u(u = v) \rightarrow \forall z(z = y)$.

3.4 Semantics

The art of interpreting (mathematical) statements presupposes a strict separation
between "language" and the mathematical "universe" of entities. The objects of language are symbols, or strings of symbols, the entities of mathematics are numbers,
sets, functions, triangles, etc. It is a matter for the philosophy of mathematics to
reflect on the universe of mathematics; here we will simply accept it as given to
us. Our requirements concerning the mathematical universe are, at present, fairly
modest. For example, ordinary set theory will do very well for us. Likewise our
desiderata with respect to language are modest. We just suppose that there is an
unlimited supply of symbols.

The idea behind the semantics of predicate logic is very simple. Following Tarski,
we assume that a statement σ is true in a structure, if σ actually is the case (the
sentence "Snow is white" is true if snow actually is white). A mathematical example:
"$\overline{2} + \overline{2} = \overline{4}$" is true in the structure of natural numbers (with addition) if $2 + 2 = 4$
(i.e. if addition of the *numbers* 2 and 2 yields the *number* 4). Interpretation is the art
of relating syntactic objects (strings of symbols) and states of affairs "in reality".

We will start by giving an example of an interpretation in a simple case. We
consider the structure $\mathfrak{A} = (\mathbb{Z}, <, +, -, 0)$, i.e. the ordered group of integers.

The language has in its alphabet:

predicate symbols: $\doteq, L$
function symbols: P, M
constant symbol: $\overline{0}$

$L(\mathfrak{A})$ has, in addition to all that, constant symbols $\overline{m}$ for all $m \in \mathbb{Z}$. We first
interpret the closed terms of $L(\mathfrak{A})$; the interpretation $t^{\mathfrak{A}}$ of a term t is an element
of $\mathbb{Z}$.

t	$t^{\mathfrak{A}}$
$\overline{m}$	m
$P(t_1, t_2)$	$t_1^{\mathfrak{A}} + t_2^{\mathfrak{A}}$
$M(t)$	$-t^{\mathfrak{A}}$

Roughly speaking, we interpret $\overline{m}$ as "its number", P as *plus*, M as *minus*. Note
that we interpret only closed terms. This stands to reason, how should one assign a
definite integer to x?

Next we interpret *sentences* of $L(\mathfrak{A})$ by assigning one of the truth values 0 or 1.
As far as the propositional connectives are concerned, we follow the semantics for
propositional logic.

$$v(\bot) = 0,$$
$$v(t \doteq s) = \begin{cases} 1 & \text{if } t^{\mathfrak{A}} = s^{\mathfrak{A}} \\ 0 & \text{else,} \end{cases}$$
$$v(L(t, s)) = \begin{cases} 1 & \text{if } t^{\mathfrak{A}} < s^{\mathfrak{A}} \\ 0 & \text{else} \end{cases}$$

$$v(\varphi \square \psi)$$
$$v(\neg \varphi)$$
$\quad$ as in Definition 2.2.1

$$v(\forall x \varphi) = \min\{v(\varphi[\overline{n}/x]) \mid n \in \mathbb{Z}\}$$
$$v(\exists x \sigma) = \max\{v(\varphi[\overline{n}/x]) \mid n \in \mathbb{Z}\}$$

A few remarks are in order.

1. In fact we have defined a function v by recursion on φ.
2. The valuation of a universally quantified formula is obtained by taking the minimum of all valuations of the individual instances, i.e. the value is 1 (true) iff all instances have the value 1. In this respect $\forall$ is a generalization of $\wedge$. Likewise $\exists$ is a generalization of $\vee$.
3. v is uniquely determined by $\mathfrak{A}$, hence $v_{\mathfrak{A}}$ would be a more appropriate notation. For convenience we will, however, stick to just v.
4. As in the semantics of propositional logic, we will write $[\![\varphi]\!]_{\mathfrak{A}}$ for $v_{\mathfrak{A}}(\varphi)$, and when no confusion arises we will drop the subscript $\mathfrak{A}$.
5. It would be tempting to make our notation really uniform by writing $[\![t]\!]_{\mathfrak{A}}$ for $t^{\mathfrak{A}}$. We will, however, keep both notations and use whichever is the most readable. The superscript notation has the drawback that it requires more brackets, but the $[\![\;]\!]$-notation does not improve readability.

Examples

1. $(P(P(\overline{2},\overline{3}), M(\overline{7})))^{\mathfrak{A}} = P(\overline{2},\overline{3})^{\mathfrak{A}} + M(\overline{7})^{\mathfrak{A}} = (\overline{2}^{\mathfrak{A}} + \overline{3}^{\mathfrak{A}}) + (-\overline{7}^{\mathfrak{A}}) = 2 + 3 + (-7) = -2$,
2. $[\![\overline{2} \doteq \overline{-1}]\!] = 0$, since $2 \neq -1$,
3. $[\![\overline{0} \doteq \overline{1} \rightarrow L(\overline{25}, \overline{10})]\!] = 1$, since $[\![\overline{0} = \overline{1}]\!] = 0$ and $[\![L(\overline{25}, \overline{10})]\!] = 0$; by the interpretation of the implication the value is 1,
4. $[\![\forall x \exists y (L(x, y))]\!] = \min_n (\max_m [\![L(\overline{n}, \overline{m})]\!])$.
   $[\![L(\overline{n}, \overline{m})]\!] = 1$ for $m > n$, so for fixed n, $\max_m [\![L(\overline{n}, \overline{m})]\!] = 1$, and hence $\min_n \max_m [\![L(\overline{n}, \overline{m})]\!] = 1$.

Let us now present a definition of interpretation for the general case. Consider $\mathfrak{A} = \langle A, R_1, \ldots, R_n, F_1, \ldots, F_m, \{c_i \mid i \in I\}\rangle$ of a given similarity type $\langle r_1, \ldots, r_n; a_1, \ldots, a_m; |I|\rangle$.

The corresponding language has predicate symbols $\overline{R}_1, \ldots, \overline{R}_n$, function symbols $\overline{F}_1, \ldots, \overline{F}_m$ and constant symbols $\overline{c}_i$. $L(\mathfrak{A})$, moreover, has constant symbols $\overline{a}$ for all $a \in |\mathfrak{A}|$.

Definition 3.4.1 An interpretation of the closed terms of $L(\mathfrak{A})$ in $\mathfrak{A}$ is a mapping $(.)^{\mathfrak{A}} : TERM_c \rightarrow |\mathfrak{A}|$ satisfying:

(i) $\overline{c}_i^{\,\mathfrak{A}} = c_i \quad (= [\![\overline{c}_i]\!]_{\mathfrak{A}})$

$\quad\; \overline{a}^{\,\mathfrak{A}} = a, \quad (= [\![\overline{a}]\!]_{\mathfrak{A}})$

(ii) $(\overline{F}_i(t_1,\ldots,t_p))^{\mathfrak{A}} = F_i(t_1^{\mathfrak{A}},\ldots,t_p^{\mathfrak{A}}),\quad (= [\![F_i(t_1,\ldots,t_p)]\!]_{\mathfrak{A}}$

where $p = a_i$ $\qquad\qquad\qquad\qquad\qquad = \overline{F}_i([\![t_1]\!]_{\mathfrak{A}},\ldots,[\![t_p]\!]_{\mathfrak{A}}))$

There is also a *valuation notation* using Scott brackets; we have indicated in the above definition how these brackets are to be used. The following definition is exclusively in terms of valuations.

Definition 3.4.2 An interpretation of the sentences φ of $L(\mathfrak{A})$ in $\mathfrak{A}$ is a mapping $[\![.]\!]_{\mathfrak{A}} : SENT \to \{0, 1\}$, satisfying:

(i) $[\![\bot]\!]_{\mathfrak{A}} := 0$,

  $[\![R]\!]_{\mathfrak{A}} := R$ (i.e. 0 or 1).

(ii) $[\![\overline{R}_i(t_1,\ldots,t_p)]\!]_{\mathfrak{A}} := \begin{cases} 1 \text{ if } \langle t_1^{\mathfrak{A}},\ldots,t_p^{\mathfrak{A}}\rangle \in R_i, & \text{where } p = r_i, \\ 0 & \text{else.} \end{cases}$

  $[\![t_1 = t_2]\!]_{\mathfrak{A}}, := \begin{cases} 1 & \text{if } t_1^{\mathfrak{A}} = t_2^{\mathfrak{A}} \\ 0 & \text{else.} \end{cases}$

(iii) $[\![\varphi \wedge \psi]\!]_{\mathfrak{A}} := \min([\![\varphi]\!]_{\mathfrak{A}}, [\![\psi]\!]_{\mathfrak{A}})$,

  $[\![\varphi \vee \psi]\!]_{\mathfrak{A}} := \max([\![\varphi]\!]_{\mathfrak{A}}, [\![\psi]\!]_{\mathfrak{A}})$,

  $[\![\varphi \to \psi]\!]_{\mathfrak{A}} := \max(1 - [\![\varphi]\!]_{\mathfrak{A}}, [\![\psi]\!]_{\mathfrak{A}})$,

  $[\![\varphi \leftrightarrow \psi]\!]_{\mathfrak{A}} := 1 - |[\![\varphi]\!]_{\mathfrak{A}} - [\![\psi]\!]_{\mathfrak{A}}|$,

  $[\![\neg\varphi]\!]_{\mathfrak{A}} := 1 - [\![\varphi]\!]_{\mathfrak{A}}$.

(iv) $[\![\forall x\varphi]\!]_{\mathfrak{A}} := \min\{[\![\varphi[\overline{a}/x]]\!]_{\mathfrak{A}} \mid a \in |\mathfrak{A}|\}$,

  $[\![\exists x\varphi]\!]_{\mathfrak{A}} := \max\{[\![\varphi[\overline{a}/x]]\!]_{\mathfrak{A}} \mid a \in |\mathfrak{A}|\}$.

Convention: from now on we will assume that all structures and languages have the appropriate similarity type, so that we don't have to specify the types all the time.

In predicate logic there is a popular and convenient alternative for the valuation notation:

$\mathfrak{A} \models \varphi$ stands for $[\![\varphi]\!]_{\mathfrak{A}} = 1$. We say that "$\varphi$ is true, valid, in $\mathfrak{A}$" if $\mathfrak{A} \models \varphi$. The relation $\models$ is called the *satisfaction relation*.

Note that the same notation is available in propositional logic—there the role of $\mathfrak{A}$ is taken by the valuation, so one could very well write $v \models \varphi$ for $[\![\varphi]\!]_v = 1$.

So far we have only defined truth for sentences of $L(\mathfrak{A})$. In order to extend $\models$ to arbitrary formulas we introduce a new notation.

Definition 3.4.3 Let $FV(\varphi) = \{z_1,\ldots,z_k\}$, then $Cl(\varphi) := \forall z_1\ldots z_k\varphi$ is the *universal closure* of φ (we assume the order of variables z_i to be fixed in some way).

Definition 3.4.4

(i) $\mathfrak{A} \models \varphi$ iff $\mathfrak{A} \models Cl(\varphi)$,

(ii) $\models \varphi$ iff $\mathfrak{A} \models \varphi$ for all $\mathfrak{A}$ (of the appropriate type),

(iii) $\mathfrak{A} \models \Gamma$ iff $\mathfrak{A} \models \psi$ for all $\psi \in \Gamma$,

(iv) $\Gamma \models \varphi$ iff $(\mathfrak{A} \models \Gamma \Rightarrow \mathfrak{A} \models \varphi)$, where $\Gamma \cup \{\varphi\}$ consists of sentences.

If $\mathfrak{A} \models \sigma$, we call $\mathfrak{A}$ a *model* of σ. In general: if $\mathfrak{A} \models \Gamma$, we call $\mathfrak{A}$ a *model* of Γ. We say that φ is *true* if $\models \varphi$, φ is a *semantic consequence* of Γ if $\Gamma \models \varphi$, i.e. φ holds in each model of Γ. Note that this is all a straightforward generalization of Definition 2.2.4.

If φ is a formula with free variables, say $FV(\varphi) = \{z_1, \ldots, z_k\}$, then we say that φ is *satisfied by* $a_1, \ldots, a_k \in |\mathfrak{A}|$ if $\mathfrak{A} \models \varphi[\overline{a}_1, \ldots, \overline{a}_k / z_1, \ldots, z_k]$, φ is called *satisfiable in* $\mathfrak{A}$ if there are $a_1, \ldots, a_k$ such that φ is satisfied by $a_1, \ldots, a_k$ and φ is called *satisfiable* if it is satisfiable in some $\mathfrak{A}$. Note that φ is satisfiable in $\mathfrak{A}$ iff $\mathfrak{A} \models \exists z_1 \ldots z_k \varphi$.

The properties of the satisfaction relation are in understandable and convenient correspondence with the intuitive meaning of the connectives.

Lemma 3.4.5 *If we restrict ourselves to sentences, then*

(i) $\mathfrak{A} \models \varphi \wedge \psi \Leftrightarrow \mathfrak{A} \models \varphi$ *and* $\mathfrak{A} \models \psi$,

(ii) $\mathfrak{A} \models \varphi \vee \psi \Leftrightarrow \mathfrak{A} \models \varphi$ *or* $\mathfrak{A} \models \psi$,

(iii) $\mathfrak{A} \models \neg\varphi \Leftrightarrow \mathfrak{A} \not\models \varphi$,

(iv) $\mathfrak{A} \models \varphi \rightarrow \psi \Leftrightarrow (\mathfrak{A} \models \varphi \Rightarrow \mathfrak{A} \models \psi)$,

(v) $\mathfrak{A} \models \varphi \leftrightarrow \psi \Leftrightarrow (\mathfrak{A} \models \varphi \Leftrightarrow \mathfrak{A} \models \psi)$,

(vi) $\mathfrak{A} \models \forall x\varphi \Leftrightarrow \mathfrak{A} \models \varphi[\overline{a}/x]$, *for all* $a \in |\mathfrak{A}|$,

(vii) $\mathfrak{A} \models \exists x\varphi \Leftrightarrow \mathfrak{A} \models \varphi[\overline{a}/x]$, *for some* $a \in |\mathfrak{A}|$.

Proof Immediate from Definition 3.4.2. We will do two cases.

(iv) $\mathfrak{A} \models \varphi \rightarrow \psi \Leftrightarrow [\![\varphi \rightarrow \psi]\!]_{\mathfrak{A}} = \max(1 - [\![\varphi]\!]_{\mathfrak{A}}, [\![\psi]\!]_{\mathfrak{A}}) = 1$. Suppose $\mathfrak{A} \models \varphi$, i.e. $[\![\varphi]\!]_{\mathfrak{A}} = 1$, then clearly $[\![\psi]\!]_{\mathfrak{A}} = 1$, or $\mathfrak{A} \models \psi$.
Conversely, let $\mathfrak{A} \models \varphi \Rightarrow \mathfrak{A} \models \psi$, and suppose $\mathfrak{A} \not\models \varphi \rightarrow \psi$, then $[\![\varphi \rightarrow \psi]\!]_{\mathfrak{A}} = \max(1 - [\![\varphi]\!]_{\mathfrak{A}}, [\![\psi]\!]_{\mathfrak{A}}) = 0$. Hence $[\![\psi]\!]_{\mathfrak{A}} = 0$ and $[\![\varphi]\!]_{\mathfrak{A}} = 1$. Contradiction.

(vii) $\mathfrak{A} \models \exists x\varphi(x) \Leftrightarrow \max\{[\![\varphi(\overline{a})]\!]_{\mathfrak{A}} | a \in |\mathfrak{A}|\} = 1 \Leftrightarrow$ there is an $a \in |\mathfrak{A}|$ such that $[\![\varphi(\overline{a})]\!]_{\mathfrak{A}} = 1 \Leftrightarrow$ there is an $a \in |\mathfrak{A}|$ such that $\mathfrak{A} \models \varphi(\overline{a})$. $\qquad\square$

Lemma 3.4.5 tells us that the interpretation of sentences in $\mathfrak{A}$ runs parallel to the construction of the sentences by means of the connectives. In other words, we replace the connectives by their analogues in the meta-language and interpret the atoms by checking the relations in the structure.

For example, consider our example of the ordered additive group of integers: $\mathfrak{A} \models \neg\forall x \exists y (x \doteq P(y, y)) \Leftrightarrow$ It is not the case that for each number n there exists an m such that $n = 2m \Leftrightarrow$ not every number can be halved in $\mathfrak{A}$. This clearly is correct, take for instance $n = 1$.

Let us reflect for a moment on the valuation of proposition symbols; an 0-ary relation is a subset of $A^\emptyset = \{\emptyset\}$, i.e. it is $\emptyset$ or $\{\emptyset\}$ and these are, considered as ordinals, 0 or 1. So $[\![\overline{P}]\!]_\mathfrak{A} = P$, and P is a truth value. This makes our definition perfectly reasonable. Indeed, without aiming for a systematic treatment, we may observe that formulas correspond to subsets of A^k, where k is the number of free variables. For example, let $FV(\varphi) = \{z_1, \ldots, z_k\}$, then we could put $[\![\varphi]\!]_\mathfrak{A} = \{\langle a_1, \ldots, a_k \rangle | \mathfrak{A} \models \varphi(\overline{a}_1, \ldots, \overline{a}_k)\}(= \{\langle a_1, \ldots, a_n \rangle | [\![\varphi(\overline{a}_1, \ldots, \overline{a}_k)]\!]_\mathfrak{A} = 1\})$, thus stretching the meaning of $[\![\varphi]\!]_\mathfrak{A}$ a bit. It is immediately clear that applying quantifiers to φ reduces the "dimension". For example, $[\![\exists x\, P(x, y)]\!]_\mathfrak{A} = \{a | \mathfrak{A} \models P(\overline{b}, \overline{a}) \text{ for some } b\}$, which is the projection of $[\![P(x, y)]\!]_\mathfrak{A}$ onto the y-axis.

Exercises

1. Let $\mathfrak{N} = \langle N, +, \cdot, S, 0 \rangle$, and L a language of type $\langle -; 2, 2, 1; 1 \rangle$.
   (i) Give two distinct terms t in L such that $t^\mathfrak{N} = 5$.
   (ii) Show that for each natural number $n \in N$ there is a term t such that $t^\mathfrak{N} = n$.
   (iii) Show that for each $n \in N$ there are infinitely many terms t such that $t^\mathfrak{N} = n$.
2. Let $\mathfrak{A}$ be the structure of Exercise 1 (v) of Sect. 3.2. Evaluate $(((\overline{1} \to \overline{0}) \to \neg\overline{0}) \wedge (\neg\overline{0}) \to (\overline{1} \to \overline{0}))^\mathfrak{A}$, $(\overline{1} \to \neg(\neg\overline{0} \vee \overline{1}))^\mathfrak{A}$.
3. Let $\mathfrak{A}$ be the structure of Exercise 1 (viii) of Sect. 3.2. Evaluate $(|(\sqrt{3})^2 - \overline{5}|)^\mathfrak{A}$, $(\overline{1} - (|\overline{(-2)}| - (\overline{5} - \overline{(-2)})))^\mathfrak{A}$.
4. Which cases of Lemma 3.4.5 remain correct if we consider formulas in general?
5. For sentences σ we have $\mathfrak{A} \models \sigma$ or $\mathfrak{A} \models \neg\sigma$. Show that this does not hold for φ with $FV(\varphi) \neq \emptyset$. Show that not even for sentences $\models \sigma$ or $\models \neg\sigma$ holds in general.
6. Show for closed terms t and formulas φ (in $L(\mathfrak{A})$):
   $\mathfrak{A} \models t = \overline{[\![t]\!]_\mathfrak{A}}$,
   $\mathfrak{A} \models \varphi(t) \leftrightarrow \varphi(\overline{[\![t]\!]}_\mathfrak{A})$. (We will also obtain this as a corollary to the Substitution Theorem, Corollary 3.5.9.)
7. Show that $\mathfrak{A} \models \varphi \Rightarrow \mathfrak{A} \models \psi$ for all $\mathfrak{A}$, implies $\models \varphi \Rightarrow \models \psi$, but not vice versa.

3.5 Simple Properties of Predicate Logic

Our definition of validity (truth) was a straightforward extension of the valuation definition of propositional logic. As a consequence formulas which are instances of tautologies are true in all structures $\mathfrak{A}$ (Exercise 1). So we can copy many results from Sects. 2.2 and 2.3. We will use these results with a simple reference to propositional logic.

The specific properties concerning quantifiers will be treated in this section. First we consider the generalizations of De Morgan's laws.

Theorem 3.5.1

   (i) $\models \neg\forall x\varphi \leftrightarrow \exists x\neg\varphi$,
   (ii) $\models \neg\exists x\varphi \leftrightarrow \forall x\neg\varphi$,

(iii) $\models \forall x\varphi \leftrightarrow \neg\exists x\neg\varphi$,

(iv) $\models \exists x\varphi \leftrightarrow \neg\forall x\neg\varphi$.

Proof If there are no free variables involved, then the above equivalences are almost trivial. We will do one general case.

(i) Let $FV(\forall x\varphi) = \{z_1, \ldots, z_k\}$, then we must show $\mathfrak{A} \models \forall z_1 \ldots z_k(\neg\forall x\varphi(x, z_1, \ldots, z_k) \leftrightarrow \exists x\neg\varphi(x, z_1, \ldots, z_k))$, for all $\mathfrak{A}$.

So we have to show $\mathfrak{A} \models \neg\forall x\varphi(x, \overline{a}_1, \ldots, \overline{a}_k) \leftrightarrow \exists x\neg\varphi(x, \overline{a}_1, \ldots, \overline{a}_k)$ for arbitrary $a_1, \ldots, a_k \in |\mathfrak{A}|$. We apply the properties of $\models$ as listed in Lemma 3.4.5: $\mathfrak{A} \models \neg\forall x\varphi(x, \overline{a}_1, \ldots, \overline{a}_k) \Leftrightarrow \mathfrak{A} \not\models \forall x\varphi(x, \overline{a}_1, \ldots, \overline{a}_k) \Leftrightarrow$ not for all $b \in |\mathfrak{A}|\ \mathfrak{A} \models \varphi(\overline{b}, \overline{a}_1, \ldots, \overline{a}_k) \Leftrightarrow$ there is a $b \in |\mathfrak{A}|$ such that $\mathfrak{A} \models \neg\varphi(\overline{b}, \overline{a}_1, \ldots, \overline{a}_k) \Leftrightarrow \mathfrak{A} \models \exists x\neg\varphi(x, \overline{a}_1, \ldots, \overline{a}_k)$.

(ii) is similarly dealt with,

(iii) can be obtained from (i), (ii),

(iv) can be obtained from (i), (ii). $\qquad\square$

The order of quantifiers of the same sort is irrelevant, and quantification over a variable that does not occur can be deleted.

Theorem 3.5.2

(i) $\models \forall x\forall y\varphi \leftrightarrow \forall y\forall x\varphi$,

(ii) $\models \exists x\exists y\varphi \leftrightarrow \exists y\exists x\varphi$,

(iii) $\models \forall x\varphi \leftrightarrow \varphi$ *if* $x \notin FV(\varphi)$,

(iv) $\models \exists x\varphi \leftrightarrow \varphi$ *if* $x \notin FV(\varphi)$.

Proof Left to the reader. $\qquad\square$

We have already observed that $\forall$ and $\exists$ are, in a way, generalizations of $\wedge$ and $\vee$. Therefore it is not surprising that $\forall$ (resp. $\exists$) distributes over $\wedge$ (resp. $\vee$). $\forall$ (and $\exists$) distributes over $\vee$ (resp. $\wedge$) only if a certain condition is met.

Theorem 3.5.3

(i) $\models \forall x(\varphi \wedge \psi) \leftrightarrow \forall x\varphi \wedge \forall x\psi$,

(ii) $\models \exists x(\varphi \vee \psi) \leftrightarrow \exists x\varphi \vee \exists x\psi$,

(iii) $\models \forall x(\varphi(x) \vee \psi) \leftrightarrow \forall x\varphi(x) \vee \psi$ *if* $x \notin FV(\psi)$,

(iv) $\models \exists x(\varphi(x) \wedge \psi) \leftrightarrow \exists x\varphi(x) \wedge \psi$ *if* $x \notin FV(\psi)$.

Proof (i) and (ii) are immediate.

(iii) Let $FV(\forall x(\varphi(x) \vee \psi)) = \{z_1, \ldots, z_k\}$. We must show that $\mathfrak{A} \models \forall z_1 \ldots z_k$ $[\forall x(\varphi(x) \vee \psi) \leftrightarrow \forall x\varphi(x) \vee \psi]$ for all $\mathfrak{A}$, so we show, using Lemma 3.4.5, that $\mathfrak{A} \models \forall x[\varphi(x, \overline{a}_1, \ldots, \overline{a}_k) \vee \psi(\overline{a}_1, \ldots, \overline{a}_k)] \Leftrightarrow \mathfrak{A} \models \forall x\varphi(x, \overline{a}_1, \ldots, \overline{a}_k) \vee \psi(\overline{a}_1, \ldots, \overline{a}_k)$ for all $\mathfrak{A}$ and all $a_1, \ldots, a_k \in |\mathfrak{A}|$.

Note that in the course of the argument $a_1, \ldots, a_k$ remain fixed, so in the future we will no longer write them down.

$\Leftarrow$: $\mathfrak{A} \models \forall x \varphi(x, \text{\textemdash}) \vee \psi(\text{\textemdash}) \Leftrightarrow \mathfrak{A} \models \forall x \varphi(x, \text{\textemdash})$ or $\mathfrak{A} \models \psi(\text{\textemdash})$
$\Leftrightarrow \mathfrak{A} \models \varphi(\overline{b}, \text{\textemdash})$ for all b or $\mathfrak{A} \models \psi(\text{\textemdash})$.
If $\mathfrak{A} \models \psi(\text{\textemdash})$, then also $\mathfrak{A} \models \varphi(\overline{b}, \text{\textemdash}) \vee \psi(\text{\textemdash})$ for all b, and so
$\mathfrak{A} \models \forall x[\varphi(x, \text{\textemdash}) \vee \psi(\text{\textemdash})]$. If for all b $\mathfrak{A} \models \varphi(\overline{b}, \text{\textemdash})$ then
$\mathfrak{A} \models \varphi(\overline{b}, \text{\textemdash}) \vee \psi(\text{\textemdash})$ for all b, so $\mathfrak{A} \models \forall x(\varphi(x, \text{\textemdash}) \vee \psi(\text{\textemdash}))$.
In both cases we get the desired result.
$\Rightarrow$: We know that for each $b \in |\mathfrak{A}|$ $\mathfrak{A} \models \varphi(\overline{b}, \text{\textemdash}) \vee \psi(\text{\textemdash})$.
If $\mathfrak{A} \models \psi(\text{\textemdash})$, then also $\mathfrak{A} \models \forall x \varphi(x, \text{\textemdash}) \vee \psi(\text{\textemdash})$, so we are done.
If $\mathfrak{A} \not\models \psi(\text{\textemdash})$ then necessarily $\mathfrak{A} \models \varphi(\overline{b}, \text{\textemdash})$ for all b, so
$\mathfrak{A} \models \forall x \varphi(x, \text{\textemdash})$ and hence $\mathfrak{A} \models \forall x \varphi(x, \text{\textemdash}) \vee \psi(\text{\textemdash})$.

(iv) is similar. $\square$

In the proof above we have demonstrated a technique for dealing with the extra
free variables $z_1, \ldots, z_k$, that do not play an actual role. One chooses an arbitrary
string of elements $a_1, \ldots, a_k$ to substitute for the z_i's and keeps them fixed during
the proof. So in the future we will mostly ignore the extra variables.

WARNING

$\forall x (\varphi(x) \vee \psi(x)) \rightarrow \forall x \varphi(x) \vee \forall x \psi(x)$, and
$\exists x \varphi(x) \wedge \exists x \psi(x) \rightarrow \exists x (\varphi(x) \wedge \psi(x))$ are *not* true.

One of the Cinderella tasks in logic is the bookkeeping of substitution, keeping track of things in iterated substitution, etc. We will provide a number of useful
lemmas; none of them is difficult—it is a mere matter of clerical labor.

A word of advice to the reader: none of these syntactical facts is hard to prove, nor
is there a great deal to be learned from the proofs (unless one is after very specific
goals, such as complexity of certain predicates); the best procedure is to give the
proofs directly and only to look at the proofs in the book in case of emergency.

Lemma 3.5.4

(i) *Let x and y be distinct variables such that $x \notin FV(r)$, then $(t[s/x])[r/y] = (t[r/y])[s[r/y]/x]$,*

(ii) *Let x and y be distinct variables such that $x \notin FV(s)$ and let t and s be free for x and y in φ, then $(\varphi[t/x])[s/y] = (\varphi[s/y])[t[s/y]/x]$,*

(iii) *Let ψ be free for \$ in φ, and let t be free for x in φ and ψ, then $(\varphi[\psi/\$])[t/x] = (\varphi[t/x])[\psi[t/x]/\$]$,*

(iv) *Let φ, ψ be free for $\$_1, \$_2$ in σ, let ψ be free for $\$_2$ in φ, and let $\$_1$ not occur in ψ, then $(\sigma[\varphi/\$_1])[\psi/\$_2] = (\sigma[\psi/\$_2])[\varphi[\psi/\$_2]/\$_1]$.*

Proof (i) Induction on t.

- $t = c$, trivial.
- $t = x$. Then $t[s/x] = s$ and $(t[s/x])[r/y] = s[r/y]$; $(t[r/y])[s[r/y]/x] = x[s[r/y]/x] = s[r/y]$.

- $t = y$. Then $(t[s/x])[r/y] = y[r/y] = r$ and $(t[r/y])[s[r/y]/x] = r(s[r/y]/x) = r$, since $x \notin FV(r)$.
- $t = z$, where $z \neq x, y$, trivial.
- $t = f(t_1, \ldots, t_n)$. Then $(t[s/x])[r/y] = (f(t_1[s/x], \ldots))[r/y] = f((t_1[s/x])[r/y], \ldots) \overset{\text{i.h.}}{=} f((t_1[r/y])[s[r/y]/x], \ldots) = f(t_1[r/y], \ldots)[s[r/y]/x] = (t[r/y])[s[r/y]/x].$[1]

(ii) Induction on φ. Left to the reader.

(iii) Induction on φ.

- $\varphi = \bot$ or P distinct from \$. Trivial.
- $\varphi = \$$. Then $(\$[\psi/\$])[t/x] = \psi[t/x]$ and $(\$[t/x])[\psi[t/x]/\$] = \$[\psi[t/x]/\$] = \psi[t/x]$.
- $\varphi = \varphi_1 \square \varphi_2, \neg \varphi_1$. Trivial.
- $\varphi = \forall y \varphi_1$. Then $(\forall y \cdot \varphi_1[\psi/\$])[t/x] = (\forall y \cdot \varphi_1[\psi/\$])[t/x] = \forall y \cdot ((\varphi_1[\psi/\$])[t/x]) \overset{\text{i.h.}}{=} \forall y((\varphi_1[t/x])[\psi[t/x]/\$]) = ((\forall y \varphi_1)[t/x])[\psi[t/x]/\$]. \; \varphi = \exists y \varphi_1$. Idem.

(iv) Induction on σ. Left to the reader. $\qquad \square$

We immediately get the following corollary.

Corollary 3.5.5

(i) If $z \notin FV(t)$, then $t[\overline{a}/x] = (t[z/x])[\overline{a}/z]$,

(ii) If $z \notin FV(\varphi)$ and z is free for x in φ, then $\varphi[\overline{a}/x] = (\varphi[z/x])[\overline{a}/z]$.

It is possible to pull out quantifiers from a formula. The trick is well known in analysis: the bound variable in an integral may be changed, e.g. $\int x \, dx + \int \sin y \, dy = \int x \, dx + \int \sin x \, dx = \int (x + \sin x) \, dx$. In predicate logic we have a similar phenomenon.

Theorem 3.5.6 (Change of Bound Variables) *If x, y are free for z in φ and $x, y \notin FV(\varphi)$, then $\models \exists x \varphi[x/z] \leftrightarrow \exists y \varphi[y/z], \models \forall x \varphi[x/z] \leftrightarrow \forall y \varphi[y/z]$.*

Proof It suffices to consider φ with $FV(\varphi) \subseteq \{z\}$. We have to show $\mathfrak{A} \models \exists x \varphi[x/z] \Leftrightarrow \mathfrak{A} \models \exists y \varphi[y/z]$ for any $\mathfrak{A}$. $\mathfrak{A} \models \exists x \varphi[x/z] \Leftrightarrow \mathfrak{A} \models (\varphi[x/z])[\overline{a}/x]$ for some $a \Leftrightarrow \mathfrak{A} \models \varphi[\overline{a}/z]$ for some $a \Leftrightarrow \mathfrak{A} \models (\varphi[y/z])[\overline{a}/y]$ for some $a \Leftrightarrow \mathfrak{A} \models \exists y \varphi[y/z]$.

The universal quantifier is handled completely similarly. $\qquad \square$

The upshot of this theorem is that one can always replace a bound variable by a "fresh" one, i.e. one that did not occur in the formula. Now one easily concludes the following.

Corollary 3.5.7 *Every formula is equivalent to one in which no variable occurs both free and bound.*

---

[1] "i.h." indicates the use of the induction hypothesis.

We now can pull out quantifiers: $\forall x \varphi(x) \vee \forall x \psi(x) \leftrightarrow \forall x \varphi(x) \vee \forall y \psi(y)$ and $\forall x \varphi(x) \vee \forall y \psi(y) \leftrightarrow \forall x y (\varphi(x) \vee \psi(y))$, for a suitable y.

In order to handle predicate logic in an algebraic way we need the technique of substituting equivalents for equivalents.

Theorem 3.5.8 (Substitution Theorem)

  (i) $\models t_1 = t_2 \rightarrow s[t_1/x] = s[t_2/x]$,
 (ii) $\models t_1 = t_2 \rightarrow (\varphi[t_1/x] \leftrightarrow \varphi[t_2/x])$,
(iii) $\models (\varphi \leftrightarrow \psi) \rightarrow (\sigma[\varphi/\$] \leftrightarrow \sigma[\psi/\$])$.

Proof It is no restriction to assume that the terms and formulas are closed. We tacitly assume that the substitutions satisfy the "free for" conditions.

  (i) Let $\mathfrak{A} \models t_1 = t_2$, i.e. $t_1^{\mathfrak{A}} = t_2^{\mathfrak{A}}$. Now use induction on s.

- s is a constant or a variable. Trivial.
- $s = \overline{F}(s_1, \ldots, s_k)$. Then $s[t_i/x] = \overline{F}(s_1[t_i/x], \ldots)$ and $(s[t_i/x])^{\mathfrak{A}} = F((s_1[t_i])^{\mathfrak{A}}/x, \ldots)$. Induction hypothesis: $(s_j[t_1/x])^{\mathfrak{A}} = (s_j[t_2/x])^{\mathfrak{A}}$, $1 \leq j \leq k$. So $(s[t_1/x])^{\mathfrak{A}} = F((s_1[t_1/x])^{\mathfrak{A}}, \ldots) = F((s_1[t_2/x])^{\mathfrak{A}}, \ldots) = (s[t_2/x])^{\mathfrak{A}}$. Hence $\mathfrak{A} \models s[t_1/x] = s[t_2/x]$.

 (ii) Let $\mathfrak{A} \models t_1 = t_2$, so $t_1^{\mathfrak{A}} = t_2^{\mathfrak{A}}$. We show $\mathfrak{A} \models \varphi[t_1/x] \Leftrightarrow \mathfrak{A} \models \varphi[t_2/x]$ by induction on φ.

- φ is atomic. The case of a propositional symbol (including $\bot$) is trivial. So consider $\varphi = \overline{P}(s_1, \ldots, s_k)$. $\mathfrak{A} \models \overline{P}(s_1, \ldots, s_k)[t_1/x] \Leftrightarrow \mathfrak{A} \models \overline{P}(s_1[t_1/x], \ldots) \Leftrightarrow \langle (s_1[t_1/x])^{\mathfrak{A}}, \ldots, (s_k[t_1/x])^{\mathfrak{A}} \in P$. By (i) $(s_j[t_1/x])^{\mathfrak{A}} = (s_j[t_2/x])^{\mathfrak{A}}$, $j = 1, \ldots, k$.
  So we get $\langle (s_1[t_1/x])^{\mathfrak{A}}, \ldots \rangle \in P \Leftrightarrow \cdots \Leftrightarrow \mathfrak{A} \models \overline{P}(s_1, \ldots)[t_2/x]$.
- $\varphi = \varphi_1 \vee \varphi_2$, $\varphi_1 \wedge \varphi_2$, $\varphi_1 \rightarrow \varphi_2$, $\neg \varphi_1$. We consider the disjunction: $\mathfrak{A} \models (\varphi_1 \vee \varphi_2)[t_1/x] \Leftrightarrow \mathfrak{A} \models \varphi_1[t_1/x]$ or $\mathfrak{A} \models \varphi_2[t_1/x] \overset{i.h.}{\Leftrightarrow} \mathfrak{A} \models \varphi_1[t_2/x]$ or $\mathfrak{A} \models \varphi_2[t_2/x] \Leftrightarrow \mathfrak{A} \models (\varphi_1 \vee \varphi_2)[t_2/x]$.
  The remaining connectives are treated similarly.
- $\varphi = \exists y \psi$, $\varphi = \forall y \psi$.
  We consider the existential quantifier. $\mathfrak{A} \models (\exists y \psi)[t_1/x] \Leftrightarrow \mathfrak{A} \models \exists y (\psi[t_1/x]) \Leftrightarrow \mathfrak{A} \models \psi[t_1/x][\overline{a}/y]$ for some a.
  By Lemma 3.5.4 $\mathfrak{A} \models \psi[t_1/x][\overline{a}/y] \Leftrightarrow \mathfrak{A} \models (\psi[\overline{a}/y])[t_1[\overline{a}/y]/x]$. Apply the induction hypothesis to $\psi[\overline{a}/y]$ and the terms $t_1[\overline{a}/y], t_2[\overline{a}/y]$. Observe that t_1 and t_2 are closed, so $t_1[\overline{a}/y] = t_1$ and $t_2 = t_2[\overline{a}/y]$. We get $\mathfrak{A} \models \psi[t_2/x][\overline{a}/y]$, and hence $\mathfrak{A} \models \exists y \psi[t_2/x]$. The other implication is similar, and so is the case of the universal quantifier.

(iii) Let $\mathfrak{A} \models \varphi \Leftrightarrow \mathfrak{A} \models \psi$. We show $\mathfrak{A} \models \sigma[\varphi/\$] \Leftrightarrow \mathfrak{A} \models \sigma[\psi/\$]$ by induction on σ.

- σ is atomic. Both cases $\sigma = \$$ and $\sigma \neq \$$ are trivial.
- $\sigma = \sigma_1 \Box \sigma_2$ (or $\neg \sigma_1$). Left to the reader.

- $\sigma = \forall x \cdot \tau$. Observe that φ and ψ are closed, but even if they were not then x could not occur free in φ, ψ.

  $\mathfrak{A} \models (\forall x \cdot \tau)[\varphi/\$] \Leftrightarrow \mathfrak{A} \models \forall x (\tau[\varphi/\$])$. Pick an $a \in |\mathfrak{A}|$, then $\mathfrak{A} \models (\tau[\varphi/\$])[\overline{a}/x] \overset{3.5.4}{\Leftrightarrow} \mathfrak{A} \models (\tau[\overline{a}/x])[\varphi[\overline{a}/x]/\$] \Leftrightarrow \mathfrak{A} \models (\tau[\overline{a}/x])[\varphi/\$] \overset{i.h.}{\Leftrightarrow} \mathfrak{A} \models \tau[\overline{a}/x][\psi/\$] \Leftrightarrow \mathfrak{A} \models \tau[\overline{a}/x][\psi[\overline{a}/x]/\$] \Leftrightarrow \mathfrak{A} \models (\tau[\psi/\$])[\overline{a}/x]$. Hence $\mathfrak{A} \models \sigma[\varphi/\$] \Leftrightarrow \mathfrak{A} \models \sigma[\psi/\$]$. The existential quantifier is treated similarly. $\qquad\square$

Observe that in the above proof we have applied induction to "$\sigma[\varphi/\$]$ for all φ", because the substitution formula changed during the quantifier case.

Note that also the σ changed, so properly speaking we are applying induction to the rank (or we have to formulate the induction principle 3.3.4 a bit more liberally).

Corollary 3.5.9

(i) $[\![s[t/x]]\!] = [\![s[\overline{[\![t]\!]}/x]]\!]$,

(ii) $[\![\varphi[t/x]]\!] = [\![\varphi[\overline{[\![t]\!]}/x]]\!]$.

Proof We apply the Substitution Theorem. Consider an arbitrary $\mathfrak{A}$. Note that $[\![\overline{[\![t]\!]}]\!] = [\![t]\!]$ (by definition), so $\mathfrak{A} \models \overline{[\![t]\!]} = t$. Now (i) and (ii) follow immediately. $\qquad\square$

In a more relaxed notation, we can write (i) and (ii) as $[\![s(t)]\!] = [\![s(\overline{[\![t]\!]})]\!]$, or $\mathfrak{A} \models s(t) = s(\overline{[\![t]\!]})$ and $[\![\varphi(t)]\!] = [\![\varphi(\overline{[\![t]\!]})]\!]$, or $\mathfrak{A} \models \varphi(t) \leftrightarrow \varphi(\overline{[\![t]\!]})$.

Observe that $[\![t]\!] (= [\![t]\!]_{\mathfrak{A}})$ is just another way to write $t^{\mathfrak{A}}$.

Proofs involving detailed analysis of substitution are rather dreary but, unfortunately, unavoidable. The reader may simplify the above and other proofs by supposing the formulas involved to be closed. There is no real loss in generality, since we only introduce a number of constants from $L(\mathfrak{A})$ and check that the result is valid for all choices of constants.

Now we really can manipulate formulas in an algebraic way. Again, write $\varphi \approx \psi$ for $\models \varphi \leftrightarrow \psi$.

Examples

1. $\forall x \varphi(x) \to \psi \approx \neg \forall x \varphi(x) \vee \psi \approx \exists x (\neg \varphi(x)) \vee \psi \approx \exists x (\neg \varphi(x) \vee \psi) \approx \exists x (\varphi(x) \to \psi)$, where $x \notin FV(\psi)$.

2. $\forall x \varphi(x) \to \exists x \varphi(x) \approx \neg \forall x \varphi(x) \vee \exists x \varphi(x) \approx \exists x (\neg \varphi(x) \vee \varphi(x))$. The formula in the scope of the quantifier is true (already by propositional logic), so the formula itself is true.

Definition 3.5.10 A formula φ is in *prenex (normal) form* if φ consists of a (possibly empty) string of quantifiers followed by an open (i.e. quantifier free) formula. We also say that φ is a prenex formula.

Examples $\exists x \forall y \exists z \exists v(x = z \lor y = z \to v < y)$, $\forall x \forall y \exists z(P(x, y) \land Q(y, x) \to P(z, z))$.

By pulling out quantifiers we can reduce each formula to a formula in prenex form.

Theorem 3.5.11 *For each φ there is a prenex formula ψ such that $\models \varphi \leftrightarrow \psi$.*

Proof First eliminate $\to$ and $\leftrightarrow$. Use induction on the resulting formula φ'.

For atomic φ' the theorem is trivial. If $\varphi' = \varphi_1 \lor \varphi_2$ and φ_1, φ_2 are equivalent to prenex ψ_1, ψ_2 then

$$\psi_1 = (Q_1 y_1) \cdots (Q_n y_n)\psi^1,$$
$$\psi_2 = (Q'_1 z_1) \cdots (Q'_m z_m)\psi^2,$$

where Q_i, Q'_j are quantifiers and ψ^1, ψ^2 open. By Theorem 3.5.6 we can choose all bound variables distinct, taking care that no variable is both free and bound. Applying Theorem 3.5.3 we find

$$\models \varphi' \leftrightarrow (Q_1 y_1) \cdots (Q_n y_n)(Q'_1 z_1) \cdots (Q'_m z_m)(\psi^1 \lor \psi^2),$$

so we are done.

The remaining cases are left to the reader.   □

In ordinary mathematics it is usually taken for granted that the benevolent reader can guess the intentions of the author, not only the explicit ones, but also the ones that are tacitly handed down generations of mathematicians. Take for example the definition of convergence of a sequence: $\forall \varepsilon > 0 \exists n \forall m(|a_n - a_{n+m}| < \varepsilon)$. In order to make sense out of this expression one has to add: the variables n, m range over natural numbers. Unfortunately our syntax does not allow for variables of different sorts. So how do we incorporate expressions of the above kind? The answer is simple: we add predicates of the desired sort and indicate inside the formula the "nature" of the variable.

Example Let $\mathfrak{A} = \langle R, Q, < \rangle$ be the structure of the reals with the set of rational numbers singled out, provided with the natural order. The sentence $\sigma := \forall x y(x < y \to \exists z(Q(z) \land x < z \land z < y))$ can be interpreted in $\mathfrak{A} : \mathfrak{A} \models \sigma$, and it tells us that the rationals are dense in the reals (in the natural ordering). We find this mode of expression, however, rather cumbersome. Therefore we introduce the notion of *relativized quantifiers*. Since it does not matter whether we express informally "x is rational" by $x \in Q$ or $Q(x)$, we will suit ourselves and at any time choose the notation which is most convenient. We use $(\exists x \in Q)$ and $(\forall x \in Q)$ as informal notation for "there exists an x in Q" and "for all x in Q". Now we can write σ as $\forall x y(x < y \to \exists z \in Q(x < z \land z < y))$. Note that we do *not* write $(\forall x y \in R)(\text{———})$, since: (1) there is no relation R in $\mathfrak{A}$, (2) variables automatically range over $|\mathfrak{A}| = R$.

Let us now define the relativization of a quantifier properly.

Definition 3.5.12 If P is a unary predicate symbol, then $(\forall x \in P)\varphi :=$ $\forall x(P(x) \to \varphi)$, $(\exists x \in P)\varphi := (\exists x)(P(x) \wedge \varphi)$.

This notation has the intended meaning, as appears from $\mathfrak{A} \models (\forall x \in P)\varphi \Leftrightarrow$ for all $a \in P^{\mathfrak{A}}$ $\mathfrak{A} \models \varphi[\bar{a}/x]$, $\mathfrak{A} \models (\exists x \in P)\varphi \Leftrightarrow$ there exists an $a \in P^{\mathfrak{A}}$ such that $\mathfrak{A} \models \varphi[\bar{a}/x]$. The proof is immediate.

We will often use informal notation, such as $(\forall x > 0)$ or $(\exists y \neq 1)$, which can be cast into the above form. The meaning of such notation will always be evident. One can restrict *all* quantifiers to the same set (predicate), which amounts to passing to a restricted universe (cf. Exercise 11).

It is a common observation that by strengthening a part of a conjunction (disjunction) the whole formula is strengthened, but that by strengthening φ in $\neg\varphi$ the whole formula is weakened. This phenomenon has a syntactic origin, and we will introduce a bit of terminology to handle it smoothly. We inductively define that a subformula occurrence φ is positive (negative) in σ.

Definition 3.5.13 Sub^+ and Sub^- are defined simultaneously by

$$Sub^+(\varphi) = \{\varphi\}$$
$$Sub^-(\varphi) = \emptyset \quad \text{for atomic } \varphi$$
$$Sub^+(\varphi_1 \square \varphi_2) = Sub^+(\varphi_1) \cup Sub^+(\varphi_2) \cup \{\varphi_1 \square \varphi_2\}$$
$$Sub^-(\varphi_1 \square \varphi_2) = Sub^-(\varphi_1) \cup Sub^-(\varphi_2) \quad \text{for } \square \in \{\wedge, \vee\}$$
$$Sub^+(\varphi_1 \to \varphi_2) = Sub^+(\varphi_2) \cup Sub^-(\varphi_1) \cup \{\varphi_1 \to \varphi_2\}$$
$$Sub^-(\varphi_1 \to \varphi_2) = Sub^+(\varphi_1) \cup Sub^-(\varphi_2)$$
$$Sub^+(Qx.\varphi) = Sub^+(\varphi) \cup \{Qx.\varphi\}$$
$$Sub^-(Qx.\varphi) = Sub^-(\varphi) \quad \text{for } Q \in \{\forall, \exists\}.$$

If $\varphi \in Sub^+(\psi)$, then we say that φ *occurs positively* in ψ (similarly for negative occurrences).

We could have restricted ourselves to $\wedge$, $\to$ and $\forall$, but it does not ask much extra space to handle the other connectives.

The following theorem makes the basic intuition clear: if a positive part of a formula increases in truth value then the formula increases in truth value (better: does not decrease in truth value). We express this role of positive and negative subformulas as follows.

Theorem 3.5.14 *Let* φ (ψ) *not occur negatively (not positively) in* σ, *then:*

  (i) $[\![\varphi_1]\!] \leq [\![\varphi_2]\!] \Rightarrow [\![\sigma[\varphi_1/\varphi]]\!] \leq [\![\sigma[\varphi_2/\varphi]]\!]$,
  (ii) $[\![\psi_1]\!] \leq [\![\psi_2]\!] \Rightarrow [\![\sigma[\psi_1/\psi]]\!] \geq [\![\sigma[\psi_2/\psi]]\!]$,
  (iii) $\mathfrak{A} \models (\varphi_1 \to \varphi_2) \to (\sigma[\varphi_1/\varphi] \to \sigma[\varphi_2/\varphi])$,
  (iv) $\mathfrak{A} \models (\psi_1 \to \psi_2) \to (\sigma[\psi_2/\psi] \to \sigma[\psi_1/\psi])$.

Proof Induction on σ.   □

Exercises

1. Show that all propositional tautologies are true in all structures (of the right similarity type).
2. Let $x \notin FV(\psi)$. Show
   (i) $\models (\forall x \varphi \to \psi) \leftrightarrow \exists x (\varphi \to \psi)$,
   (ii) $\models (\exists x \varphi \to \psi) \leftrightarrow \forall x (\varphi \to \psi)$,
   (iii) $\models (\psi \to \exists x \varphi) \leftrightarrow \exists x (\psi \to \varphi)$,
   (iv) $\models (\psi \to \forall x \varphi) \leftrightarrow \forall x (\psi \to \varphi)$.
3. Show that the condition on $FV(\psi)$ in Exercise 2 is necessary.
4. Show $\not\models \forall x \exists y \varphi \leftrightarrow \exists y \forall x \varphi$.
5. Show $\models \varphi \Rightarrow \models \forall x \varphi$ and $\models \exists x \varphi$.
6. Show $\not\models \exists x \varphi \to \forall x \varphi$.
7. Show $\not\models \exists x \varphi \wedge \exists x \psi \to \exists x (\varphi \wedge \psi)$.
8. Show that the condition on x, y in Theorem 3.5.6 is necessary.
9. Show
   (i) $\models \forall x (\varphi \to \psi) \to (\forall x \varphi \to \forall x \psi)$;
   (ii) $\models (\exists x \varphi \to \exists x \psi) \to \exists x (\varphi \to \psi)$;
   (iii) $\models \forall x (\varphi \leftrightarrow \psi) \to (\forall x \varphi \leftrightarrow \forall x \psi)$;
   (iv) $\models (\forall x \varphi \to \exists x \psi) \leftrightarrow \exists x (\varphi \to \psi)$;
   (v) $\models (\exists x \varphi \to \forall x \psi) \to \forall x (\varphi \to \psi)$.
10. Show that the converses of Exercises 9 (i)–(iii) and (v) do not hold.
11. Let L have a unary predicate P. Define the relativization σ^P of σ by

$$\sigma^P := \sigma \quad \text{for atomic } \varphi,$$
$$(\varphi \square \psi)^P := \varphi^P \square \psi^P,$$
$$(\neg \varphi)^P := \neg \varphi^P,$$
$$(\forall x \varphi)^P := \forall x \big(P(x) \to \varphi^P \big),$$
$$(\exists x \varphi)^P := \exists x \big(P(x) \wedge \varphi^P \big).$$

Let $\mathfrak{A}$ be a structure without functions and constants. Consider the structure $\mathfrak{B}$ with universe $P^{\mathfrak{A}}$ and relations which are restrictions of the relations of $\mathfrak{A}$, where $P^{\mathfrak{A}} \neq \emptyset$. Show $\mathfrak{A} \models \sigma^P \Leftrightarrow \mathfrak{B} \models \sigma$ for sentences σ. Why are only relations allowed in $\mathfrak{A}$?

12. Let S be a binary predicate symbol. Show $\models \neg \exists y \forall x (S(y, x) \leftrightarrow \neg S(x, x))$. (Think of "$y$ shaves x" and recall Russell's barber paradox.)
13.   (i) Show that the "free for" conditions cannot be dropped from Theorem 3.5.8.
      (ii) Show $\models t = s \Rightarrow \models \varphi[t/x] \leftrightarrow \varphi[s/x]$.
      (iii) Show $\models \varphi \leftrightarrow \psi \Rightarrow \models \sigma[\varphi/\$] \leftrightarrow \sigma[\psi/\$]$.
14. Find prenex normal forms for
   (a) $\neg((\neg \forall x \varphi(x) \vee \forall x \psi(x)) \wedge (\exists x \sigma(x) \to \forall x \tau(x)))$,
   (b) $\forall x \varphi(x) \leftrightarrow \exists x \psi(x)$,
   (c) $\neg(\exists x \varphi(x, y) \wedge (\forall y \psi(y) \to \varphi(x.x)) \to \exists x \forall y \sigma(x, y))$,
   (d) $((\forall x \varphi(x) \to \exists y \psi(x, y)) \to \psi(x, x)) \to \exists x \forall y \sigma(x, y)$.
15. Show $\models \exists x (\varphi(x) \to \forall y \varphi(y))$. (It is instructive to think of $\varphi(x)$ as "x drinks".)

3.6 Identity

We have limited ourselves in this book to the consideration of structures with identity (see Definition 3.4.2), and hence of languages with identity. Therefore we classified "=" as a logical symbol, rather than a mathematical one. We can, however, not treat = as just some binary predicate, since identity satisfies a number of characteristic axioms, listed below.

I_1 $\forall x (x = x)$,

I_2 $\forall xy (x = y \rightarrow y = x)$,

I_3 $\forall xyz (x = y \wedge y = z \rightarrow x = z)$,

I_4 $\forall x_1 \ldots x_n y_1 \ldots y_n (\bigwedge_{i \leq n} x_i = y_i \rightarrow t(x_1, \ldots, x_n) = t(y_1, \ldots, y_n))$,

   $\forall x_1 \ldots x_n y_1 \ldots y_n (\bigwedge_{i \leq n} x_i = y_i \rightarrow (\varphi(x_1, \ldots, x_n) \rightarrow \varphi(y_1, \ldots, y_n)))$.

One simply checks that I_1, I_2, I_3 are true in all structures $\mathfrak{A}$. For I_4, observe that we can suppose the formulas to be closed. Otherwise we add quantifiers for the remaining variables and add dummy identities, e.g. $\forall z_1 \ldots z_k x_1 \ldots x_n y_1 \ldots y_n$ $(\bigwedge_{i \leq n} x_i = y_i \wedge \bigwedge_{i \leq k} z_i = z_i \rightarrow t(x_1, \ldots, x_n) = t(y_1, \ldots, y_n))$. Now $(t(\overline{a}_1, \ldots, \overline{a}_n))^{\mathfrak{A}}$ defines a function $t^{\mathfrak{A}}$ on $|\mathfrak{A}|^n$, obtained from the given functions of $\mathfrak{A}$ by various substitutions, hence $a_i = b_i (i \leq n) \Rightarrow (t(\overline{a}_1, \ldots, \overline{a}_n))^{\mathfrak{A}} = (t(\overline{b}_1, \ldots, \overline{b}_n))^{\mathfrak{A}}$. This establishes the first part of I_4.

The second part is proved by induction on φ (using the first part): e.g. consider the universal quantifier case and let $a_i = b_i$ for all $i \leq n$. $\mathfrak{A} \models \forall u \varphi(u, \overline{a}_1, \ldots, \overline{a}_n)$ $\Leftrightarrow$ $\mathfrak{A} \models \varphi(\overline{c}, \overline{a}_1, \ldots, \overline{a}_n)$ for all $c \overset{\text{i.h.}}{\Leftrightarrow} \mathfrak{A} \models \varphi(\overline{c}, \overline{b}_1, \ldots, \overline{b}_n)$ for all $c \Leftrightarrow \mathfrak{A} \models \forall u \varphi(u, \overline{b}_1, \ldots, \overline{b}_n)$. So $\mathfrak{A} \models (\bigwedge_{i \leq n} \overline{a}_i = \overline{b}_i) \Rightarrow \mathfrak{A} \models \forall u \varphi(u, \overline{a}_1, \ldots, \overline{a}_n) \rightarrow \forall u \varphi(u, \overline{b}_1, \ldots \overline{b}_n)$. This holds for all $a_1, \ldots, a_n, \ b_1, \ldots, b_n$, hence $\mathfrak{A} \models \forall x_1, \ldots x_n y_1 \ldots y_n (\bigwedge_{i \leq n} x_i = y_i \rightarrow (\forall u \varphi(u, x_1, \ldots, x_n) \rightarrow \forall u \varphi(u, y_1, \ldots, y_n)))$.

Note that φ (respectively t) in I_4 can be any formula (respectively term), so I_4 stands for infinitely many axioms. We call such an "instant axiom" an *axiom schema*.

The first three axioms state that identity is an equivalence relation. I_4 states that identity is a congruence with respect to all (definable) relations.

It is important to realize that from the axioms alone, we cannot determine the precise nature of the interpreting relation. We explicitly adopt the convention that "=" will always be interpreted by real equality. Inequality, $\neg x = y$, is abbreviated as $x \neq y$.

Exercises

1. Show $\models \forall x \exists y (x = y)$.
2. Show $\models \forall x (\varphi(x) \leftrightarrow \exists y (x = y \wedge \varphi(y)))$ and $\models \forall x (\varphi(x) \leftrightarrow \forall y (x = y \rightarrow \varphi(y)))$, where y does not occur in $\varphi(x)$.
3. Show that $\models \varphi(t) \leftrightarrow \forall x (x = t \rightarrow \varphi(x))$ if $x \notin FV(t)$.
4. Show that the conditions in Exercises 2 and 3 are necessary.
5. Consider $\sigma_1 = \forall x (x \sim x)$, $\sigma_2 = \forall xy (x \sim y \rightarrow y \sim x)$, $\sigma_3 = \forall xyz (x \sim y \wedge y \sim z \rightarrow x \sim z)$. Show that if $\mathfrak{A} \models \sigma_1 \wedge \sigma_2 \wedge \sigma_3$, where $\mathfrak{A} = \langle A, R \rangle$, then R is an equivalence relation. N.B. $x \sim y$ is a suggestive notation for the atom $\overline{R}(x, y)$.

6. Let $\sigma_4 = \forall xyz(x \sim y \wedge x \sim z \rightarrow y \sim z)$. Show that $\sigma_1, \sigma_4 \models \sigma_2 \wedge \sigma_3$.
7. Consider the schema $\sigma_5 : x \sim y \rightarrow (\varphi[x/z] \rightarrow \varphi[y/z])$. Show that $\sigma_1, \sigma_5 \models \sigma_2 \wedge \sigma_3$. N.B. If σ is a schema, then $\bigwedge \cup \{\sigma\} \models \varphi$ stands for $\bigwedge \cup \Sigma \models \varphi$, where Σ consists of all instances of σ.
8. Derive the term version of I_4 from the formula version.

3.7 Examples

We will consider languages for some familiar kinds of structures. Since all languages are built in the same way, we shall not list the logical symbols. All structures are supposed to satisfy the identity axioms $I_1 - I_4$.

For a refinement see Lemma 3.10.2.

1. *The language of identity.* Type: $\langle -; -; 0 \rangle$.

   Alphabet.
   Predicate symbol: $=$

The structures of this type are of the form $\mathfrak{A} = \langle A \rangle$, and satisfy I_1, I_2, I_3. (In this language I_4 follows from I_1, I_2, I_3, cf. Sect. 3.10 Exercise 5.)

In an identity structure there is so little "structure", that all one can virtually do is look for the number of elements (cardinality). There are sentences λ_n and μ_n saying that there are at least (or at most) n elements (Exercise 3, Sect. 4.1)

$$\lambda_n := \exists y_1 \ldots y_n \bigwedge_{i \neq j} y_i \neq y_j \quad (n > 1),$$

$$\mu_n := \forall y_0 \ldots y_n \bigvee_{i \neq j} y_i = y_j \quad (n > 0).$$

So $\mathfrak{A} \models \lambda_n \wedge \mu_n$ iff $|\mathfrak{A}|$ has exactly n elements. Since universes are not empty $\models \exists x(x = x)$ always holds.

We can also formulate *"there exists a unique x such that ..."*.

Definition 3.7.1 $\exists! x \varphi(x) := \exists x(\varphi(x) \wedge \forall y(\varphi(y) \rightarrow x = y))$, where y does not occur in $\varphi(x)$.

Note that $\exists! x \varphi(x)$ is an (informal) abbreviation.

2. *The language of partial order.* Type: $\langle 2; -; 0 \rangle$.

   Alphabet.
   Predicate symbols: $=, \leq$

   Abbreviations

$$x \neq y := \neg x = y, \qquad\qquad x < y := x \leq y \wedge x \neq y,$$

$$x > y := y < x, \qquad\qquad x \geq y := y \leq x,$$

$$x \leq y \leq z := x \leq y \wedge y \leq z.$$

Definition 3.7.2 $\mathfrak{A}$ is a partially ordered set (poset) if $\mathfrak{A}$ is a model of

$$\forall xyz(x \leq y \leq z \rightarrow x \leq z),$$

$$\forall xy(x \leq y \leq x \leftrightarrow x = y).$$

The notation may be misleading, since one usually introduces the relation $\leq$ (e.g. on the reals) as a disjunction: $x < y$ or $x = y$. In our alphabet the relation is primitive; another symbol might have been preferable, but we chose to observe the tradition. Note that the relation is reflexive: $x \leq x$.

Partially ordered sets are very basic in mathematics, appearing in many guises. It is often convenient to visualize posets by means of diagrams, where $a \leq b$ is represented as equal or above (respectively to the right). One of the traditions in logic is to keep objects and their names apart. Thus we speak of function symbols which are interpreted as functions, etc. However, in practice this is a bit cumbersome. We prefer to use the same notation for the syntactic objects and their interpretations, e.g. if $\mathfrak{R} = \langle \mathbb{R}, \leq \rangle$ is the partially ordered set of reals, then $\mathfrak{R} \models \forall x \exists y(x \leq y)$, whereas it should be something like $\forall x \exists y(x \widetilde{\leq} y)$ to distinguish the symbol from the relation.

The "$\leq$" in $\mathfrak{R}$ stands for the actual relation and the "$\leq$" in the sentence stands for the predicate symbol. The reader is urged to distinguish symbols in their various guises.

We show some diagrams of posets.

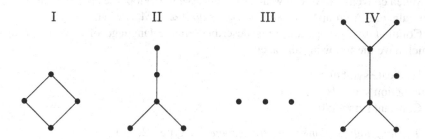

From the diagrams we can easily read off a number of properties. For example, $\mathfrak{A}_1 \models \exists x \forall y(x \leq y)$ ($\mathfrak{A}_i$ is the structure with the diagram of Fig. I), i.e. $\mathfrak{A}_1$ has a least element (a *minimum*). $\mathfrak{A}_3 \models \forall x \neg \exists y(x < y)$, i.e. in $\mathfrak{A}_3$ no element is strictly less than another element.

Definition 3.7.3

(i) $\mathfrak{A}$ is a (*linearly* or *totally*) *ordered set* if it is a poset and $\mathfrak{A} \models \forall xy(x \leq y \lor y \leq x)$ (each two elements are comparable).
(ii) $\mathfrak{A}$ is *densely ordered* if $\mathfrak{A} \models \forall xy(x < y \rightarrow \exists z(x < z \land z < y))$ (between any two elements there is a third one).

It is a moderately amusing exercise to find sentences that distinguish between structures and vice versa. For example, we can distinguish $\mathfrak{A}_3$ and $\mathfrak{A}_4$ (from the diagram above) as follows: in $\mathfrak{A}_4$ there is precisely *one* element that is incomparable with all other elements, in $\mathfrak{A}_3$ there are more such elements. Put $\sigma(x) :=$ $\forall y(y \neq x \to \neg y \leq x \land \neg x \leq y)$. Then $\mathfrak{A}_4 \models \forall xy(\sigma(x) \land \sigma(y) \to x = y)$, but $\mathfrak{A}_3 \models \neg \forall xy(\sigma(x) \land \sigma(y) \to x = y)$.

3. *The language of groups.* Type: $\langle -; 2, 1; 1 \rangle$.

   Alphabet.
   Predicate symbol: $=$
   Function symbols: $\cdot, ^{-1}$
   Constant symbol: e

Notation In order to conform with practice we write $t \cdot s$ and t^{-1} instead of $\cdot (t, s)$ and $^{-1}(t)$.

Definition 3.7.4 $\mathfrak{A}$ is a *group* if it is a model of

$$\forall xyz((x \cdot y) \cdot z = x \cdot (y \cdot z)),$$
$$\forall x(x \cdot e = x \land e \cdot x = x),$$
$$\forall x(x \cdot x^{-1} = e \land x^{-1} \cdot x = e).$$

When convenient, we will write ts for $t \cdot s$; we will adopt the bracket conventions from algebra. A group $\mathfrak{A}$ is *commutative* or *abelian* if $\mathfrak{A} \models \forall xy(xy = yx)$.

Commutative groups are often described in the language of *additive* groups, which have the following alphabet:

   Predicate symbol: $=$
   Function symbols: $+, -$
   Constant symbol: 0

4. *The language of plane projective geometry.* Type: $\langle 2; -; 0 \rangle$

The structures one considers are projective planes, which are usually taken to consist of *points* and *lines* with an *incidence relation*. In this approach the type would be $\langle 1, 1, 2; -; 0 \rangle$. We can, however, use a more simple type, since a point can be defined as something that is incident with a line, and a line as something for which we can find a point which is incident with it. Of course this requires a non-symmetric incidence relation.

We will now list the axioms, which deviate somewhat from the traditional set. It is a simple exercise to show that the system is equivalent to the standard sets.

   Alphabet.
   Predicate symbols: $I, =$

We introduce the following abbreviations:

$$\Pi(x) := \exists y(xIy), \qquad \Lambda(y) := \exists x(xIy).$$

Definition 3.7.5 $\mathfrak{A}$ is a *projective plane* if it satisfies

$\gamma_0 : \forall x(\Pi(x) \leftrightarrow \neg \Lambda(x))$,

$\gamma_1 : \forall xy(\Pi(x) \wedge \Pi(y) \to \exists z(xIz \wedge yIz))$,

$\gamma_2 : \forall uv(\Lambda(u) \wedge \Lambda(v) \to \exists x(xIu \wedge xIv))$,

$\gamma_3 : \forall xyuv(xIu \wedge yIu \wedge xIv \wedge yIv \to x = y \vee u = v)$,

$\gamma_4 : \exists x_0 x_1 x_2 x_3 u_0 u_1 u_2 u_3 \left(\bigwedge x_i I u_i \wedge \bigwedge_{j=i-1(\mathrm{mod}\ 3)} x_i I u_j \wedge \bigwedge_{\substack{j \neq i-1(\mathrm{mod}\ 3) \\ i \neq j}} \neg x_i I u_j \right)$.

γ_0 tells us that in a projective plane everything is either a point or a line, γ_1 and γ_2 tell us that "any two lines intersect in a point" and "any two points can be joined by a line", by γ_3 this point (or line) is unique if the given lines (or points) are distinct. Finally γ_4 makes projective planes non-trivial, in the sense that there are enough points and lines.

$\Pi^{\mathfrak{A}} = \{a \in |\mathfrak{A}| \, | \, \mathfrak{A} \models \Pi(\bar{a})\}$ and $\Lambda^{\mathfrak{A}} = \{b \in |\mathfrak{A}| \, | \, \mathfrak{A} \models \Lambda(\bar{b})\}$ are the sets of *points* and *lines* of $\mathfrak{A}$; $I^{\mathfrak{A}}$ is the *incidence relation* on $\mathfrak{A}$.

The above formalization is rather awkward. One usually employs a two-sorted formalism, with $P, Q, R, \ldots$ varying over points and $\ell, m, n \ldots$ varying over lines. The first axiom is then suppressed by convention. The remaining axioms become

$\gamma_1' : \forall PQ \exists \ell(PI\ell \wedge QI\ell)$,

$\gamma_2' : \forall \ell m \exists P(PI\ell \wedge PIm)$,

$\gamma_3' : \forall PQ\ell m(PI\ell \wedge QI\ell \wedge PIm \wedge QIm \to P = Q \vee \ell = m)$,

$\gamma_4' : \exists P_0 P_1 P_2 P_3 \ell_0 \ell_1 \ell_2 \ell_3 \left(\bigwedge P_i I \ell_i \wedge \bigwedge_{j=i-1(mod 3)} P_i I \ell_j \wedge \bigwedge_{\substack{j \neq i-1(\mathrm{mod}\ 3) \\ i \neq j}} \neg P_i I \ell_j \right)$.

The translation from one language to the other presents no difficulty. The above axioms are different from the ones usually given in the course in projective geometry. We have chosen these particular axioms because they are easy to formulate and also because the *duality principle* follows immediately (cf. Sect. 3.10, Exercise 8). The fourth axiom is an existence axiom, it merely says that certain things exist; it can be paraphrased differently: there are four points no three of which are collinear (i.e. on a line). Such an existence axiom is merely a precaution to make sure that trivial models are excluded. In this particular case, one would not do much geometry if there were only one triangle!

5. *The language of rings with unity.* Type: $\langle -; 2, 2, 1; 2 \rangle$

Alphabet.
Predicate symbol: $=$
Function symbols: $+, \cdot, -$
Constant symbols: $0, 1$

Definition 3.7.6 $\mathfrak{A}$ is a *ring* (with unity) if it is a model of

$$\forall xyz((x+y)+z=x+(y+z)),$$
$$\forall xy(x+y=y+x),$$
$$\forall xyz((xy)z=x(yz)),$$
$$\forall xyz(x(y+z)=xy+xz),$$
$$\forall xyz((x+y)z=xz+yz),$$
$$\forall x(x+0=x),$$
$$\forall x(x+(-x)=0),$$
$$\forall x(1\cdot x=x \wedge x\cdot 1=x), 0\neq 1.$$

A ring $\mathfrak{A}$ is *commutative* if $\mathfrak{A}\models \forall xy(xy=yx)$.
A ring $\mathfrak{A}$ is a *division ring* if $\mathfrak{A}\models \forall x(x\neq 0 \rightarrow \exists y(xy=1))$.
A commutative division ring is called a *field*.

Actually it is more convenient to have an inverse function symbol available in the language of fields, which therefore has type $\langle -; 2, 2, 1, 1; 2\rangle$.

Therefore we add to the above list the sentences $\forall x(x\neq 0 \rightarrow x\cdot x^{-1}=1 \wedge x^{-1}\cdot x=1)$ and $0^{-1}=1$.

Note that we must somehow "fix the value of 0^{-1}", the reason will appear in Sect. 3.10, Exercise 2.

6. *The language of arithmetic.* Type $\langle -; 2, 2, 1; 1\rangle$.

Alphabet.
Predicate symbol: $=$
Function symbols: $+, \cdot, S$
Constant symbol: 0
(S stands for the successor function $n \mapsto n+1$).

Historically, the language of arithmetic was introduced by Peano with the intention to describe the natural numbers with plus, times and successor up to an isomorphism. This is in contrast to, e.g. the theory of groups, in which one tries to capture a large class of non-isomorphic structures. It has turned out, however, that Peano's axioms characterize a large class of structures, which we will call (lacking a current term) *Peano structures*. Whenever confusion threatens we will use the official notation for the zero symbol: $\overline{0}$, but mostly we will trust the good sense of the reader.

Definition 3.7.7 A *Peano structure* $\mathfrak{A}$ is a model of

$$\forall x(0\neq S(x)),$$
$$\forall xy(S(x)=S(y)\rightarrow x=y),$$
$$\forall x(x+0=x),$$

$$\forall xy(x + S(y) = S(x + y)),$$
$$\forall x(x \cdot 0 = 0),$$
$$\forall xy(x \cdot S(y) = x \cdot y + x),$$
$$\varphi(0) \wedge \forall x(\varphi(x) \rightarrow \varphi(S(x))) \rightarrow \forall x \varphi(x).$$

The last axiom schema is called the *induction schema* or the *principle of mathematical induction*.

It will prove convenient to introduce some notation. We define:

$$\overline{1} := S(\overline{0}), \qquad \overline{2} := S(\overline{1}), \quad \text{and in general } \overline{n+1} := S(\overline{n}),$$
$$x < y := \exists z(x + Sz = y),$$
$$x \leq y := x < y \vee x = y.$$

There is one particular Peano structure which is the intended model of arithmetic, namely the structure of the ordinary natural numbers, with the ordinary addition, multiplication and successor (e.g. the finite ordinals in set theory). We call this Peano structure the *standard model* $\mathfrak{N}$, and the ordinary natural numbers are called the *standard numbers*.

One easily checks that $\overline{n}^{\mathfrak{N}} = n$ and $\mathfrak{N} \models \overline{n} < \overline{m} \Leftrightarrow n < m$: by definition of interpretation we have $\overline{0}^{\mathfrak{N}} = 0$. Assume $\overline{n}^{\mathfrak{N}} = n$, then $\overline{n+1}^{\mathfrak{N}} = (S(\overline{n}))^{\mathfrak{N}} = \overline{n}^{\mathfrak{N}} + 1 = n + 1$. We now apply mathematical induction in the meta-language, and obtain $\overline{n}^{\mathfrak{N}} = n$ for all n. For the second claim see Exercise 13. In $\mathfrak{N}$ we can define all kinds of sets, relations and numbers. To be precise we say that a k-ary relation R in $\mathfrak{N}$ is defined by φ if $\langle a_1, \ldots, a_k \rangle \in R \Leftrightarrow \mathfrak{N} \models \varphi(\overline{a_1}, \ldots, \overline{a_k})$. An element $a \in |\mathfrak{N}|$ is defined in $\mathfrak{N}$ by φ if $\mathfrak{N} \models \varphi(\overline{b}) \Leftrightarrow b = a$, or $\mathfrak{N} \models \forall x(\varphi(x) \leftrightarrow x = \overline{a})$.

Examples

(a) The set of even numbers is defined by $E(x) := \exists y(x = y + y)$.
(b) The divisibility relation is defined by $x|y := \exists z(xz = y)$.
(c) The set of prime numbers is defined by $P(x) := \forall yz(x = yz \rightarrow y = 1 \vee z = 1) \wedge x \neq 1$.

We say that we have introduced predicates $E, |$ and P by (explicit) definition.

7. *The language of graphs*

We usually think of graphs as geometric figures consisting of vertices and edges connecting certain of the vertices. A suitable language for the theory of graphs is obtained by introducing a predicate R which expresses the fact that two vertices are connected by an edge. Hence, we don't need variables or constants for edges.

Alphabet.
Predicate symbols: $R, =$

Definition 3.7.8 A *graph* is a structure $\mathfrak{A} = \langle A, R \rangle$ satisfying the following axioms:

$$\forall xy(R(x, y) \to R(y, x))$$
$$\forall x \neg R(x, x).$$

This definition is in accordance with the geometric tradition. There are elements, called vertices, of which some are connected by edges. Note that two vertices are connected by at most one edge. Furthermore there is no (need for an) edge from a vertex to itself. This is geometrically inspired; however, from the point of view of the numerous applications of graphs it appears that more liberal notions are required.

Examples

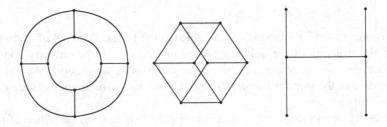

We can also consider graphs in which the edges are directed. A *directed graph* $\mathfrak{A} = \langle A, R \rangle$ satisfies only $\forall x \neg R(x, x)$.

Examples

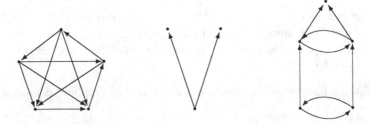

If we drop the condition of irreflexivity then a "graph" is just a set with a binary relation. We can generalize the notion even further, so that more edges may connect a pair of vertices.

In order to treat those generalized graphs we consider a language with two unary predicates V, E and one ternary predicate C. Think of $V(x)$ as "x is a vertex", $E(x)$

as "x is an edge", and $C(x, z, y)$ as "z connects x and y". A directed *multigraph* is a structure $= \langle A, V, E, C \rangle$ satisfying the following axioms:

$$\forall x(V(x) \leftrightarrow \neg E(x)),$$
$$\forall xyz(C(x, z, y) \rightarrow V(x) \wedge V(y) \wedge E(z)).$$

The edges can be seen as arrows. By adding the symmetry condition, $\forall xyz(C(x, z, y) \rightarrow C(y, z, x))$ one obtains plain multigraphs.

Examples

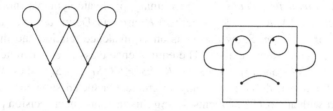

Remark The nomenclature in graph theory is not very uniform. We have chosen our formal framework such that it lends itself to treatment in first-order logic.

For the purpose of describing multigraphs a two-sorted language (cf. geometry) is well suited. The reformulation is left to the reader.

Exercises

1. Consider the language of partial order. Define predicates for (a) x is the *maximum*; (b) x is *maximal*; (c) there is no element between x and y; (d) x is an *immediate successor* (respectively *predecessor*) of y; (e) z is the *infimum* of x and y.
2. Give a sentence σ such that $\mathfrak{A}_2 \models \sigma$ and $\mathfrak{A}_4 \models \neg\sigma$ (for $\mathfrak{A}_i$ associated to the diagrams of p. 79).
3. Let $\mathfrak{A}_1 = \langle \mathbb{N}, \leq \rangle$ and $\mathfrak{A}_2 = \langle \mathbb{Z}, \leq \rangle$ be the ordered sets of natural, respectively integer, numbers. Give a sentence σ such that $\mathfrak{A}_1 \models \sigma$ and $\mathfrak{A}_2 \models \neg\sigma$. Do the same for $\mathfrak{A}_2$ and $\mathfrak{B} = \langle \mathbb{Q}, \leq \rangle$ (the ordered set of rationals). N.B. σ is in the language of posets; in particular, you may not add extra constants, function symbols, etc., defined abbreviations are of course harmless.
4. Let $\sigma = \exists x \forall y(x \leq y \vee y \leq x)$. Find posets $\mathfrak{A}$ and $\mathfrak{B}$ such that $\mathfrak{A} \models \sigma$ and $\mathfrak{B} \models \neg\sigma$.
5. Do the same for $\sigma = \forall xy \exists z[(x \leq z \wedge y \leq z) \vee (z \leq x \wedge z \leq y)]$.
6. Using the language of identity structures give an (infinite) set Γ such that $\mathfrak{A}$ is a model of Γ iff $\mathfrak{A}$ is infinite.
7. Consider the language of groups. Define the properties: (a) x is idempotent; (b) x belongs to the center.

8. Let $\mathfrak{A}$ be a ring, give a sentence σ such that $\mathfrak{A} \models \sigma \Leftrightarrow \mathfrak{A}$ is an integral domain (has no divisors of zero).

9. Give a formula $\sigma(x)$ in the language of rings such that $\mathfrak{A} \models \sigma(\overline{a}) \Leftrightarrow$ the principal ideal (a) is prime (in $\mathfrak{A}$).

10. Define in the language of arithmetic: (a) x and y are relatively prime; (b) x is the smallest prime greater than y; (c) x is the greatest number with $2x < y$.

11. $\sigma := \forall x_1 \ldots x_n \exists y_1 \ldots y_m \varphi$ and $\tau := \exists y_1 \ldots y_m \psi$ are sentences in a language without identity, function symbols and constants, where φ and ψ are quantifier free. Show: $\models \sigma \Leftrightarrow \sigma$ holds in all structures with n elements. $\models \tau \Leftrightarrow \tau$ holds in all structures with 1 element.

12. *Monadic predicate calculus* has only unary predicate symbols (no identity). Consider $\mathfrak{A} = \langle A, R_1, \ldots, R_n \rangle$ where all R_i are sets. Define $a \sim b := a \in R_i \Leftrightarrow b \in R_i$ for all $i \leq n$. Show that $\sim$ is an equivalence relation and that $\sim$ has at most 2^n equivalence classes. The equivalence class of a is denoted by $[a]$. Define $B = A/\sim$ and $[a] \in S_i \Leftrightarrow a \in R_i$, $\mathfrak{B} = \langle B, S_1, \ldots, S_n \rangle$. Show $\mathfrak{A} \models \sigma \Leftrightarrow \mathfrak{B} \models \sigma$ for all σ in the corresponding language. For such σ show $\models \sigma \Leftrightarrow \mathfrak{A} \models \sigma$ for all $\mathfrak{A}$ with at most 2^n elements. Using this fact, outline a decision procedure for truth in monadic predicate calculus.

13. Let $\mathfrak{N}$ be the standard model of arithmetic. Show $\mathfrak{N} \models \overline{n} < \overline{m} \Leftrightarrow n < m$.

14. Let $\mathfrak{A} = \langle \mathbb{N}, < \rangle$ and $\mathfrak{B} = \langle \mathbb{N}, \triangle \rangle$, where $n \triangle m$ iff (i) $n < m$ and n, m both even or both odd, or (ii) if n is even and m odd. Give a sentence σ such that $\mathfrak{A} \models \sigma$ and $\mathfrak{B} \models \neg\sigma$.

15. If $\langle A, R \rangle$ is a projective plane, then $\langle A, \check{R} \rangle$ is also a projective plane (*the dual plane*), where $\check{R}$ is the converse of the relation R. Formulated in the two-sorted language: if $\langle A_P, A_L, I \rangle$ is a projective plane, then so is $\langle A_L, A_P, \check{I} \rangle$.

3.8 Natural Deduction

We extend the system of Sect. 2.5 to predicate logic. For reasons similar to the ones mentioned in Sect. 2.5 we consider a language with connectives $\wedge, \rightarrow, \perp$ and $\forall$. The existential quantifier is left out, but will be considered later.

We adopt all the rules of propositional logic and we add

$$\forall I \frac{\varphi(x)}{\forall x \varphi(x)} \qquad \forall E \frac{\forall x \varphi(x)}{\varphi(t)}$$

where in $\forall I$ the variable x may not occur free in any hypothesis on which $\varphi(x)$ depends, i.e. an uncanceled hypothesis in the derivation of $\varphi(x)$. In $\forall E$ we, of course, require t to be free for x.

$\forall I$ has the following intuitive explanation: if an arbitrary object x has the property φ, then every object has the property φ. The problem is that none of the objects we know in mathematics can be considered "arbitrary". So instead of looking for the "arbitrary object" in the real world (as far as mathematics is concerned), let us try to find a syntactic criterion. Consider a variable x (or a constant) in a derivation, are there reasonable grounds for calling x "arbitrary"? Here is a plausible suggestion: in the context of the derivations we shall call x arbitrary if nothing has been assumed concerning x. In more technical terms, x is arbitrary at its particular occurrence in a derivation if the part of the derivation above it contains no hypotheses containing x free.

We will demonstrate the necessity of the above restrictions, keeping in mind that the system at least has to be *sound*, i.e. that derivable statements should be true. Restriction on $\forall I$:

$$\frac{\dfrac{\dfrac{\dfrac{[x = 0]}{\forall x(x = 0)}}{x = 0 \to \forall x(x = 0)}}{\forall x(x = 0 \to \forall x(x = 0))}}{0 = 0 \to \forall x(x = 0)}$$

The $\forall$ introduction at the first step was illegal.

So $\vdash 0 = 0 \to \forall x(x = 0)$, but clearly $\nvDash 0 = 0 \to \forall x(x = 0)$ (take any structure containing more than just 0). Restriction on $\forall E$:

$$\frac{\dfrac{[\forall x \neg \forall y(x = y)]}{\neg \forall y(y = y)}}{\forall x \neg \forall y(x = y) \to \neg \forall y(y = y)}$$

The $\forall$ elimination at the first step was illegal.

Note that y is not free for x in $\neg \forall y(x = y)$. The derived sentence is clearly not true in structures with at least two elements.

We now give some examples of derivations. We assume that the reader has enough experience by now in canceling hypotheses, so we will no longer indicate the cancellations by encircled numbers.

$$\frac{\dfrac{\dfrac{\dfrac{\dfrac{\dfrac{[\forall x \forall y \varphi(x, y)]}{\forall y \varphi(x, y)} \forall E}{\varphi(x, y)} \forall E}{\forall x \varphi(x, y)} \forall I}{\forall y \forall x (\varphi(x, y))} \forall I}{\forall x \forall y \varphi(x, y) \to \forall y \forall x \varphi(x, y)} \to I$$

Let $x \notin FV(\varphi)$

$$\frac{\dfrac{[\forall x(\varphi(x) \wedge \psi(x))]}{\varphi(x) \wedge \psi(x)}}{\dfrac{\varphi(x)}{\forall x \varphi(x)}} \qquad \frac{\dfrac{[\forall x(\varphi(x) \wedge \psi(x)))]}{\varphi(x) \wedge \psi(x)}}{\dfrac{\psi(x)}{\forall x \psi(x)}}$$

$$\frac{\forall x \varphi(x) \wedge \forall x \psi(x)}{\forall x(\varphi \wedge \psi) \to \forall x \varphi \wedge \forall x \psi}$$

$$\frac{[\forall x(\varphi \to \psi(x)))]}{\varphi \to \psi(x)} \forall E$$

$$\frac{\dfrac{\dfrac{\dfrac{}{\varphi \to \psi(x)} \quad [\varphi]}{\psi(x)} \to E}{\forall x \psi(x)} \forall I}{\varphi \to \forall x \psi(x)} \to I$$

$$\frac{[\varphi]}{\forall x \varphi} \forall I \qquad \frac{[\forall x \varphi]}{\varphi} \forall E$$

$$\varphi \leftrightarrow \forall x \varphi$$

$$\forall x(\varphi \to \psi(x)) \to (\varphi \to \forall x \psi(x))$$

In the right-hand derivation $\forall I$ is allowed, since $x \notin FV(\varphi)$, and $\forall E$ is applicable.

Note that $\forall I$ in the bottom left derivation is allowed because $x \notin FV(\varphi)$, for at that stage φ is still (part of) a hypothesis.

The reader will have grasped the technique behind the quantifier rules: reduce a $\forall x \varphi$ to φ and reintroduce $\forall$ later, if necessary. Intuitively, one makes the following step: to show "for all $x \ldots x \ldots$" it suffices to show "$\ldots x \ldots$" for an arbitrary x. The latter statement is easier to handle. Without going into fine philosophical distinctions, we note that the distinction "for all $x \ldots x \ldots$" – "for an arbitrary $x \ldots x \ldots$" is embodied in our system by means of the distinction. "quantified statement" – "free variable statement".

The reader will also have observed that under a reasonable derivation strategy, roughly speaking, elimination precedes introduction. There is a sound explanation for this phenomenon, its proper treatment belongs to *proof theory*, where *normal derivations* (derivations without superfluous steps) are considered. See Chap. 7. For the moment the reader may accept the above mentioned fact as a convenient rule of thumb.

We can formulate the derivability properties of the universal quantifier in terms of the relation $\vdash$:

$$\Gamma \vdash \varphi(x) \Rightarrow \Gamma \vdash \forall x \varphi(x) \quad \text{if } x \notin FV(\psi) \text{ for all } \psi \in \Gamma$$

$$\Gamma \vdash \forall x \varphi(x) \Rightarrow \Gamma \vdash \varphi(t) \quad \text{if } t \text{ is free for } x \text{ in } \varphi.$$

The above implications follow directly from $(\forall I)$ and $(\forall E)$.

Our next goal is the correctness of the system of natural deduction for predicate logic. We first extend the definition of $\models$.

Definition 3.8.1 Let Γ be a set of formulas and let $\{x_{i_1}, x_{i_2}, \ldots\} = \bigcup\{FV(\psi) | \psi \in \Gamma \cup \{\sigma\}\}$. If $\mathbf{a}$ is a sequence $(a_1, a_2, \ldots)$ of elements (repetitions allowed) of $|\mathfrak{A}|$, then $\Gamma(\mathbf{a})$ is obtained from Γ by replacing simultaneously in all formulas of Γ the x_{i_j} by $\bar{a}_j (j \geq 1)$ (for $\Gamma = \{\psi\}$ we write $\psi(\mathbf{a})$). We now define

(i) $\mathfrak{A} \models \Gamma(\mathbf{a})$ if $\mathfrak{A} \models \psi$ for all $\psi \in \Gamma(\mathbf{a})$,

(ii) $\Gamma \models \sigma$ if $\mathfrak{A} \models \Gamma(\mathbf{a}) \Rightarrow \mathfrak{A} \models \sigma(\mathbf{a})$ for all $\mathfrak{A}, \mathbf{a}$.

In case only sentences are involved, the definition can be simplified:

$\Gamma \models \sigma$ if $\mathfrak{A} \models \Gamma \Rightarrow \mathfrak{A} \models \sigma$ for all $\mathfrak{A}$.

If $\Gamma = \emptyset$, we write $\models \sigma$.

We can paraphrase this definition as: $\Gamma \models \sigma$, if for all structures $\mathfrak{A}$ and all choices of $\mathbf{a}$, $\sigma(\mathbf{a})$ is true in $\mathfrak{A}$ if all hypotheses of $\Gamma(\mathbf{a})$ are true in $\mathfrak{A}$.

Now we can formulate the following lemma.

Lemma 3.8.2 (Soundness) $\Gamma \vdash \sigma \Rightarrow \Gamma \models \sigma$.

Proof By definition of $\Gamma \vdash \sigma$ it suffices to show that for each derivation $\mathcal{D}$ with hypothesis set Γ and conclusion σ $\Gamma \models \sigma$. We use induction on $\mathcal{D}$ (cf. Lemma 2.5.1 and Exercise 2).

Since we have cast our definition of satisfaction in terms of valuations, which evidently contains the propositional logic as a special case, we can copy the cases of (1) the one-element derivation, and (2) the derivations with a propositional rule as the last step, from Lemma 2.6.1 (please check this claim).

So we have to treat derivations with ($\forall I$) or ($\forall E$) as the final step.

$(\forall I)$ $\quad \mathcal{D} \quad$ $\mathcal{D}$ has its hypotheses in Γ and x is not free in Γ.

$\dfrac{\varphi(x)}{\forall x \varphi(x)}$ Induction hypothesis: $\Gamma \models \varphi(x)$, i.e. $\mathfrak{A} \models \Gamma(\mathbf{a}) \Rightarrow$
$\mathfrak{A} \models (\varphi(x))(\mathbf{a})$ for all $\mathfrak{A}$ and all $\mathbf{a}$.

It is no restriction to suppose that x is the first of the free variables involved (why?). So we can substitute $\overline{a}_1$ for x in φ. Put $\mathbf{a} = (a_1, \mathbf{a}')$. Now we have:

for all a_1 and $\mathbf{a}' = (a_2, \ldots)$ $\mathfrak{A} \models \Gamma(\mathbf{a}') \Rightarrow \mathfrak{A} \models \varphi(\overline{a_1})(\mathbf{a}')$, so
for all $\mathbf{a}'$ $\mathfrak{A} \models \Gamma(\mathbf{a}') \Rightarrow \mathfrak{A} \models (\varphi(\overline{a}_1))(\mathbf{a}')$ for all a_1, so
for all $\mathbf{a}'$ $\mathfrak{A} \models \Gamma(\mathbf{a}') \Rightarrow \mathfrak{A} \models (\forall x \varphi(x))(\mathbf{a}')$.

This shows $\Gamma \models \forall x \varphi(x)$. (Note that in this proof we used $\forall x(\sigma \to \tau(x)) \to (\sigma \to \forall x \tau(x))$, where $x \notin FV(\sigma)$, in the meta-language. Of course we may use sound principles on the meta-level.)

$(\forall E)$ $\quad \mathcal{D} \quad$ Induction hypothesis: $\Gamma \models \forall x \varphi(x)$,

$\dfrac{\forall x \varphi(x)}{\varphi(t)}$ i.e. $\mathfrak{A} \models \Gamma(\mathbf{a}) \Rightarrow \mathfrak{A} \models (\forall x \varphi(x))(\mathbf{a})$,
for all $\mathbf{a}$ and $\mathfrak{A}$.

So let $\mathfrak{A} \models \Gamma(\mathbf{a})$, then $\mathfrak{A} \models \varphi(\overline{b})(\mathbf{a})$ for all $b \in |\mathfrak{A}|$. In particular we may take $t[\overline{\mathbf{a}}/\mathbf{z}]$ for $\overline{b}$, where we slightly abuse the notation; since there are finitely many variables $z_1, \ldots, z_n$, we only need finitely many of the a_i's, and therefore we consider it an ordinary simultaneous substitution.

$\mathfrak{A} \models (\varphi[\mathbf{a}/\mathbf{z}])[t[\mathbf{a}/\mathbf{z}]/x]$, hence by Lemma 3.5.4, $\mathfrak{A} \models (\varphi[t/x])[\mathbf{a}/\mathbf{z}]$, or $\mathfrak{A} \models (\varphi(t))(\mathbf{a})$. $\qquad \square$

Having established the soundness of our system, we can easily get non-derivability results.

Examples

1. $\not\vdash \forall x \exists y \varphi \to \exists y \forall x \varphi$.
   Take $\mathfrak{A} = \langle \{0, 1\}, \{\langle 0, 1\rangle, \langle 1, 0\rangle\} \rangle$ (type $\langle 2; -; 0\rangle$) and consider $\varphi := P(x, y)$, the predicate interpreted in $\mathfrak{A}$.

$\mathfrak{A} \models \forall x \exists y P(x, y)$, since for 0 we have $\langle 0, 1 \rangle \in P$ and for 1 we have $\langle 1, 0 \rangle \in P$. But, $\mathfrak{A} \not\models \exists y \forall x P(x, y)$, since for 0 we have $\langle 0, 0 \rangle \notin P$ and for 1 we have $\langle 1, 1 \rangle \notin P$.

2. $\forall x \varphi(x, x), \forall x y (\varphi(x, y) \rightarrow \varphi(y, x)) \not\vdash \forall x y z (\varphi(x, y) \wedge \varphi(y, z) \rightarrow \varphi(x, z))$.
   Consider $\mathfrak{B} = \langle \mathbb{R}, P \rangle$ with $P = \{\langle a, b \rangle \mid |a - b| \leq 1\}$.

Although variables and constants are basically different, they share some properties. Both constants and free variables may be introduced in derivations through $\forall E$, but only free variables can be subjected to $\forall I$,—that is free variables can disappear in derivations by other than propositional means. It follows that a variable can take the place of a constant in a derivation but in general not vice versa. We make this precise as follows.

Theorem 3.8.3 *Let x be a variable not occurring in Γ or φ.*

(i) $\Gamma \vdash \varphi \Rightarrow \Gamma[x/c] \vdash \varphi[x/c]$.
(ii) *If c does not occur in Γ, then $\Gamma \vdash \varphi(c) \Rightarrow \Gamma \vdash \forall x \varphi(x)$.*

Proof (ii) follows immediately from (i) by $\forall I$. (i) Induction on the derivation of $\Gamma \vdash \varphi$. Left to the reader.   □

Observe that the result is rather obvious, changing c to x is just as harmless as coloring c red—the derivation remains intact.

Exercises

1. Show:
   (i) $\vdash \forall x (\varphi(x) \rightarrow \psi(x)) \rightarrow (\forall x \varphi(x) \rightarrow \forall x \psi(x))$,
   (ii) $\vdash \forall x \varphi(x) \rightarrow \neg \forall x \neg \varphi(x)$,
   (iii) $\vdash \forall x \varphi(x) \rightarrow \forall z \varphi(z)$ if z does not occur in $\varphi(x)$,
   (iv) $\vdash \forall x \forall y \varphi(x, y) \rightarrow \forall y \forall x \varphi(x, y)$,
   (v) $\vdash \forall x \forall y \varphi(x, y) \rightarrow \forall x \varphi(x, x)$,
   (vi) $\vdash \forall x (\varphi(x) \wedge \psi(x)) \leftrightarrow \forall x \varphi(x) \wedge \forall x \psi(x)$,
   (vii) $\vdash \forall x (\varphi \rightarrow \psi(x)) \leftrightarrow (\varphi \rightarrow \forall x \psi(x))$, where $x \notin FV(\varphi)$.
2. Extend the definition of derivation to the present system (cf. Definition 2.4.1).
3. Show $(s(t)[\overline{a}/x])^{\mathfrak{A}} = (s((t[\overline{a}/x])^{\mathfrak{A}})[\overline{a}/x])^{\mathfrak{A}}$.
4. Show the inverse implications of Theorem 3.8.3.
5. Assign to each atom $P(t_1, \ldots, t_n)$ a proposition symbol, denoted by P. Now define a translation † from the language of predicate logic into the language of propositional logic by

$$(P(t_1, \ldots, t_n))^{\dagger} := P \quad \text{and} \quad \perp^{\dagger} := \perp,$$
$$(\varphi \square \psi)^{\dagger} := \varphi^{\dagger} \square \psi^{\dagger},$$
$$(\neg \varphi)^{\dagger} := \neg \varphi^{\dagger},$$
$$(\forall x \varphi)^{\dagger} := \varphi^{\dagger}.$$

Show $\Gamma \vdash \varphi \Rightarrow \Gamma^\dagger \vdash^\dagger \varphi^\dagger$, where $\vdash^\dagger$ stands for "derivable without using $(\forall I)$ or $(\forall E)$" (does the converse hold?)

Conclude the consistency of predicate logic.

Show that predicate logic is conservative over propositional logic (cf. Definition 4.1.5).

3.9 Adding the Existential Quantifier

Let us introduce $\exists x \varphi$ as an abbreviation for $\neg \forall x \neg \varphi$ (Theorem 3.5.1 tells us that there is a good reason for doing so). We can prove the following:

Lemma 3.9.1

(i) $\varphi(t) \vdash \exists x \varphi(x)$ (t free for x in φ)

(ii) $\Gamma, \varphi(x) \vdash \psi \Rightarrow \Gamma, \exists x \varphi(x) \vdash \psi$ if x is not free in ψ or any formula of Γ.

Proof (i)

$$\cfrac{\cfrac{\dfrac{[\forall x \neg \varphi(x)]}{\neg \varphi(t)} \forall E \qquad \varphi(t)}{\bot} \to E}{\neg \forall x \neg \varphi(x)} \to I$$

so $\varphi(t) \vdash \exists x \varphi(x)$

(ii)

$$\cfrac{\neg \forall x \neg \varphi(x) \qquad \cfrac{\cfrac{\cfrac{\dfrac{[\varphi(x)]}{\mathcal{D}} \\ \psi \qquad [\neg \psi]}{\bot} \to E}{\neg \varphi(x)} \to I}{\forall x \neg \varphi(x)} \forall I}{\cfrac{\bot}{\psi} \text{ RAA}} \to E \qquad \square$$

Explanation The subderivation top left is the given one; its hypotheses are in $\Gamma \cup \{\varphi(x)\}$ (only $\varphi(x)$ is shown). Since $\varphi(x)$ (that is, all occurrences of it) is canceled and x does not occur free in Γ or ψ, we may apply $\forall I$. From the derivation we conclude that $\Gamma, \exists x \varphi(x) \vdash \psi$.

We can compress the last derivation into an elimination rule for $\exists$:

$$\frac{\begin{array}{c}[\varphi[y/x]]\\ \vdots\\ \end{array}}{}$$

$$\frac{\exists x\varphi(x) \qquad \psi}{\psi}\; \exists E$$

with the conditions: y is not free in ψ, or in a hypothesis of the subderivation of ψ, other than $\varphi(x)$.

This is easily seen to be correct since we can always fill in the missing details, as shown in the preceding derivation.

By (i) we also have an introduction rule: $\frac{\varphi(t)}{\exists x\,\varphi(x)}\exists I$ for t free for x in φ.

Examples of derivations

$$\frac{\dfrac{[\forall x(\varphi(x)\to\psi)]^3}{\varphi(x)\to\psi}\;\forall E \qquad [\varphi(x)]^1}{\cfrac{\dfrac{[\exists x\varphi(x)]^2 \qquad \dfrac{\psi}{}}{\psi}\;\exists E_1}{\cfrac{\cfrac{\psi}{\exists x\varphi(x)\to\psi}\;\to I_2}{\forall x(\varphi(x)\to\psi)\to(\exists x\varphi(x)\to\psi)}\;\to I_3}}\;\to E}\qquad x\notin FV(\psi)$$

$$\frac{[\exists x(\varphi(x)\vee\psi(x))]^3 \qquad \dfrac{[\varphi(x)\vee\psi(x)]^2 \quad \dfrac{\dfrac{[\varphi(x)]^1}{\exists x\varphi(x)}\;\; \dfrac{[\psi(x)]^1}{\exists x\psi(x)}}{\exists x\varphi(x)\vee\exists x\psi(x)}\;\vee E_1}{\exists x\varphi(x)\vee\exists x\psi(x)}\;\exists E_2}{\cfrac{\exists x\varphi(x)\vee\exists x\psi(x)}{\exists x(\varphi(x)\vee\psi(x))\to\exists x\varphi(x)\vee\exists x\psi(x)}\;\to I_3}$$

We will also sketch the alternative approach, that of enriching the language.

Theorem 3.9.2 *Consider predicate logic with the full language and rules for all connectives, then* $\vdash \exists x\varphi(x) \leftrightarrow \neg\forall x\neg\varphi(x)$.

Proof Compare Theorem 2.6.3.   □

It is time now to state the rules for $\forall$ and $\exists$ with more precision. We want to allow substitution of terms for some occurrences of the quantified variable in $(\forall E)$ and $(\exists E)$. The following example motivates this.

$$\frac{\dfrac{\dfrac{\forall x(x=x)}{x=x}\;\forall E}{\exists y(x=y))}\;\exists I}{}$$

The result would not be derivable if we could only make substitutions for *all* occurrences at the same time. Yet, the result is evidently true.

The proper formulation of the rules now is:

$$\forall I \ \frac{\varphi}{\forall x \varphi} \qquad \forall E \ \frac{\forall x \varphi}{\varphi[t/x]}$$

$$[\varphi]$$

$$\vdots$$

$$\exists I \ \frac{\varphi[t/x]}{\exists x \varphi} \qquad \exists E \ \frac{\exists x \varphi \quad \psi}{\psi}$$

with the appropriate restrictions.

Exercises

1. $\vdash \exists x (\varphi(x) \wedge \psi) \leftrightarrow \exists x \varphi(x) \wedge \psi$ if $x \notin FV(\psi)$,
2. $\vdash \forall x (\varphi(x) \vee \psi) \leftrightarrow \forall x \varphi(x) \vee \psi$ if $x \notin FV(\psi)$,
3. $\vdash \forall x \varphi(x) \leftrightarrow \neg \exists x \neg \varphi(x)$,
4. $\vdash \neg \forall x \varphi(x) \leftrightarrow \exists x \neg \varphi(x)$,
5. $\vdash \neg \exists x \varphi(x) \leftrightarrow \forall x \neg \varphi(x)$,
6. $\vdash \exists x (\varphi(x) \rightarrow \psi) \leftrightarrow (\forall x \varphi(x) \rightarrow \psi)$ if $x \notin FV(\psi)$,
7. $\vdash \exists x (\varphi \rightarrow \psi(x)) \leftrightarrow (\varphi \rightarrow \exists x \psi(x))$ if $x \notin FV(\varphi)$,
8. $\vdash \exists x \exists y \varphi \leftrightarrow \exists y \exists x \varphi$,
9. $\vdash \exists x \varphi \leftrightarrow \varphi$ if $x \notin FV(\varphi)$.

3.10 Natural Deduction and Identity

We will give rules, corresponding to the axioms $I_1 - I_4$ of Sect. 3.6.

$$\frac{}{x = x} \ RI_1$$

$$\frac{x = y}{y = x} \ RI_2$$

$$\frac{x = y \quad y = z}{x = z} \ RI_3$$

$$\frac{x_1 = y_1, \ldots, x_n = y_n}{t(x_1, \ldots, x_n) = t(y_1, \ldots, y_n)} \ RI_4$$

$$\frac{x_1 = y_1, \ldots, x_n = y_n \quad \varphi(x_1, \ldots, x_n)}{\varphi(y_1, \ldots, y_n)} \ RI_4$$

where $y_1, \ldots, y_n$ are free for $x_1, \ldots, x_n$ in φ. Note that we want to allow substitution of the variable y_i ($i \leq n$) for *some* and not necessarily all occurrences of the variable x_i. We can express this by formulating RI_4 in the precise terms of the simultaneous substitution operator:

$$\frac{x_1 = y_1, \ldots, x_n = y_n}{t[x_1, \ldots, x_n/z_1, \ldots, z_n] = t[y_1, \ldots, y_n/z_1, \ldots, z_n]}$$

$$\frac{x_1 = y_1, \ldots, x_n = y_n \quad \varphi[x_1, \ldots, x_n/z_1, \ldots, z_n]}{\varphi[y_1, \ldots, y_n/z_1, \ldots, z_n]}$$

Example

$$\frac{x = y \quad x^2 + y^2 > 12x}{2y^2 > 12x}$$

$$\frac{x = y \quad x^2 + y^2 > 12x}{x^2 + y^2 > 12y}$$

$$\frac{x = y \quad x^2 + y^2 > 12x}{2y^2 > 12y}$$

The above are three legitimate applications of RI_4 having three different conclusions.

The rule RI_1 has no hypotheses, which may seem surprising, but which certainly is not forbidden.

The rules RI_4 have many hypotheses; as a consequence the derivation trees can look a bit more complicated. Of course one can get all the benefits from RI_4 by a restricted rule, allowing only one substitution at the time.

Lemma 3.10.1 $\vdash I_i$ for $i = 1, 2, 3, 4$.

Proof Immediate.   □

We can weaken the rules RI_4 slightly by considering only the simplest terms and formulas.

Lemma 3.10.2 *Let L be of type $\langle r_1, \ldots, r_n; a_1, \ldots, a_m; k \rangle$. If the rules*

$$\frac{x_1 = y_1, \ldots, x_{r_i} = y_{r_i} \quad P_1(x_1, \ldots, x_{r_i})}{P_1(y_1, \ldots, y_{r_i})} \quad \textit{for all } i \leq n$$

and

$$\frac{x_1 = y_1, \ldots, x_{a_j} = y_{a_j}}{f_j(x_1, \ldots, x_{a_j}) = f_j(y_1, \ldots, y_{a_j})} \quad \textit{for all } j \leq m$$

are given, then the rules RI_4 are derivable.

Proof We consider a special case. Let L have one binary predicate symbol and one unary function symbol.

(i) We show $x = y \vdash t(x) = t(y)$ by induction on t.
   (a) $t(x)$ is a variable or a constant. Immediate.
   (b) $t(x) = f(s(x))$. Induction hypothesis: $x = y \vdash s(x) = s(y)$

$$\frac{\dfrac{\dfrac{[x = y]}{\dfrac{f(x) = f(y)}{\forall xy(x = y \to f(x) = f(y))} \forall I \; 2\times}}{s(x) = s(y) \to f(s(x)) = f(s(y))} \qquad \dfrac{\begin{array}{c} x = y \\ \mathcal{D} \end{array}}{s(x) = s(y)}}{f(s(x)) = f(s(y))}$$

   This shows $x = y \vdash f(s(x)) = f(s(y))$.
(ii) We show $\vec{x} = \vec{y}, \varphi(\vec{x}) \vdash \varphi(\vec{y})$
   (a) φ is atomic, then $\varphi = P(t, s)$. t and s may (in this example) contain at most one variable each. So it suffices to consider $x_1 = y_1$, $x_2 = y_2$, $P(t(x_1, x_2), s(x_1, x_2)) \vdash P(t(y_1, y_2), s(y_1, y_2))$ (i.e. $P(t[x_1, x_2/z_1, z_2], \ldots)$).
   Now we get, by applying $\to E$ twice, from

$$\frac{\dfrac{\dfrac{[x_1 = y_1] \quad [x_2 = y_2] \quad [P(x_1, x_2)]}{P(y_1, y_2)}}{\dfrac{x_1 = y_1 \to (x_2 = y_2 \to (P(x_1, x_2) = P(y_1, y_2)))}{\forall x_1 x_2 y_1 y_2 (x_1 = y_1 \to (x_2 = y_2 \to (P(x_1, x_2) = P(y_1, y_2))))} \forall I} \to I \; 3\times}{s(x_1, x_2) = s(y_1, y_2) \to (t(x_1, x_2) = t(y_1, y_2) \to (P(s_x, t_x) = P(s_y, t_y)))} \forall E$$

and the following two instances of (i)

$$\dfrac{\begin{array}{c} x_1 = y_1 \quad x_2 = y_2 \\ \mathcal{D} \end{array}}{s(x_1, x_2) = s(y_1, y_2)} \qquad \text{and} \qquad \dfrac{\begin{array}{c} x_1 = y_1 \quad x_2 = y_2 \\ \mathcal{D}' \end{array}}{t(x_1, x_2) = t(y_1, y_2),}$$

the required result, $(P(s_x, t_x) = P(s_y, t_y))$.
   So

$$x_1 = y_1, x_2 = y_2 \vdash P(s_x, t_x) \to P(s_y, t_y)$$

where

$$s_x = s(x_1, x_2), \quad s_y = s(y_1, y_2)$$
$$t_x = t(x_1, x_2), \quad t_y = t(y_1, y_2).$$

   (b) $\varphi = \sigma \to \tau$.
   Induction hypotheses:

$$\vec{x} = \vec{y}, \sigma(\vec{y}) \vdash \sigma(\vec{x})$$
$$\vec{x} = \vec{y}, \tau(\vec{x}) \vdash \tau(\vec{y})$$

$$\vec{x} = \vec{y} \; [\sigma(\vec{y})]$$

$$\mathcal{D}$$

$$\frac{\sigma(\vec{x}) \to \tau(\vec{x}) \qquad \sigma(\vec{x})}{\tau(\vec{x})} \qquad\qquad \vec{x} = \vec{y}$$

$$\mathcal{D}'$$

$$\frac{\tau(\vec{y})}{\sigma(\vec{y}) \to \tau(\vec{y})}$$

So $\vec{x} = \vec{y}, \sigma(\vec{x}) \to \tau(\vec{x}) \vdash \sigma(\vec{y}) \to \tau(\vec{y})$.

(c) $\varphi = \sigma \wedge \tau$, left to the reader.

(d) $\varphi = \forall z \psi(z, \vec{x})$

Induction hypothesis: $\vec{x} = \vec{y}, \psi(z, \vec{x}) \vdash \psi(z, \vec{y})$

$$\frac{\forall z \psi(z, \vec{x})}{\psi(z, \vec{x}) \qquad \vec{x} = \vec{y}}$$

$$\mathcal{D}$$

$$\frac{\psi(z, \vec{y})}{\forall z \psi(z, \vec{y})}$$

So $\vec{x} = \vec{y}, \forall z \psi(z, \vec{x}) \vdash \forall z \psi(z, \vec{y})$.

This establishes, by induction, the general rule.   □

Exercises

1. Show that $\forall x(x = x), \forall xyz(x = y \wedge z = y \to x = z) \vdash I_2 \wedge I_3$ (using predicate logic only).
2. Show $\vdash \exists x(t = x)$ for any term t. Explain why all functions in a structure are total (i.e. defined for all arguments); what is 0^{-1}?
3. Show $\vdash \forall z(z = x \to z = y) \to x = y$.
4. Show $\vdash \forall xyz(x \neq y \to x \neq z \vee y \neq z)$.
5. Show that in the language of identity $I_1, I_2, I_3 \vdash I_4$.
6. Show $\forall x(x = a \vee x = b \vee x = c) \vdash \forall x \varphi(x) \leftrightarrow (\varphi(a) \wedge \varphi(b) \wedge \varphi(c))$, where a, b, c, are constants.
7. Show
   (i) $\forall xy(f(x) = f(y) \to x = y), \forall xy(g(x) = g(y) \to x = y) \vdash \forall xy(f(g(x)) = f(g(y)) \to x = y)$,
   (ii) $\forall y \exists x(f(x) = y), \forall y \exists x(g(x) = y) \vdash \forall y \exists x(f(g(x)) = y)$.
   Which properties are expressed by this exercise?
8. Prove the following *duality principle* for projective geometry (cf. Definition 3.7.5): If $\Gamma \vdash \varphi$ then also $\Gamma \vdash \varphi^d$, where Γ is the set of axioms of projective geometry and φ^d is obtained from φ by replacing each atom $x I y$ by $y I x$. (Hint: check the effect of the translation d on the derivation of φ from Γ.)

Chapter 4
Completeness and Applications

4.1 The Completeness Theorem

Just as in the case of propositional logic we shall show that "derivability" and "semantical consequence" coincide. We will do quite a bit of work before we come to the theorem. Although the proof of the completeness theorem is not harder than, say, some proofs in analysis, we would advise the reader to read the statement of the theorem but to skip the proof at the first reading and return to it later. It is more instructive to go to the applications, and it will probably give the reader a better feeling for the subject.

The main tool in this chapter is the following lemma.

Lemma 4.1.1 (Model Existence Lemma) *If Γ is a consistent set of sentences, then Γ has a model.*

A sharper version is the following.

Lemma 4.1.2 *Let L have cardinality κ. If Γ is a consistent set of sentences, then Γ has a model of cardinality $\leq \kappa$.*

From Lemma 4.1.1 we immediately deduce Gödel's completeness theorem.

Theorem 4.1.3 (Completeness Theorem) $\Gamma \vdash \varphi \Leftrightarrow \Gamma \models \varphi$.

We will now go through all the steps of the proof of the Completeness Theorem. In this section we will consider sentences, unless we specifically mention non-closed formulas. Furthermore "$\vdash$" will stand for "derivability in predicate logic with identity".

Just as in the case of propositional logic we have to construct a model and the only thing we have is our consistent theory. This construction is a kind of Baron von Münchhausen trick; we have to pull ourselves (or rather, a model) out of the quicksand of syntax and proof rules. The most plausible idea is to make a universe

D. van Dalen, *Logic and Structure*, Universitext, DOI 10.1007/978-1-4471-4558-5_4,
© Springer-Verlag London 2013

out of the closed terms and to define relations as the sets of (tuples of) terms in the atoms of the theory. There are basically two things we have to take care of: (i) if the theory tells us that $\exists x\varphi(x)$, then the model has to make $\exists x\varphi(x)$ true, and so it has to exhibit an element (which in this case is a closed term t) such that $\varphi(t)$ is true. This means that the theory has to prove $\varphi(t)$ for a suitable closed term t. This problem is solved in Henkin theories. (ii) A model has to decide sentences, i.e. it has to say σ or $\neg\sigma$ for each sentence σ. As in propositional logic, this is handled by maximal consistent theories.

Definition 4.1.4

(i) A *theory* T is a collection of sentences with the property $T \vdash \varphi \Rightarrow \varphi \in T$ (a theory is *closed under derivability*).
(ii) A set Γ such that $T = \{\varphi | \Gamma \vdash \varphi\}$ is called an *axiom set* of the theory T. The elements of Γ are called *axioms*.
(iii) T is called a *Henkin theory* if for each sentence $\exists x\varphi(x)$ there is a constant c such that $\exists x\varphi(x) \to \varphi(c) \in T$ (such a c is called a *witness* for $\exists x\varphi(x)$).

Note that $T = \{\sigma | \Gamma \vdash \sigma\}$ is a theory. For, if $T \vdash \varphi$, then $\sigma_1, \ldots, \sigma_k \vdash \varphi$ for certain σ_i with $\Gamma \vdash \sigma_i$.

$\mathcal{D}_1 \, \mathcal{D}_2 \ldots \mathcal{D}_k$ From the derivations $\mathcal{D}_1, \ldots, \mathcal{D}_k$ of $\Gamma \vdash \sigma_1, \ldots,$
$\sigma_1 \, \sigma_2 \ldots \sigma_k$ $\Gamma \vdash \sigma_k$ and $\mathcal{D}$ of $\sigma_1, \ldots, \sigma_k \vdash \varphi$ a derivation,
$\dfrac{\mathcal{D}}{\varphi}$ of $\Gamma \vdash \varphi$ is obtained, as indicated.

Definition 4.1.5 Let T and T' be theories in the languages L and L'.

(i) T' is an *extension* of T if $T \subseteq T'$,
(ii) T' is a *conservative extension* of T if $T' \cap L = T$ (i.e. all theorems of T' in the language L are already theorems of T).

Example of a conservative extension: consider propositional logic P' in the language L with $\to, \wedge, \bot, \leftrightarrow, \neg$. Then Exercise 2, Sect. 2.6, tells us that P' is conservative over P.

Our first task is the construction of *Henkin extensions* of a given theory T, that is to say: extensions of T which are Henkin theories.

Definition 4.1.6 Let T be a theory with language L. The language L^* is obtained from L by adding a constant c_φ for each sentence of the form $\exists x\varphi(x)$. T^* is the theory with axiom set $T \cup \{\exists x\varphi(x) \to \varphi(c_\varphi) | \exists x\varphi(x)$ closed, with witness $c_\varphi\}$.

Lemma 4.1.7 T^* *is conservative over* T.

Proof (a) Let $\exists x\varphi(x) \to \varphi(c)$ be one of the new axioms. Suppose $\Gamma, \exists x\varphi(x) \to \varphi(c) \vdash \psi$, where ψ does not contain c and where Γ is a set of sentences, none of which contains the constant c. We show $\Gamma \vdash \psi$ in a number of steps.

1. $\Gamma \vdash (\exists x \varphi(x) \rightarrow \varphi(c)) \rightarrow \psi$.
2. $\Gamma \vdash (\exists x \varphi(x) \rightarrow \varphi(y)) \rightarrow \psi$, where y is a variable that does not occur in the associated derivation. 2 follows from 1 by Theorem 3.8.3.
3. $\Gamma \vdash \forall y[(\exists x \varphi(x) \rightarrow \varphi(y)) \rightarrow \psi]$. This application of $(\forall I)$ is correct, since c did not occur in Γ.
4. $\Gamma \vdash \exists y (\exists x \varphi(x) \rightarrow \varphi(y)) \rightarrow \psi$ (cf. example of Sect. 3.9).
5. $\Gamma \vdash (\exists x \varphi(x) \rightarrow \exists y \varphi(y)) \rightarrow \psi$ (Sect. 3.9 Exercise 7).
6. $\vdash \exists x \varphi(x) \rightarrow \exists y \varphi(y)$.
7. $\Gamma \vdash \psi$ (from 5,6).

(b) Let $T^* \vdash \psi$ for a $\psi \in L$. By the definition of derivability $T \cup \{\sigma_1, \ldots, \sigma_n\} \vdash \psi$, where the σ_i are the new axioms of the form $\exists x \varphi(x) \rightarrow \varphi(c)$. We show $T \vdash \psi$ by induction on n. For $n = 0$ we are done. Let $T \cup \{\sigma_1, \ldots, \sigma_{n+1}\} \vdash \psi$. Put $T' = T \cup \{\sigma_1, \ldots, \sigma_n\}$, then $T', \sigma_{n+1} \vdash \psi$ and we may apply (a). Hence $T \cup \{\sigma_1, \ldots, \sigma_n\} \vdash \psi$. Now by the induction hypothesis $T \vdash \psi$. $\qquad\square$

Although we have added a large number of witnesses to T, there is no evidence that T^* is a Henkin theory, since by enriching the language we also add new existential statements $\exists x \tau(x)$ which may not have witnesses. In order to overcome this difficulty we iterate the above process countably many times.

Lemma 4.1.8 *Define* $T_0 := T$; $T_{n+1} := (T_n)^*$; $T_\omega := \bigcup\{T_n | n \geq 0\}$. *Then* T_ω *is a Henkin theory and it is conservative over* T.

Proof Call the language of T_n (resp. T_ω) L_n (resp. L_ω).

(i) T_n is conservative over T. Induction on n.
(ii) T_ω is a theory. Suppose $T_\omega \vdash \sigma$, then $\varphi_0, \ldots, \varphi_n \vdash \sigma$ for certain $\varphi_0, \ldots, \varphi_n \in T_\omega$. For each $i \leq n$ $\varphi_i \in T_{m_i}$ for some m_i. Let $m = \max\{m_i | i \leq n\}$. Since $T_k \subseteq T_{k+1}$ for all k, we have $T_{m_i} \subseteq T_m (i \leq n)$. Therefore $T_m \vdash \sigma$. T_m is (by definition) a theory, so $\sigma \in T_m \subseteq T_\omega$.
(iii) T_ω is a Henkin theory. Let $\exists x \varphi(x) \in L_\omega$, then $\exists x \varphi(x) \in L_n$ for some n. By definition $\exists x \varphi(x) \rightarrow \varphi(c) \in T_{n+1}$ for a certain c. So $\exists x \varphi(x) \rightarrow \varphi(c) \in T_\omega$.
(iv) T_ω is conservative over T. Observe that $T_\omega \vdash \sigma$ if $T_n \vdash \sigma$ for some n and apply (i). $\qquad\square$

As a corollary we get: T_ω is consistent if T is so. For suppose T_ω inconsistent, then $T_\omega \vdash \perp$. As T_ω is conservative over T (and $\perp \in L$) $T \vdash \perp$. Contradiction.

Our next step is to extend T_ω as far as possible, just as we did in propositional logic (2.5.7). We state a general principle.

Lemma 4.1.9 (Lindenbaum) *Each consistent theory is contained in a maximally consistent theory.*

Proof We give a straightforward application of *Zorn's lemma*. Let T be consistent. Consider the set A of all consistent extensions T' of T, partially ordered by inclusion. Claim: A has a maximal element.

1. Each chain, i.e. linearly ordered set, in A has an upper bound. Let $\{T_i | i \in I\}$ be a chain. Then $T' = \bigcup T_i$ is a consistent extension of T containing all T_i's (Exercise 2). So T' is an upper bound.
2. Therefore A has a maximal element T_m (Zorn's lemma).
3. T_m is a maximally consistent extension of T. We only have to show: $T_m \subseteq T'$ and $T' \in A$, then $T_m = T'$. But this is trivial as T_m is maximal in the sense of $\subseteq$. Conclusion: T is contained in the maximally consistent theory T_m. □

Note that in general T has many maximally consistent extensions. The above existence is far from unique (as a matter of fact the proof of its existence essentially uses the axiom of choice). Note, however, that if the language is countable, one can mimic the proof of Lemma 2.5.7 and dispense with Zorn's lemma.

We now combine the construction of a Henkin extension with a maximally consistent extension. Fortunately the property of being a Henkin theory is preserved under taking a maximally consistent extension.

Lemma 4.1.10 *An extension of a Henkin theory in the same language is again a Henkin theory.*

Proof If T' extends the Henkin theory T in the same language, then for any φ we have $\exists x \varphi(x) \to \varphi(c) \in T \Rightarrow \exists x \varphi(x) \to \varphi(c) \in T'$. □

We now get to the proof of our main result.

Lemma 4.1.11 (Model Existence Lemma) *If Γ is consistent, then Γ has a model.*

Proof Let $T = \{\sigma | \Gamma \vdash \sigma\}$ be the theory given by Γ. Any model of T is, of course, a model of Γ.

Let T_m be a maximally consistent Henkin extension of T (which exists by the preceding lemmas), with language L_m.

We will construct a model of T_m using T_m itself. At this point the reader should realize that a language is, after all, a set, that is a set of strings of symbols. So, we will exploit this set to build the universe of a suitable model.

1. $A = \{t \in L_m | t \text{ is closed}\}$.
2. For each function symbol $\overline{f}$ we define a function $\hat{f} : A^k \to A$ by $\hat{f}(t_1, \ldots, t_k) := f(t_1, \ldots, t_k)$.
3. For each predicate symbol $\overline{P}$ we define a relation $\hat{P} \subseteq A^p$ by $\langle t_1, \ldots, t_p \rangle \in \hat{P} \Leftrightarrow T_m \vdash P(t_1, \ldots, t_p)$.
4. For each constant symbol c we define a constant $\hat{c} := c$.

Although it looks as if we have created the required model, we have to improve the result, because "=" is not interpreted as the real equality; we still have to equate many non-identical terms. Think of $3 + 4 = 2 + 5$. We can only assert the following.

   (a) The relation $t \sim s$ defined by $T_m \vdash t = s$ for $t, s \in A$ is an equivalence relation. By Lemma 3.10.1, I_1, I_2, I_3 are theorems of T_m, so $T_m \vdash \forall x(x = x)$, and hence (by $\forall E$) $T_m \vdash t = t$, or $t \sim t$. Symmetry and transitivity follow in the same way.

(b) $t_i \sim s_i (i \leq p)$ and $\langle t_1, \ldots, t_p \rangle \in \hat{P} \Rightarrow \langle s_1, \ldots, s_p \rangle \in \hat{P}$.

$t_i \sim s_i (i \leq k) \Rightarrow \hat{f}(t_1, \ldots, t_k) \sim \hat{f}(s_1, \ldots, s_k)$ for all symbols P and f.

The proof is simple: use $T_m \vdash I_4$ (Lemma 3.10.1).

Once we have an equivalence relation, which, moreover, is a congruence with respect to the basic relations and functions, it is natural to introduce the quotient structure.

Denote the equivalence class of t under $\sim$ by $[t]$.

Define $\mathfrak{A} := \langle A/\sim, \tilde{P}_1, \ldots, \tilde{P}_n, \tilde{f}_1, \ldots, \tilde{f}_m, \{\tilde{c}_i | i \in I\}\rangle$, where

$$\tilde{P}_i := \{\langle [t_1], \ldots, [t_{r_i}]\rangle | \langle t_1, \ldots, t_{r_i}\rangle \in \hat{P}_i\}$$
$$\tilde{f}_j([t_1], \ldots, [t_{a_j}]) = [\hat{f}_j(t_1, \ldots, t_{a_j})]$$
$$\tilde{c}_i := [\hat{c}_i].$$

One has to show that the relations and functions on $A/\sim$ are well defined, but that is taken care of by (b) above.

Closed terms lead a kind of double life. On the one hand they are syntactical objects, on the other hand they are the stuff that elements of the universe are made from. The two things are related by $t^{\mathfrak{A}} = [t]$. This is shown by induction on t.

(i) $t = c$, then $t^{\mathfrak{A}} = \tilde{c} = [\hat{c}] = [c] = [t]$,

(ii) $t = f(t_1, \ldots, t_k)$, then

$$t^{\mathfrak{A}} = \tilde{f}(t_1{}^{\mathfrak{A}}, \ldots, t_k{}^{\mathfrak{A}}) \overset{\text{i.h.}}{=} \tilde{f}([t_1], \ldots, [t_k]) = [\hat{f}(t_1, \ldots, t_k)] = [f(t_1, \ldots, t_k)].$$

Furthermore we have $\mathfrak{A} \models \varphi(t) \Leftrightarrow \mathfrak{A} \models \varphi(\overline{[t]})$, by the above and by Exercise 6, Sect. 3.4.

Claim. $\mathfrak{A} \models \varphi(t) \Leftrightarrow T_m \vdash \varphi(t)$ for all sentences φ in the language L_m of T_m which, by the way, is also $L(\mathfrak{A})$, since each element of $A/\sim$ has a name in L_m. We prove the claim by induction on φ.

(i) φ is atomic. $\mathfrak{A} \models P(t_1, \ldots, t_p) \Leftrightarrow \langle t_1^{\mathfrak{A}}, \ldots, t_p^{\mathfrak{A}}\rangle \in \tilde{P} \Leftrightarrow \langle [t_1], \ldots, [t_p]\rangle \in \tilde{P} \Leftrightarrow$

$\langle t_1, \ldots, t_p\rangle \in \hat{P} \Leftrightarrow T_m \vdash P(t_1, \ldots, t_p)$. The case $\varphi = \perp$ is trivial.

(ii) $\varphi = \sigma \wedge \tau$. Trivial.

(iii) $\varphi = \sigma \rightarrow \tau$. We recall that, by Lemma 2.5.9, $T_m \vdash \sigma \rightarrow \tau \Leftrightarrow (T_m \vdash \sigma \Rightarrow T_m \vdash \tau)$. Note that we can copy this result, since its proof only uses propositional logic, and hence remains correct in predicate logic.

$\mathfrak{A} \models \varphi \rightarrow \tau \Leftrightarrow (\mathfrak{A} \models \sigma \Rightarrow \mathfrak{A} \models \tau) \overset{\text{i.h.}}{\Leftrightarrow} (T_m \vdash \sigma \Rightarrow T_m \vdash \tau) \Leftrightarrow T_m \vdash \sigma \rightarrow \tau$.

(iv) $\varphi = \forall x \psi(x)$. $\mathfrak{A} \models \forall x \psi(x) \Leftrightarrow \mathfrak{A} \not\models \exists x \neg \psi(x) \Leftrightarrow \mathfrak{A} \not\models \neg \psi(\bar{a})$, for all $a \in |\mathfrak{A}| \Leftrightarrow$ for all $a \in |\mathfrak{A}|$ $\mathfrak{A} \models \psi(\bar{a})$. Assuming $\mathfrak{A} \models \forall x \psi(x)$, we get in particular $\mathfrak{A} \models \psi(c)$ for the witness c belonging to $\exists x \neg \psi(x)$. By the induction hypothesis: $T_m \vdash \psi(c)$. $T_m \vdash \exists x \neg \psi(x) \rightarrow \neg \psi(c)$, so $T_m \vdash \psi(c) \rightarrow \neg \exists \neg \psi(x)$. Hence $T_m \vdash \forall x \varphi(x)$.

Conversely: $T_m \vdash \forall x \psi(x) \Rightarrow T_m \vdash \psi(t)$, so $T_m \vdash \psi(t)$ for all closed t, and therefore by the induction hypothesis, $\mathfrak{A} \models \psi(t)$ for all closed t. Hence $\mathfrak{A} \models \forall x \psi(x)$.

Now we see that $\mathfrak{A}$ is a model of Γ, as $\Gamma \subseteq T_m$. $\square$

The model constructed above goes by various names; it is sometimes called the *canonical model* or the *(closed) term model*. In logic programming the set of closed terms of any language is called the *Herbrand universe* or *domain* and the canonical model is called the *Herbrand model*.

In order to get an estimation of the cardinality of the model we have to compute the number of closed terms in L_m. As we did not change the language going from T_ω to T_m, we can look at the language L_ω. We will indicate how to get the required cardinalities, given the alphabet of the original language L. We will use the axiom of choice freely, in particular in the form of absorption laws (i.e. $\kappa + \lambda = \kappa \cdot \lambda = \max(\kappa, \lambda)$ for infinite cardinals). Say L has type $\langle r_1, \ldots, r_n; a_1, \ldots, a_m; \kappa \rangle$.

1. Define
   $TERM_0 := \{c_i | i \in I\} \cup \{x_j | j \in N\}$
   $TERM_{n+1} := TERM_n \cup \{f_j(t_1, \ldots, t_{a_j}) | j \leq m, t_k \in TERM_n \text{ for } k \leq a_j\}$.
   Then $TERM = \bigcup \{TERM_n | n \in N\}$ (Exercise 5)
   $|TERM_0| = \max(\kappa, \aleph_0) = \mu$.
   Suppose $|TERM_n| = \mu$. Then $|\{f_j(t_1, \ldots, t_{a_j}) | t_1, \ldots, t_{a_j} \in TERM_n\}| = |TERM_n|^{a_j} = \mu^{a_j} = \mu$. So $|TERM_{n+1}| = \mu + \mu + \cdots + \mu$ ($m + 1$ times) $= \mu$. Finally $|TERM| = \sum_{n \in N} |TERM_n| = \aleph_0 \cdot \mu = \mu$.

2. Define
   $FORM_0 := \{P_i(t_1, \ldots, t_{r_i}) | i \leq n, t_k \in TERM\} \cup \{\bot\}$
   $FORM_{n+1} := FORM_n \cup \{\varphi \square \psi \mid \square \in \{\wedge, \rightarrow\}, \varphi, \psi \in FORM_n\}$
   $\cup \{\forall x_i \varphi | i \in N, \varphi \in FORM_n\}$.
   Then $FORM = \bigcup \{FORM_n | n \in N\}$ (Exercise 5)
   As in 1, one shows $|FORM| = \mu$.

3. The set of sentences of the form $\exists x \varphi(x)$ has cardinality μ. It trivially is $\leq \mu$. Consider $A = \{\exists x_j (x_j = c_i) | j \in \mathbb{N}, i \in I\}$. Clearly $|A| = \kappa \cdot \aleph_0 = \mu$. Hence the cardinality of the existential statements is μ.

4. L_1 has the constant symbols of L, plus the witnesses. By 3 the cardinality of the set of constant symbols is μ. Using 1 and 2 we find L_0 has μ terms and μ formulas. By induction on n each L_n has μ terms and μ formulas. Therefore L_ω has $\aleph_0 \cdot \mu = \mu$ terms and formulas. L_ω is also the language of T_m.

5. L_ω has at most μ closed terms. Since L_1 has μ witnesses, L_ω has at least μ, and hence exactly μ closed terms.

6. The set of closed terms has $\leq \mu$ equivalence classes under $\sim$, so $\|\mathfrak{A}\| \leq \mu$.

All this adds up to the strengthened version of the Model Existence Lemma.

Lemma 4.1.12 Γ *is consistent* $\leftrightarrow$ Γ *has a model of cardinality at most the cardinality of the language.*

Note the following facts:

- If L has finitely many constants, then L is countable.
- If L has $\kappa \geq \aleph_0$ constants, then $|L| = \kappa$.

The completeness theorem for predicate logic raises the same question as the completeness theorem for propositional logic: can we effectively find a derivation of φ if φ is true? The problem is that we don't have much to go on; φ is true in all structures (of the right similarity type). Even though (in the case of a countable language) we can restrict ourselves to countable structures, the fact that φ is true in all those structures does not give the combinatorial information necessary to construct a derivation for φ. At this stage the matter is beyond us. A treatment of the problem belongs to proof theory.

In the case of predicate logic there are certain improvements on the completeness theorem. One can, for example, ask how complicated the model is that we constructed in the Model Existence Lemma. The proper setting for those questions is found in recursion theory. We can, however, have a quick look at a simple case.

Let T be a *decidable* theory with a countable language, i.e. we have an effective method to test membership (or, which comes to the same, we can test $\Gamma \vdash \varphi$ for a set of axioms of T). Consider the Henkin theory T_ω introduced in Lemma 4.1.8; $\sigma \in T_\omega$ if $\sigma \in T_n$ for a certain n. This number n can be read off from σ by inspection of the witnesses occurring in σ. From the witnesses we can also determine which axioms of the form $\exists x \varphi(x) \to \varphi(c)$ are involved. Let $\{\tau_1, \ldots, \tau_n\}$ be the set of axioms required for the derivation of σ, then $T \cup \{\tau_1, \ldots, \tau_n\} \vdash \sigma$. By the rules of logic this reduces to $T \vdash \tau_1 \land \cdots \land \tau_n \to \sigma$. Since the constants c_i are new with respect to T, this is equivalent to $T \vdash \forall z_1, \ldots, z_k (\tau_1' \land \cdots \land \tau_n' \to \sigma')$ for suitable variables $z_1, \ldots, z_k$, where $\tau_1', \ldots, \tau_n', \sigma'$ are obtained by substitution. Thus we see that $\sigma \in T_\omega$ is decidable. The next step is the formation of a maximal extension T_m.

Let $\varphi_0, \varphi_1, \varphi_2, \ldots$ be an enumeration of all sentences of T_ω. We add sentences to T_ω in steps. Falle:

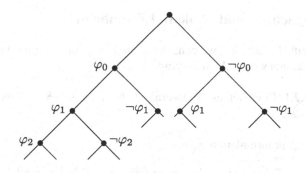

Step 0: $T_0 = \begin{cases} T_\omega \cup \{\varphi_0\} & \text{if } T_\omega \cup \{\varphi_0\} \text{ is consistent,} \\ T_\omega \cup \{\neg\varphi_0\} & \text{else.} \end{cases}$

Step $n + 1$: $T_{n+1} = \begin{cases} T_n \cup \{\varphi_{n+1}\} & \text{if } T_n \cup \{\varphi_{n+1}\} \text{ is consistent,} \\ T_n \cup \{\neg\varphi_{n+1}\} & \text{else.} \end{cases}$

$T^\circ = \bigcup T_n$ (T° is given by a suitable infinite path in the tree). It is easily seen that T° is maximally consistent. Moreover, T° is decidable. To test $\varphi_n \in T^\circ$ we have to test if $\varphi_n \in T_n$ or $T_{n-1} \cup \{\varphi_n\} \vdash \bot$, which is decidable. So T° is decidable.

The model $\mathfrak{A}$ constructed in Lemma 4.1.11 is therefore also decidable in the following sense: the operations and relations of $\mathfrak{A}$ are decidable, which means that $\langle [t_1], \ldots, [t_p] \rangle \in \tilde{P}$ and $\tilde{f}([t_1], \ldots, [t_k]) = [t]$ are decidable.

Summing up, we say that a decidable consistent theory has a decidable model (this can be made more precise by replacing "decidable" by "recursive").

Exercises

1. Consider the language of groups. $T = \{\sigma | \mathfrak{A} \models \sigma\}$, where $\mathfrak{A}$ is a fixed non-trivial group. Show that T is not a Henkin theory.
2. Let $\{T_i | i \in I\}$ be a set of theories, linearly ordered by inclusion. Show that $T = \bigcup \{T_i | i \in I\}$ is a theory which extends each T_i. If each T_i is consistent, then T is consistent.
3. Show that $\lambda_n \vdash \sigma \Leftrightarrow \sigma$ holds in all models with at least n elements, $\mu_n \vdash \sigma \Leftrightarrow \sigma$ holds in all models with at most n elements, $\lambda_n \wedge \mu_n \vdash \sigma \Leftrightarrow \sigma$ holds in all models with exactly n elements, $\{\lambda_n | n \in \mathbb{N}\} \vdash \sigma \Leftrightarrow \sigma$ holds in all infinite models (for a definition of λ_n, μ_n cf. Sect. 3.7).
4. Show that $T = \{\sigma | \lambda_2 \vdash \sigma\} \cup \{c_1 \neq c_2\}$ in a language with $=$ and two constant symbols c_1, c_2, is a Henkin theory.
5. Show $TERM = \bigcup \{TERM_n | n \in \mathbb{N}\}$, $FORM = \bigcup \{FORM_n | n \in \mathbb{N}\}$ (cf. Lemma 2.1.5).
6. T is a theory in the language L. Extend L to L' by adding a set of new constants. $T' = \{\varphi | T \vdash \varphi \in L'\}$. Show that T' is conservative over T.

4.2 Compactness and Skolem–Löwenheim

Unless specified otherwise, we consider sentences in this section. From the Model Existence Lemma we get the following.

Theorem 4.2.1 (Compactness Theorem) Γ *has a model* $\Leftrightarrow$ *each finite subset* Δ *of* Γ *has a model.*

An equivalent formulation is:

$$\Gamma \text{ has no model} \Leftrightarrow \text{some finite } \Delta \subseteq \Gamma \text{ has no model.}$$

Proof We consider the second version.

$\Leftarrow$: Trivial.

$\Rightarrow$: Suppose Γ has no model, then by the Model Existence Lemma Γ is inconsistent, i.e. $\Gamma \vdash \bot$. Therefore there are $\sigma_1, \ldots, \sigma_n \in \Gamma$ such that $\sigma_1, \ldots, \sigma_n \vdash \bot$. This shows that $\Delta = \{\sigma_1, \ldots, \sigma_n\}$ has no model. $\qquad \square$

Let us introduce a bit of notation: $Mod(\Gamma) = \{\mathfrak{A} | \mathfrak{A} \models \sigma \text{ for all } \sigma \in \Gamma\}$. For convenience we will often write $\mathfrak{A} \models \Gamma$ for $\mathfrak{A} \in Mod(\Gamma)$. We write $Mod(\varphi_1, \ldots, \varphi_2)$ instead of $Mod(\{\varphi_1, \ldots, \varphi_n\})$.

In general $Mod(\Gamma)$ is not a set (in the technical sense of set theory: $Mod(\Gamma)$ is most of the time a proper class). We will not worry about that since the notation is only used as an abbreviation.

Conversely, let $\mathcal{K}$ be a class of structures (we have fixed the similarity type), then $Th(\mathcal{K}) = \{\sigma \,|\, \mathfrak{A} \models \sigma \text{ for all } \mathfrak{A} \in \mathcal{K}\}$. We call $Th(\mathcal{K})$ the *theory of* $\mathcal{K}$.

We adopt the convention (already used in Sect. 3.7) not to include the identity axioms in a set Γ; these will always be satisfied.

Examples

1. $Mod(\forall xy(x \leq y \wedge y \leq x \leftrightarrow x = y), \forall xyz(x \leq y \wedge y \leq z \rightarrow x \leq z))$ is the class of posets.
2. Let $\mathcal{G}$ be the class of all groups. $Th(\mathcal{G})$ is the theory of groups.

We can consider the set of integers with the usual additive group structure, but also with the ring structure, so there are two structures $\mathfrak{A}$ and $\mathfrak{B}$, of which the first one is in a sense a part of the second (category theory uses a forgetful functor to express this). We say that $\mathfrak{A}$ is a *reduct* of $\mathfrak{B}$, or $\mathfrak{B}$ is an *expansion* of $\mathfrak{A}$.

In general, we have the following definition.

Definition 4.2.2 $\mathfrak{A}$ is a *reduct* of $\mathfrak{B}$ ($\mathfrak{B}$ an *expansion* of $\mathfrak{A}$) if $|\mathfrak{A}| = |\mathfrak{B}|$ and moreover all relations, functions and constants of $\mathfrak{A}$ occur also as relations, functions and constants of $\mathfrak{B}$.

Notation $(\mathfrak{A}, S_1, \ldots, S_n, g_1, \ldots, g_m, \{a_j \,|\, j \in J\})$ is the expansion of $\mathfrak{A}$ with the indicated extras.

In the early days of logic (before "model theory" was introduced) Skolem (1920) and Löwenheim (1915) studied the possible cardinalities of models of consistent theories. The following generalization follows immediately from the preceding results.

Theorem 4.2.3 (Downward Skolem–Löwenheim Theorem) *Let Γ be a set of sentences in a language of cardinality κ, and let $\kappa < \lambda$. If Γ has a model of cardinality λ, then Γ has a model of cardinality κ', for all κ' with $\kappa \leq \kappa' < \lambda$.*

Proof Add to the language L of Γ a set of fresh constants (not occurring in the alphabet of L) $\{c_i \,|\, i \in I\}$ of cardinality κ', and consider $\Gamma' = \Gamma \cup \{c_i \neq c_j \,|\, i, j \in I, i \neq j\}$. Claim: $Mod(\Gamma') \neq \emptyset$.

Consider a model $\mathfrak{A}$ of Γ of cardinality λ. We expand $\mathfrak{A}$ to $\mathfrak{A}'$ by adding κ' distinct constants (this is possible: $|\mathfrak{A}|$ contains a subset of cardinality κ'). $\mathfrak{A}' \in Mod(\Gamma)$ (cf. Exercise 3) and $\mathfrak{A}' \models c_i \neq c_j (i \neq j)$. Consequently $Mod(\Gamma') \neq \emptyset$. The cardinality of the language of Γ' is κ'. By the Model Existence Lemma Γ' has a model $\mathfrak{B}'$ of cardinality $\leq \kappa'$, but, by the axioms $c_i \neq c_j$, the cardinality is also $\geq \kappa'$. Hence $\mathfrak{B}'$ has cardinality κ'. Now take the reduct $\mathfrak{B}$ of $\mathfrak{B}'$ in the language of Γ, then $\mathfrak{B} \in Mod(\Gamma)$ (Exercise 3). $\qquad\square$

Examples

1. The theory of real numbers, $Th(\mathbb{R})$, in the language of fields, has a countable model.
2. Consider Zermelo–Fraenkel's set theory ZF. If $Mod(ZF) \neq \emptyset$, then ZF has a countable model. This fact was discovered by Skolem. Because of its baffling nature, it was called *Skolem's paradox*. One can prove in ZF the existence of un-countable sets (e.g. the continuum); how can ZF then have a countable model? The answer is simple: countability as seen from outside and from inside the model is not the same. To establish countability one needs a bijection to the nat-ural numbers. Apparently a model can be so poor that it misses some bijections which do exist outside the model.

Theorem 4.2.4 (Upward Skolem–Löwenheim Theorem) *Let Γ have a language L of cardinality κ, and $\mathfrak{A} \in Mod(\Gamma)$ with cardinality $\lambda \geq \kappa$. For each $\mu > \lambda$ Γ has a model of cardinality μ.*

Proof Add μ fresh constants $c_i, i \in I$ to L and consider $\Gamma' = \Gamma \cup \{c_i \neq c_j | i \neq j, i, j \in I\}$. Claim: $Mod(\Gamma') \neq \emptyset$. We apply the Compactness Theorem.

Let $\Delta \subseteq \Gamma'$ be finite. Say Δ contains new axioms with constants $c_{i_0}, \ldots, c_{i_k}$, then $\Delta \subseteq \Gamma \cup \{c_{i_p} \neq c_{i_q} | p, q \leq k\} = \Gamma_0$. Clearly each model of Γ_0 is a model of Δ (Exercise 1 (i)).

Now take $\mathfrak{A}$ and expand it to $\mathfrak{A}' = (\mathfrak{A}, a_1, \ldots, a_k)$, where the a_i are distinct.

Then obviously $\mathfrak{A}' \in Mod(\Gamma_0)$, so $\mathfrak{A}' \in Mod(\Delta)$. By the Compactness Theorem there is a $\mathfrak{B}' \in Mod(\Gamma')$. The reduct $\mathfrak{B}$ of $\mathfrak{A}'$ to the (type of the) language L is a model of Γ. From the extra axioms in Γ' it follows that $\mathfrak{B}'$, and hence $\mathfrak{B}$, has cardinality $\geq \mu$.

We now apply the Downward Skolem–Löwenheim Theorem and obtain the ex-istence of a model of Γ of cardinality μ.   □

We now list a number of applications.

Application I. Non-standard Models of PA

Corollary 4.2.5 *Peano arithmetic has non-standard models.*

Let $\mathcal{P}$ be the class of all Peano structures. Put $\mathbf{PA} = Th(\mathcal{P})$. By the Complete-ness Theorem $\mathbf{PA} = \{\sigma | \Sigma \vdash \sigma\}$ where Σ is the set of axioms listed in Sect. 3.7, Example 6. $\mathbf{PA}$ has a model of cardinality $\aleph_0$ (the standard model $\mathfrak{N}$), so by the Upward Skolem–Löwenheim Theorem it has models of every cardinality $\kappa > \aleph_0$. These models are clearly not isomorphic to $\mathfrak{N}$. For more see page 114.

Application II. Finite and Infinite Models

Lemma 4.2.6 *If Γ has arbitrarily large finite models, then Γ has an infinite model.*

Proof Put $\Gamma' = \Gamma \cup \{\lambda_n | n > 1\}$, where λ_n expresses the sentence "there are at least n distinct elements", cf. Sect. 3.7, Example 1. Apply the Compactness Theorem. Let $\Delta \subseteq \Gamma'$ be finite, and let λ_m be the sentence λ_n in Δ with the largest index n. Verify that $Mod(\Delta) \supseteq Mod(\Gamma \cup \{\lambda_m\})$. Now Γ has arbitrarily large finite models, so Γ has a model $\mathfrak{A}$ with at least m elements, i.e. $\mathfrak{A} \in Mod(\Gamma \cup \{\lambda_m\})$. So $Mod(\Delta) \neq \emptyset$.

By compactness $Mod(\Gamma') \neq \emptyset$, but by virtue of the axioms λ_m, a model of Γ' is infinite. Hence Γ', and therefore Γ, has an infinite model. $\qquad\square$

We get the following simple corollary.

Corollary 4.2.7 *If $\mathcal{K}$ contains arbitrarily large finite models, then, in the language of the class, there is no set Σ of sentences, such that $\mathfrak{A} \in Mod(\Sigma) \Leftrightarrow \mathfrak{A}$ is finite and $\mathfrak{A} \in \mathcal{K}$.*

Proof Immediate. $\qquad\square$

We can paraphrase the result as follows: the class of finite structures in such a class $\mathcal{K}$ is not axiomatizable in first-order logic.

We all know that finiteness can be expressed in a language that contains variables for sets or functions (e.g. Dedekind's definition), so the inability to characterize the notion of finite is a specific defect of first-order logic. We say that *finiteness is not a first-order property.*

The corollary applies to numerous classes, e.g. groups, rings, fields, posets and sets (identity structures).

Application III. Axiomatizability and Finite Axiomatizability

Definition 4.2.8 A class $\mathcal{K}$ of structures is (finitely) *axiomatizable* if there is a (finite) set Γ such that $\mathcal{K} = Mod(\Gamma)$. We say that Γ *axiomatizes* $\mathcal{K}$; the sentences of Γ are called axioms (cf. Definition 4.1.4).

Examples of axiom sets Γ for the classes of posets, ordered sets, groups and rings are listed in Sect. 3.7.

The following fact is very useful.

Lemma 4.2.9 *If $\mathcal{K} = Mod(\Gamma)$ and $\mathcal{K}$ is finitely axiomatizable, then $\mathcal{K}$ is axiomatizable by a finite subset of Γ.*

Proof Let $\mathcal{K} = Mod(\Delta)$ for a finite Δ, then $\mathcal{K} = Mod(\sigma)$, where σ is the conjunction of all sentences of Δ (Exercise 4). Then $\sigma \models \psi$ for all $\psi \in \Gamma$ and $\Gamma \models \sigma$, hence also $\Gamma \vdash \sigma$. Thus there are finitely many $\psi_1, \ldots, \psi_k \in \Gamma$ such that $\psi_1, \ldots, \psi_k \vdash \sigma$. Claim $\mathcal{K} = Mod(\psi_1, \ldots, \psi_k)$.

(i) $\{\psi_1, \ldots, \psi_k\} \subseteq \Gamma$ so $Mod(\Gamma) \subseteq Mod(\psi_1, \ldots, \psi_k)$.
(ii) From $\psi_1, \ldots, \psi_k \vdash \sigma$ it follows that $Mod(\psi_1, \ldots, \psi_k) \subseteq Mod(\sigma)$.

Using (i) and (ii) we conclude that $Mod(\psi_1, \ldots, \psi_k) = \mathcal{K}$. $\qquad\square$

The following lemma is instrumental in proving non-finite axiomatizability results. We need one more fact.

Lemma 4.2.10 $\mathcal{K}$ *is finitely axiomatizable* $\Leftrightarrow$ $\mathcal{K}$ *and its complement* $\mathcal{K}^c$ *are both axiomatizable.*

Proof $\Rightarrow$. Let $\mathcal{K} = Mod(\varphi_1, \ldots, \varphi_n)$, then $\mathcal{K} = Mod(\varphi_1 \wedge \cdots \wedge \varphi_k)$. $\mathfrak{A} \in \mathcal{K}^c$ (complement of $\mathcal{K}$) $\Leftrightarrow$ $\mathfrak{A} \not\models \varphi_1 \wedge \cdots \wedge \varphi_n \Leftrightarrow \mathfrak{A} \models \neg(\varphi_1 \wedge \cdots \wedge \varphi_n)$. So $\mathcal{K}^c = Mod(\neg(\varphi_1 \wedge \cdots \wedge \varphi_n))$.

$\Leftarrow$. Let $\mathcal{K} = Mod(\Gamma)$, $\mathcal{K}^c = Mod(\Delta)$. $K \cap \mathcal{K}^c = Mod(\Gamma \cup \Delta) = \emptyset$ (Exercise 1). By compactness, there are $\varphi_1, \ldots, \varphi_n \in \Gamma$ and $\psi_1, \ldots, \psi_m \in \Delta$ such that $Mod(\varphi_1, \ldots, \varphi_n, \psi_1, \ldots, \psi_m) = \emptyset$, or

$$Mod(\varphi_1, \ldots, \varphi_n) \cap Mod(\psi_1, \ldots, \psi_m) = \emptyset, \quad (1)$$

$$\mathcal{K} = Mod(\Gamma) \subseteq Mod(\varphi_1, \ldots, \varphi_n), \quad (2)$$

$$\mathcal{K}^c = Mod(\Delta) \subseteq Mod(\psi_1, \ldots, \psi_m), \quad (3)$$

$(1), (2), (3) \Rightarrow K = Mod(\varphi_1, \ldots, \varphi_n)$. $\square$

We now get a number of corollaries.

Corollary 4.2.11 *The class of all infinite sets* (*identity structures*) *is axiomatizable, but not finitely axiomatizable.*

Proof $\mathfrak{A}$ is infinite $\Leftrightarrow$ $\mathfrak{A} \in Mod(\{\lambda_n | n \in \mathbb{N}\})$. So the axiom set is $\{\lambda_n | n \in \mathbb{N}\}$. On the other hand the class of finite sets is not axiomatizable, so, by Lemma 4.2.10, the class of infinite sets is not finitely axiomatizable. $\square$

Corollary 4.2.12

  (i) *The class of fields of characteristic* $p(> 0)$ *is finitely axiomatizable.*
 (ii) *The class of fields of characteristic* 0 *is axiomatizable but not finitely axiomatizable.*
(iii) *The class of fields of positive characteristic is not axiomatizable.*

Proof (i) The theory of fields has a finite set Δ of axioms. $\Delta \cup \{\overline{p} = 0\}$ axiomatizes the class $\mathcal{F}_p$ of fields of characteristic p (where $\overline{p}$ stands for $1 + 1 + \cdots + 1, (p\times)$).

(ii) $\Delta \cup \{\overline{2} \neq 0, \overline{3} \neq 0, \ldots, \overline{p} \neq 0, \ldots\}$ axiomatizes the class $\mathcal{F}_0$ of fields of characteristic 0. Suppose $\mathcal{F}_0$ was finitely axiomatizable, then by Lemma 4.2.9 $\mathcal{F}_0$ was axiomatizable by $\Gamma = \Delta \cup \{\overline{p}_1 \neq 0, \ldots, \overline{p}_k \neq 0\}$, where $p_1, \ldots, p_k$ are primes (not necessarily the first k ones). Let q be a prime greater than all p_i (Euclid). Then $\mathbb{Z}/(q)$ (the integers modulo q) is a model of Γ, but $\mathbb{Z}/(q)$ is not a field of characteristic 0. Contradiction.

(iii) Follows immediately from (ii) and Lemma 4.2.10. $\square$

Corollary 4.2.13 *The class A_c of all algebraically closed fields is axiomatizable, but not finitely axiomatizable.*

Proof Let $\sigma_n = \forall y_1 \ldots y_n \exists x (x^n + y_1 x^{n-1} + \cdots + y_{n-1} x + y_n = 0)$. Then $\Gamma = \Delta \cup \{\sigma_n | n \geq 1\}$ (Δ as in Corollary 4.2.12) axiomatizes A_c. To show non-finite axiomatizability, apply Lemma 4.2.9 to Γ and find a field in which a certain polynomial does not factorize.   □

Corollary 4.2.14 *The class of all torsion-free abelian groups is axiomatizable, but not finitely axiomatizable.*

Proof Exercise 15.   □

Remark In Lemma 4.2.9 we used the Completeness Theorem and in Lemma 4.2.10 the Compactness Theorem. The advantage of using only the Compactness Theorem is that one avoids the notion of provability altogether. The reader might object that this advantage is rather artificial since the Compactness Theorem is a corollary to the Completeness Theorem. This is true in our presentation; one can, however, derive the Compactness Theorem by purely model theoretic means (using ultraproducts, cf. Chang–Keisler), so there are situations where one has to use the Compactness Theorem. For the moment the choice between using the Completeness Theorem or the Compactness Theorem is largely a matter of taste or convenience.

By way of illustration we will give an alternative proof of Lemma 4.2.9 using the Compactness Theorem.

Again we have $Mod(\Gamma) = Mod(\sigma)(*)$. Consider $\Gamma' = \Gamma \cup \{\neg\sigma\}$.

$$\mathfrak{A} \in Mod(\Gamma') \quad \Leftrightarrow \quad \mathfrak{A} \in Mod(\Gamma) \text{ and } \mathfrak{A} \models \neg\sigma,$$
$$\Leftrightarrow \quad \mathfrak{A} \in Mod(\Gamma) \text{ and } \mathfrak{A} \notin Mod(\sigma).$$

In view of $(*)$ we have $Mod(\Gamma') = \emptyset$.

By the Compactness Theorem there is a finite subset Δ of Γ' with $Mod(\Delta) = \emptyset$. It is no restriction to suppose that $\neg\sigma \in \Delta$, hence $Mod(\psi_1, \ldots, \psi_k, \neg\sigma) = \emptyset$. It now easily follows that $Mod(\psi_1, \ldots, \psi_k) = Mod(\sigma) = Mod(\Gamma)$.

Application IV. Ordering Sets   One easily shows that each finite set can be ordered; for infinite sets this is harder. A simple trick is presented below.

Theorem 4.2.15 *Each infinite set can be ordered.*

Proof Let $|X| = \kappa \geq \aleph_0$. Consider Γ, the set of axioms for linear order (3.7.3). Γ has a countable model, e.g. $\mathbb{N}$. By the Upward Skolem–Löwenheim Theorem Γ has a model $\mathfrak{A} = \langle A, < \rangle$ of cardinality κ. Since X and A have the same cardinality there is a bijection $f : X \to A$. Define $x <_X x' := f(x) < f(x')$. Evidently, $<_X$ is a linear order.   □

In the same way one gets: Each infinite set can be densely ordered. The same trick works for axiomatizable classes in general.

Exercises

1. Show:
   (i) $\Gamma \subseteq \Delta \Rightarrow Mod(\Delta) \subseteq Mod(\Gamma)$,
   (ii) $\mathcal{K}_1 \subseteq \mathcal{K}_2 \Rightarrow Th(\mathcal{K}_2) \subseteq Th(\mathcal{K}_1)$,
   (iii) $Mod(\Gamma \cup \Delta) = Mod(\Gamma) \cap Mod(\Delta)$,
   (iv) $Th(\mathcal{K}_1 \cup \mathcal{K}_2) = Th(\mathcal{K}_1) \cap Th(\mathcal{K}_2)$,
   (v) $\mathcal{K} \subseteq Mod(\Gamma) \Leftrightarrow \Gamma \subseteq Th(\mathcal{K})$,
   (vi) $Mod(\Gamma \cap \Delta) \supseteq Mod(\Gamma) \cup Mod(\Delta)$,
   (vii) $Th(\mathcal{K}_1 \cap \mathcal{K}_2) \supseteq Th(\mathcal{K}_1) \cup Th(\mathcal{K}_2)$.
   Show that in (vi) and (vii) $\supseteq$ cannot be replaced by $=$.
2. (i) $\Gamma \subseteq Th(Mod(\Gamma))$,
   (ii) $\mathcal{K} \subseteq Mod(Th(\mathcal{K}))$,
   (iii) $Th(Mod(\Gamma))$ is a theory with axiom set Γ.
3. If $\mathfrak{A}$ with language L is a reduct of $\mathfrak{B}$, then $\mathfrak{A} \models \sigma \Leftrightarrow \mathfrak{B} \models \sigma$ for $\sigma \in L$.
4. $Mod(\varphi_1, \ldots, \varphi_n) = Mod(\varphi_1 \wedge \cdots \wedge \varphi_n)$.
5. $\Gamma \models \varphi \Rightarrow \Delta \models \varphi$ for a finite subset $\Delta \subseteq \Gamma$. (Give one proof using completeness, another proof using compactness on $\Gamma \cup \{\neg\varphi\}$.)
6. Show that *well-ordering* is not a first-order notion. Suppose that Γ axiomatizes the class of well-orderings. Add countably many constants c_i and show that $\Gamma \cup \{c_{i+1} < c_i | i \in \mathcal{N}\}$ has a model.
7. If Γ has only finite models, then there is an n such that each model has at most n elements.
8. Let L have the binary predicate symbol P. $\sigma := \forall x \neg P(x, x) \wedge \forall xyz(P(x, y) \wedge P(y, z) \rightarrow P(x, z)) \wedge \forall x \exists y P(x, y)$. Show that $Mod(\sigma)$ contains only infinite models.
9. Show that $\sigma \vee \forall xy(x = y)$ has infinite models and a finite model, but no arbitrarily large finite models (σ as in Exercise 8).
10. Let L have one unary function symbol.
    (i) Write down a sentence φ such that $\mathfrak{A} \models \varphi \Leftrightarrow f^{\mathfrak{A}}$ is a surjection.
    (ii) Idem for an injection.
    (iii) Idem for a bijection (permutation).
    (iv) Use (ii) to formulate a sentence σ such that (a) $\mathfrak{A} \models \sigma \Rightarrow \mathfrak{A}$ is infinite, (b) each infinite set can be expanded to a model of σ (Dedekind).
    (v) Show that each infinite set carries a permutation without fixed points (cf. the proof of Theorem 4.2.15).
11. Show: σ holds for fields of characteristic zero $\Rightarrow \sigma$ holds for all fields of characteristic $q > p$ for a certain p.
12. Consider a sequence of theories T_i such that $T_i \neq T_{i+1}$ and $T_i \subseteq T_{i+1}$. Show that $\bigcup\{T_i | i \in \mathcal{N}\}$ is not finitely axiomatizable.
13. If T_1 and T_2 are theories such that $Mod(T_1 \cup T_2) = \emptyset$, then there is a σ such that $T_1 \models \sigma$ and $T_2 \models \neg\sigma$.

14. (i) A group can be ordered $\Leftrightarrow$ each finitely generated subgroup can be ordered.

   (ii) An abelian group $\mathfrak{A}$ can be ordered $\Leftrightarrow$ $\mathfrak{A}$ is torsion free. (Hint: look at all closed atoms of $L(\mathfrak{A})$ true in $\mathfrak{A}$.)
15. Prove Corollary 4.2.14.
16. Show that each countable, ordered set can be embedded in the rationals.
17. Show that the class of trees cannot be axiomatized. Here we define a tree as a structure $\langle T, \leq, t \rangle$, where $\leq$ is a partial order, such that for each a the predecessors form a finite chain $a = a_n < a_{n-1} < \cdots < a_1 < a_0 = t$. Moreover, no two incomparable elements have a common successor. t is called the top.
18. A graph (with symmetric and irreflexive R) is called k-colorable if we can paint the vertices with k-different colors such that adjacent vertices have distinct colors. We formulate this by adding k unary predicates $C_1, \ldots, C_k$, plus the following axioms:

$$\forall x \bigvee_i C_i(x), \forall x \bigwedge_{i \neq j} \neg(C_i(x) \wedge C_j(x)),$$

$$\bigwedge_i \forall xy\big(C_i(x) \wedge C_i(y) \to \neg R(x, y)\big).$$

Show that a graph is k-colorable if each finite subgraph is k-colorable (De Bruijn–Erdös).

4.3 Some Model Theory

In model theory one investigates the various properties of models (structures), in particular in connection with the features of their language. One could say that algebra is a part of model theory; some parts of algebra indeed belong to model theory, other parts only in the sense of the limiting case in which the role of language is negligible. It is the interplay between language and models that makes model theory fascinating. Here we will only discuss the very beginnings of the topic.

In algebra one does not distinguish structures which are isomorphic; the nature of the objects is purely accidental. In logic we have another criterion: we distinguish between two structures by exhibiting a sentence which holds in one but not in the other. So, if $\mathfrak{A} \models \sigma \Leftrightarrow \mathfrak{B} \models \sigma$ for all σ, then we cannot (logically) distinguish $\mathfrak{A}$ and $\mathfrak{B}$.

Definition 4.3.1 (i) $f : |\mathfrak{A}| \to |\mathfrak{B}|$ is a *homomorphism* if for all P_i $\langle a_1, \ldots, a_k \rangle \in P_i^{\mathfrak{A}} \Rightarrow \langle f(a_1), \ldots, f(a_k) \rangle \in P_i^{\mathfrak{B}}$, if for all F_j $f(F_j^{\mathfrak{A}}(a_1, \ldots, a_p)) = F_j^{\mathfrak{B}}(f(a_1), \ldots, f(a_p))$ and if for all c_i $f(c_i^{\mathfrak{A}}) = c_i^{\mathfrak{B}}$.

   (ii) f is an *isomorphism* if it is a homomorphism which is bijective and satisfies $\langle a_1, \ldots, a_n \rangle \in P_i^{\mathfrak{A}} \Leftrightarrow \langle f(a_1), \ldots, f(a_n) \rangle \in P_i^{\mathfrak{B}}$, for all P_i.

We write $f : \mathfrak{A} \to \mathfrak{B}$ if f is a homomorphism from $\mathfrak{A}$ to $\mathfrak{B}$. $\mathfrak{A} \cong \mathfrak{B}$ stands for "$\mathfrak{A}$ is isomorphic to $\mathfrak{B}$", i.e. there is an isomorphism $f : \mathfrak{A} \to \mathfrak{B}$.

Definition 4.3.2 $\mathfrak{A}$ and $\mathfrak{B}$ are *elementarily equivalent* if for all sentences σ of L, $\mathfrak{A} \models \sigma \Leftrightarrow \mathfrak{B} \models \sigma$.

Notation $\mathfrak{A} \equiv \mathfrak{B}$. Note that $\mathfrak{A} \equiv \mathfrak{B} \Leftrightarrow Th(\mathfrak{A}) = Th(\mathfrak{B})$.

Lemma 4.3.3 $\mathfrak{A} \cong \mathfrak{B} \Rightarrow \mathfrak{A} \equiv \mathfrak{B}$.

Proof Exercise 2. □

Definition 4.3.4 $\mathfrak{A}$ is a *substructure* (submodel) of $\mathfrak{B}$ (of the same type) if $|\mathfrak{A}| \subseteq |\mathfrak{B}|$ and for all $P_i, F_j : P_i^{\mathfrak{B}} \cap |\mathfrak{A}|^{n_i} = P_i^{\mathfrak{A}}, F_j^{\mathfrak{B}} \upharpoonright |\mathfrak{A}|^{n_j} = F_j^{\mathfrak{A}}$ and $c_i^{\mathfrak{A}} = c_i^{\mathfrak{B}}$ (where n_i, n_j are the number of arguments of P_i, F_j).

Notation $\mathfrak{A} \subseteq \mathfrak{B}$. Note that it is not sufficient for $\mathfrak{A}$ to be contained in $\mathfrak{B}$ "as a set"; the relations and functions of $\mathfrak{B}$ have to be extensions of the corresponding ones on $\mathfrak{A}$, in the specific way indicated above.

Examples The field of rationals is a substructure of the field of reals, but not of the ordered field of reals. Let $\mathfrak{A}$ be the additive group of rationals, $\mathfrak{B}$ the multiplicative group of non-zero rationals. Although $|\mathfrak{B}| \subseteq |\mathfrak{A}|$, $\mathfrak{B}$ is not a substructure of $\mathfrak{A}$. The well-known notions of subgroups, subrings, subspaces, all satisfy the above definition.

The notion of elementary equivalence only requires that sentences (which do not refer to specific elements, except for constants) are simultaneously true in two structures. We can sharpen the notion, by considering $\mathfrak{A} \subseteq \mathfrak{B}$ and by allowing reference to elements of $|\mathfrak{A}|$.

Definition 4.3.5 $\mathfrak{A}$ is an *elementary substructure* of $\mathfrak{B}$ (or $\mathfrak{B}$ is an *elementary extension* of $\mathfrak{A}$) if $\mathfrak{A} \subseteq \mathfrak{B}$ and for all $\varphi(x_1, \ldots, x_n)$ in L and $a_1, \ldots, a_n \in |\mathfrak{A}|$, $\mathfrak{A} \models \varphi(\bar{a}_1, \ldots, \bar{a}_n) \Leftrightarrow \mathfrak{B} \models \varphi(\bar{a}_1, \ldots, \bar{a}_n)$.

Notation $\mathfrak{A} \prec \mathfrak{B}$.

We say that $\mathfrak{A}$ and $\mathfrak{B}$ have the same true sentences *with parameters in $\mathfrak{A}$*.

Fact 4.3.6 $\mathfrak{A} \prec \mathfrak{B} \Rightarrow \mathfrak{A} \equiv \mathfrak{B}$.

The converse does not hold (cf. Exercise 4).

Since we will often join all elements of $|\mathfrak{A}|$ to $\mathfrak{A}$ as constants, it is convenient to have a special notation for the enriched structure: $\hat{\mathfrak{A}} = (\mathfrak{A}, |\mathfrak{A}|)$.

If one wants to describe a certain structure $\mathfrak{A}$, one has to specify all the basic relationships and functional relations. This can be done in the language $L(\mathfrak{A})$ belonging to $\mathfrak{A}$ (which, incidentally, *is* the language of the type of $\hat{\mathfrak{A}}$).

Definition 4.3.7 *The diagram*, $Diag(\mathfrak{A})$, is the set of closed atoms and negations of closed atoms of $L(\mathfrak{A})$, which are true in $\mathfrak{A}$. The *positive diagram*, $Diag^+(\mathfrak{A})$, is the set of closed atoms φ of $L(\mathfrak{A})$ such that $\mathfrak{A} \models \varphi$.

Examples

1. $\mathfrak{A} = \langle \mathbb{N} \rangle$. $Diag(\mathfrak{A}) = \{\overline{n} = \overline{n} | n \in \mathbb{N}\} \cup \{\overline{n} \neq \overline{m} | n \neq m; n, m \in \mathbb{N}\}$.
2. $\mathfrak{B} = \langle \{1, 2, 3\}, < \rangle$. (natural order). $Diag(\mathfrak{B}) = \{\overline{1} = \overline{1},\ \overline{2} = \overline{2},\ \overline{3} = \overline{3},\ \overline{1} \neq \overline{2},$
   $\overline{1} \neq \overline{3}, \overline{2} \neq \overline{3},\ \overline{2} \neq \overline{1},\ \overline{3} \neq \overline{1},\ \overline{3} \neq \overline{2},\ \overline{1} < \overline{2},\ \overline{1} < \overline{3},\ \overline{2} < \overline{3},\ \neg\overline{2} < \overline{1}, \neg\overline{3} < \overline{2},$
   $\neg\overline{3} < \overline{1},\ \neg\overline{1} < \overline{1},\ \neg\overline{2} < \overline{2},\ \neg\overline{3} < \overline{3}\}$.

Diagrams are useful for many purposes. We demonstrate one here: we say that $\mathfrak{A}$ is *isomorphically embedded* in $\mathfrak{B}$ if there is an isomorphism f from $\mathfrak{A}$ into a substructure of $\mathfrak{B}$.

Lemma 4.3.8 $\mathfrak{A}$ *is isomorphically embedded in* $\mathfrak{B} \Leftrightarrow \hat{\mathfrak{B}}$ *is a model of* $Diag(\mathfrak{A})$.

Proof $\Rightarrow$. Let f be an isomorphic embedding of $\mathfrak{A}$ in $\mathfrak{B}$, then $\mathfrak{A} \models P_i(\overline{a}_1, \ldots, \overline{a}_n)$ $\Leftrightarrow \mathfrak{B} \models P_i(\overline{f(a_1)}, \ldots, \overline{f(a_n)})$ and $\mathfrak{A} \models t(\overline{a}_1, \ldots, \overline{a}_n) = s(\overline{a}_1, \ldots, \overline{a}_n) \Leftrightarrow$ $\mathfrak{B} \models t(\overline{f(a_1)}, \ldots) = s(\overline{f(a_1)}, \ldots)$ (cf. Exercise 2). By interpreting $\overline{a}$ as $f(a)$ in $\hat{\mathfrak{B}}$ (i.e. $\overline{a}^{\hat{\mathfrak{B}}} = f(a)$), we immediately see $\hat{\mathfrak{B}} \models Diag(\mathfrak{A})$.

$\Leftarrow$: Let $\hat{\mathfrak{B}} \models Diag(\mathfrak{A})$. Define a mapping $f : |\mathfrak{A}| \to |\mathfrak{B}|$ by $f(a) = (\overline{a})^{\mathfrak{B}}$. Then, clearly, f satisfies the conditions of Definition 4.3.1 on relations and functions (since they are given by atoms and negations of atoms). Moreover if $a_1 \neq a_2$ then $\mathfrak{A} \models \neg \overline{a}_1 = \overline{a}_2$, so $\hat{\mathfrak{B}} \models \neg \overline{a}_1 = \overline{a}_2$.

Hence $\overline{a}_1^{\mathfrak{B}} \neq \overline{a}_2^{\mathfrak{B}}$, and thus $f(a_1) \neq f(a_2)$. This shows that f is an isomorphism. $\qquad\square$

We will often identify $\mathfrak{A}$ with its image under an isomorphic embedding into $\mathfrak{B}$, so that we may consider $\mathfrak{A}$ as a substructure of $\mathfrak{B}$.

We have a similar criterion for elementary extension. We say that $\mathfrak{A}$ is *elementarily embeddable* in $\mathfrak{B}$ if $\mathfrak{A} \cong \mathfrak{A}'$ and $\mathfrak{A}' \prec \mathfrak{B}$ for some $\mathfrak{A}'$. Again, we often simplify matters by just writing $\mathfrak{A} \prec \mathfrak{B}$ when we mean "elementarily embeddable".

Lemma 4.3.9 $\mathfrak{A} \prec \mathfrak{B} \Leftrightarrow \hat{\mathfrak{B}} \models Th(\hat{\mathfrak{A}})$.

N.B. $\mathfrak{A} \prec \mathfrak{B}$ holds "up to isomorphism". $\hat{\mathfrak{B}}$ is supposed to be of a similarity type which admits at least constants for all constant symbols of $L(\mathfrak{A})$.

Proof $\Rightarrow$. Let $\varphi(\overline{a}_1, \ldots, \overline{a}_n) \in Th(\hat{\mathfrak{A}})$, then $\mathfrak{A} \models \varphi(\overline{a}_1, \ldots, \overline{a}_n)$, and hence $\hat{\mathfrak{B}} \models$ $\varphi(\overline{a}_1, \ldots, \overline{a}_n)$. So $\hat{\mathfrak{B}} \models Th(\hat{\mathfrak{A}})$.

$\Leftarrow$. By Lemma 4.3.8, $\mathfrak{A} \subseteq \mathfrak{B}$ (up to isomorphism). The reader can easily finish the proof now. $\Box$

We now give some applications.

Application I. Non-standard Models of Arithmetic   Recall that $\mathfrak{N} = \langle \mathbb{N}, +, \cdot, s, 0 \rangle$ is the *standard model* of arithmetic. We know that it satisfies Peano's axioms (cf. Example 6, Sect. 3.7). We use the abbreviations introduced in Sect. 3.7.

Let us now construct a non-standard model. Consider $T = Th(\hat{\mathfrak{N}})$. By the Skolem–Löwenheim Theorem T has an uncountable model $\mathfrak{M}$. Since $\mathfrak{M} \models Th(\hat{\mathfrak{N}})$, we have, by Lemma 4.3.9, $\mathfrak{N} \prec \mathfrak{M}$. Observe that $\mathfrak{N} \not\cong \mathfrak{M}$ (why?). Let us have a closer look at the way in which $\mathfrak{N}$ is embedded in $\mathfrak{M}$.

We note that $\mathfrak{N} \models \forall xyz(x < y \wedge y < z \rightarrow x < z)$ (1)

$$\mathfrak{N} \models \forall xyz(x < y \vee x = y \vee y < x) \quad (2)$$

$$\mathfrak{N} \models \forall x(\overline{0} \le x) \quad (3)$$

$$\mathfrak{N} \models \neg \exists x(\overline{n} < x \wedge x < \overline{n+1}) \quad (4)$$

Hence, $\mathfrak{N}$ being an elementary substructure of $\mathfrak{M}$, we have (1) and (2) for $\mathfrak{M}$, i.e. $\mathfrak{M}$ is linearly ordered. From $\mathfrak{N} \prec \mathfrak{M}$ and (3) we conclude that $\overline{0}$ is the first element of $\mathfrak{M}$. Furthermore, (4) with $\mathfrak{N} \prec \mathfrak{M}$ tells us that there are no elements of $\mathfrak{M}$ between the "standard natural numbers".

As a result we see that $\mathfrak{N}$ is an initial segment of $\mathfrak{M}$:

standard numbers   *non-standard numbers*

Remark It is important to realize that (1)–(4) are not only *true in the standard model*, but even provable in **PA**. This implies that they hold not only in elementary extensions of $\mathfrak{N}$, but in *all* Peano structures. The price one has to pay is the actual proving of (1)–(4) in **PA**, which is more cumbersome than merely establishing their validity in $\mathfrak{N}$. However, anyone who can give an informal proof of these simple properties will find out that it is just one more (tedious, but not difficult) step to formalize the proof in our natural deduction system. Step-by-step proofs are outlined in the Exercises 27, 28.

So, all elements of $|\mathfrak{M}| - |\mathfrak{N}|$, the *non-standard numbers*, come after the standard ones. Since $\mathfrak{M}$ is uncountable, there is at least one non-standard number a. Note that $n < a$ for all n, so $\mathfrak{M}$ has a *non-archimedean order* (recall that $n = 1 + 1 + \cdots + 1 (n \times)$).

We see that the successor $S(n)(= n + 1)$ of a standard number is standard. Furthermore $\mathfrak{N} \models \forall x(x \ne \overline{0} \rightarrow \exists y(y + \overline{1} = x))$, so, since $\mathfrak{N} \prec \mathfrak{M}$, also $\mathfrak{M} \models \forall x(x \ne \overline{0} \rightarrow \exists y(y + \overline{1} = x))$, i.e. in $\mathfrak{M}$ each number, distinct from zero, has a (unique) predecessor. Since a is non-standard it is distinct from zero, hence it has a predecessor, say a_1. Since successors of standard numbers are standard, a_1 is non-standard. We can repeat this procedure indefinitely and obtain an infinite descending sequence $a > a_1 > a_2 > a_3 > \cdots$ of non-standard numbers. Conclusion: $\mathfrak{M}$ is not well-ordered.

However, non-empty *definable* subsets of $\mathfrak{M}$ do possess a least element. For, such a set is of the form $\{b|\mathfrak{M} \models \varphi(\bar{b})\}$, where $\varphi \in L(\mathfrak{N})$, and we know $\mathfrak{N} \models \exists x \varphi(x) \to \exists x(\varphi(x) \wedge \forall y(\varphi(y) \to x \leq y))$. This sentence also holds in $\mathfrak{M}$ and it tells us that $\{b|\mathfrak{M} \models \varphi(\bar{b})\}$ has a least element if it is not empty.

The above construction not merely gave a non-standard Peano structure (cf. Corollary 4.2.5), but also a non-standard model of *true arithmetic*, i.e. it is a model of all sentences true in the standard model. Moreover, it is an elementary extension.

The non-standard models of **PA** that are elementary extensions of $\mathfrak{N}$ are the ones that can be handled most easily, since the facts from the standard model carry over. There are also quite a number of properties that have been established for non-standard models in general. We treat two of them here.

Theorem 4.3.10 *The set of standard numbers in a non-standard model is not definable.*

Proof Suppose there is a $\varphi(x)$ in the language of **PA**, such that: $\mathfrak{M} \models \varphi(\bar{a}) \Leftrightarrow$ "a is a standard natural number", then $\neg\varphi(x)$ defines the non-standard numbers. Since **PA** proves the *least number principle*, we have $\mathfrak{M} \models \exists x(\neg\varphi(x) \wedge \forall y < x\varphi(y))$, or there is a least non-standard number. However, as we have seen above, this is not the case. So there is no such definition.   □

A simple consequence is the following.

Lemma 4.3.11 (Overspill Lemma) *If $\varphi(\bar{n})$ holds in a non-standard model for infinitely many finite numbers n, then $\varphi(\bar{a})$ holds for at least one infinite number a.*

Proof Suppose that for no infinite a $\varphi(\bar{a})$ holds, then $\exists y(x < y \wedge \varphi(y))$ defines the set of standard natural numbers in the model. This contradicts the preceding result.   □

Our technique of constructing models yields various non-standard models of Peano arithmetic. We have at this stage no means to decide if all models of **PA** are elementarily equivalent or not. The answer to this question is provided by Gödel's incompleteness theorem, which states that there is a sentence γ such that **PA** $\nvdash \gamma$ and **PA** $\nvdash \neg\gamma$. The incompleteness of **PA** has been re-established by quite different means by Paris–Kirby–Harrington, Kripke and others. As a result we now have examples for γ, which belong to "normal mathematics", whereas Gödel's γ, although purely arithmetical, can be considered as slightly artificial, cf. *Barwise*, Handbook of Mathematical Logic, D8. **PA** has a decidable (recursive) model, namely the standard model. That, however, is the only one. By Tennenbaum's theorem all non-standard models of **PA** are undecidable (not recursive).

Application II. Non-standard Real Numbers   Similarly to the above application, we can introduce non-standard models for the real number system. We use the language of the ordered field R of real numbers, and for convenience we use the function symbol, $|\ |$, for the absolute value function. By the Skolem–Löwenheim Theorem there is a model $^*R$ of $Th(\hat{R})$ such that $^*R$ has greater cardinality than R. Applying Lemma 4.3.9, we see that $R \prec {}^*R$, so $^*R$ is an ordered field, containing the standard real numbers. For cardinality reasons there is an element $a \in |^*R| - |R|$. For the element a there are two possibilities:

(i) $|a| > |r|$ for all $r \in |R|$,
(ii) there is an $r \in |R|$ such that $|a| < r$.

In the second case $\{u \in |R| \mid u < |a|\}$ is a bounded, non-empty set, which therefore has a supremum s (in R). Since $|a|$ is a non-standard number, there is no standard number between s and $|a|$. By ordinary algebra, there is no standard number between 0 and $|\,|a| - s\,|$. Hence $||a| - s|^{-1}$ is larger than all standard numbers. So in case (ii) there is also a non-standard number greater than all standard numbers. Elements satisfying the condition (i) above, are called *infinite* and elements satisfying (ii) are called *finite* (note that the standard numbers are finite).

We now list a number of facts, leaving the (fairly simple) proofs to the reader.

1. $^*R$ has a non-archimedean order.
2. There are numbers a such that for all positive standard $r, 0 < |a| < r$. We call such numbers, including 0, *infinitesimals*.
3. a is infinitesimal $\Leftrightarrow a^{-1}$ is infinite, where $a \neq 0$.
4. For each non-standard finite number a there is a unique standard number $st(a)$ such that $a - st(a)$ is infinitesimal.

   Infinitesimals can be used for elementary calculus in the Leibnizian tradition. We will give a few examples. Consider an expansion R' of R with a predicate for N and a function v. Let $^*R'$ be the corresponding non-standard model such that $R' \prec {}^*R'$. We are actually considering two extensions at the same time. N is contained in R', i.e. singled out by a special predicate N. Hence N is extended, along with R' to $^*N$. As expected, $^*N$ is an elementary extension of N (cf. Exercise 14). Therefore we may safely operate in the traditional manner with real numbers and natural numbers. In particular we also have in $^*R'$ infinite natural numbers available. We want v to be a sequence, i.e. we are only interested in the values of v for natural number arguments. The concepts of convergence, limit, etc. can be taken from analysis.

   We will use the notation of the calculus. The reader may try to give the correct formulation.

   Here is an example: $\exists m \forall n > m(|v_n - v_m| < \epsilon)$ stands for $\exists x(N(x) \wedge \forall y(N(y) \wedge y > x \rightarrow |v(y) - v(x)| < \epsilon))$. Properly speaking we should relativize quantifiers over natural numbers (cf. 3.5.12), but it is more convenient to use variables of several sorts.
5. The sequence v (or (v_n)) converges in R' iff for all infinite natural numbers n, m $|v_n - v_m|$ is infinitesimal.

Proof (v_n) converges in R' if $R' \models \forall \epsilon > 0 \exists n \forall m > n(|v_n - v_m| < \epsilon)$. Assume that (v_n) converges. Choose for $\epsilon > 0$ an $n(\epsilon) \in |R'|$ such that $R' \models \forall m > n(|v_n - v_m| < \epsilon)$. Then also $^*R' \models \forall m > n(|v_n - v_m| < \epsilon)$. In particular, if m, m' are infinite, then $m, m' > n(\epsilon)$ for all ϵ. Hence $|v_m - v_{m'}| < 2\epsilon$ for all ϵ. This means that $|v_m - v_{m'}|$ is infinitesimal. Conversely, if $|v_n - v_m|$ is infinitesimal for all infinite n, m, then $^*R \models \forall m > n(|v_n - v_m| < \epsilon)$ where n is infinite and ϵ standard, positive. So $^*R' \models \exists n \forall m > n(|v_n - v_m| < \epsilon)$, for each standard $\epsilon > 0$. Now, since $R' \prec {}^*R'$, $R' \models \exists n \forall m > n(|v_n - v_m| < \epsilon)$ for $\epsilon > 0$, so $R' \models \forall \epsilon > 0 \exists n \forall m > n(|v_n - v_m| < \epsilon)$. Hence (v_n) converges. $\square$

6. $\lim_{n \to \infty} v_n = a \Leftrightarrow |a - v_n|$ is infinitesimal for infinite n.

Proof Similar to 5. $\square$

We have only been able to touch the surface "non-standard analysis". For an extensive treatment, see e.g. *Robinson (1965), Stroyan and Luxemburg (1976)*.

We can now strengthen the Skolem–Löwenheim Theorems.

Theorem 4.3.12 (Downward Skolem–Löwenheim) *Let the language L of $\mathfrak{A}$ have cardinality κ, and suppose $\mathfrak{A}$ has cardinality $\lambda \geq \kappa$. Then there is a structure $\mathfrak{B}$ of cardinality κ such that $\mathfrak{B} \prec \mathfrak{A}$.*

Proof See Corollary 4.4.11. $\square$

Theorem 4.3.13 (Upward Skolem–Löwenheim) *Let the language L of $\mathfrak{A}$ have cardinality κ and suppose $\mathfrak{A}$ has cardinality $\lambda \geq \kappa$. Then for each $\mu > \lambda$ there is a structure $\mathfrak{B}$ of cardinality μ, such that $\mathfrak{A} \prec \mathfrak{B}$.*

Proof Apply the old Upward Skolem–Löwenheim Theorem to $Th(\hat{\mathfrak{A}})$. $\square$

In the completeness proof we used maximally consistent theories. In model theory these are called complete theories. As a rule the notion is defined with respect to axiom sets.

Definition 4.3.14 A theory with axioms Γ in the language L, is called *complete* if for each sentence σ in L, either $\Gamma \vdash \sigma$, or $\Gamma \vdash \neg \sigma$.

A complete theory leaves, so to speak, no questions open, but it does not prima facie restrict the class of models. In the old days mathematicians tried to find for such basic theories as arithmetic axioms that would determine up to isomorphism one model, i.e. to give a set Γ of axioms such that $\mathfrak{A}, \mathfrak{B} \in Mod(\Gamma) \Rightarrow \mathfrak{A} \cong \mathfrak{B}$. The Skolem–Löwenheim Theorems have taught us that this is (barring the finite case) unattainable. There is, however, a significant notion.

Definition 4.3.15 Let κ be a cardinal. A theory is κ-categorical if it has at least one model of cardinality κ and if any two of its models of cardinality κ are isomorphic.

Categoricity in some cardinality is not as unusual as one might think. We list some examples.

1. *The theory of infinite sets (identity structures) is κ-categorical for all infinite κ.*

Proof Immediate, as here "isomorphic" means "of the same cardinality".   □

2. *The theory of densely ordered sets without endpoints is $\aleph_0$-categorical.*

Proof See any textbook on set theory. The theorem was proved by Cantor using what is called the back-and-forth method.   □

3. *The theory of divisible torsion-free abelian groups is κ-categorical for $\kappa > \aleph_0$.*

Proof Check that a divisible torsion-free abelian group is a vector space over the rationals. Use the fact that vector spaces of the same dimension (over the same field) are isomorphic.   □

4. *The theory of algebraically closed fields (of a fixed characteristic) is κ-categorical for $\kappa > \aleph_0$.*

Proof Use Steinitz's theorem: two algebraically closed fields of the same characteristic and of the same uncountable transcendence degree are isomorphic.   □

The connection between categoricity and completeness, for countable languages, is given by the following.

Theorem 4.3.16 (Vaught's Theorem)   *If T has no finite models and is κ-categorical for some $n\kappa$ not less than the cardinality of L, then T is complete.*

Proof Suppose T is not complete. Then there is a σ such that $T \nvdash \sigma$ and $T \nvdash \neg\sigma$. By the Model Existence Lemma, there are $\mathfrak{A}$ and $\mathfrak{B}$ in $Mod(T)$ such that $\mathfrak{A} \models \sigma$ and $\mathfrak{B} \models \neg\sigma$. Since $\mathfrak{A}$ and $\mathfrak{B}$ are infinite we can apply the Skolem–Löwenheim Theorem (upwards or downwards), so as to obtain $\mathfrak{A}'$ and $\mathfrak{B}'$, of cardinality κ, such that $\mathfrak{A} \equiv \mathfrak{A}'$, and $\mathfrak{B} \equiv \mathfrak{B}'$. But then $\mathfrak{A}' \cong \mathfrak{B}'$, and hence $\mathfrak{A}' \equiv \mathfrak{B}'$, so $\mathfrak{A} \equiv \mathfrak{B}$.
This contradicts $\mathfrak{A} \models \sigma$ and $\mathfrak{B} \models \neg\sigma$.   □

As a consequence we see that the following theories are complete:

1. The theory of infinite sets;
2. The theory of densely ordered sets without endpoints;
3. The theory of divisible torsion-free abelian groups;
4. The theory of algebraically closed fields of fixed characteristic.

A corollary of the last fact was known as *Lefschetz's principle: if a sentence σ, in the first-order language of fields, holds for the complex numbers, it holds for all algebraically closed fields of characteristic zero.*

This means that an "algebraic" theorem σ concerning algebraically closed fields of characteristic 0 can be obtained by devising a proof by whatsoever means (analytical, topological, ...) for the special case of the complex numbers.

Decidability   We have seen in Chap. 2 that there is an effective method to test whether a proposition is provable—by means of the truth table technique, since "truth = provability".

It would be wonderful to have such a method for predicate logic. Church has shown, however, that there is no such method (if we identify "effective" with "recursive") for general predicate logic. But there might be, and indeed there are, special theories which are decidable. A technical study of decidability belongs to recursion theory. Here we will present a few informal considerations.

If T, with language L, has a decidable set of axioms Γ, then there is an effective method for enumerating all theorems of T.

One can obtain such an enumeration as follows:

(a) Make an effective list $\sigma_1, \sigma_2, \sigma_3, \ldots$ of all axioms of T (this is possible because Γ is decidable), and a list $\varphi_1, \varphi_2, \ldots$ of all formulas of L.
(b) (1) write down all derivations of size 1, using σ_1, φ_1, with at most σ_1 uncancelled,
   (2) write down all derivations of size 2, using $\sigma_1, \sigma_2, \varphi_1 \varphi_2$, with at most σ_1, σ_2 uncancelled,

$\quad\vdots$

   (n) write down all derivations of size n, using $\sigma_1, \ldots, \sigma_n, \varphi, \ldots, \varphi_n$, with at most $\sigma_1, \ldots, \sigma_n$ uncancelled,

$\quad\vdots$

Each time we get only finitely many theorems and each theorem is eventually derived. The process is clearly effective (although not efficient).

We now observe the following.

Lemma 4.3.17 *If Γ and Γ^c (complement of Γ) are effectively enumerable, then Γ is decidable.*

Proof Generate the lists of Γ and Γ^c simultaneously. In finitely many steps we will either find σ in the list for Γ or in the list for Γ^c. So for each σ we can decide in finitely many steps whether $\sigma \in \Gamma$ or not.   □

As a corollary we get the next theorem.

Theorem 4.3.18 *If T is effectively axiomatizable and complete, then T is decidable.*

Proof Since T is complete, we have $\Gamma \vdash \sigma$ or $\Gamma \vdash \neg\sigma$ for each σ (where Γ axiomatizes T). So $\sigma \in T^c \Leftrightarrow \Gamma \nvdash \sigma \Leftrightarrow \Gamma \vdash \neg\sigma$.

From the above sketch it follows that T and T^c are effectively enumerable. By the lemma T is decidable.   □

Application The following theories are decidable:

1. The theory of infinite sets;
2. The theory of densely ordered sets without endpoints;
3. The theory of divisible, torsion-free abelian groups;
4. The theory of algebraically closed fields of fixed characteristic.

Proof See the consequences of Vaught's Theorem (4.3.16). The effective enumerating is left to the reader (the simplest case is, of course, that of a finitely axiomatizable theory, e.g. (1), (2). □

We will finally present one more application of the non-standard approach, by giving a non-standard proof of the following lemma.

Lemma 4.3.19 (König's Lemma) *An infinite, finitary tree has an infinite path.*

A finitary tree, or *fan*, has the property that each node has only finitely many immediate successors ("zero successors" is included). By contraposition one obtains from König's Lemma the *fan theorem* (which was actually discovered first).

Theorem 4.3.20 *If in a fan all paths are finite then the length of the paths is bounded.*

Note that if one considers the tree as a topological space, with its canonical topology (basic open set "are" nodes), then König's Lemma is the Bolzano–Weierstrass theorem, and the fan theorem states the compactness.

We will now provide a non-standard proof of König's Lemma.

Let T be a fan, and let T^* be a proper elementary extension (use Theorem 4.3.13).

(1) the relation "...is an immediate successor of ..." can be expressed in the language of partial order:
$$x <_i y := x < y \land \forall z(x \leq z \leq y \to x = z \lor y = z)$$ where, as usual, $x < y$ stands for $x \leq y \land x \neq y$.

(2) If a is standard, then its immediate successors in T^* are also standard. Since T is finitary, we can indicate $a_1, \ldots, a_n$ such that $T \models \forall x(x <_i \overline{a} \leftrightarrow \bigvee_{1 \leq k \leq n} \overline{a}_k = x)$. By $T \prec T^*$, we also have $T^* \models \forall x(x <_i \overline{a} \leftrightarrow \bigvee_{1 \leq k \leq n} \overline{a}_k = x)$, so if b is an immediate successor of a in T^*, then $b = a_k$ for some $k \leq n$, i.e. b is standard.

Note that a node without successors in T has no successors in T^* either, for $T \models \forall x(x \leq \overline{a} \leftrightarrow x = \overline{a}) \Leftrightarrow T^* \models \forall x(x \leq \overline{a} \leftrightarrow x = \overline{a})$.

(3) In T we have that a successor of a node is an immediate successor of that node or a successor of an immediate successor, i.e.

$$T \models \forall xy\big(x < y \to \exists z(x \leq z <_i y)\big). \tag{*}$$

This is the case since for nodes a and b with $a < b$, b must occur in the finite chain of all predecessors of a. So let $a = a_n < a_{n-1} < \cdots < a_i = b < a_{i-1} < , \ldots$, then $a \leq a_{i+1} <_i b$.

Since the desired property is expressed by a first-order sentence (∗), (3) also holds for T^*.

(4) Let a^* be a non-standard element of T^*. We claim that $P = \{a \in |T| \,|\, a^* < a\}$ is an infinite path (i.e. a chain).

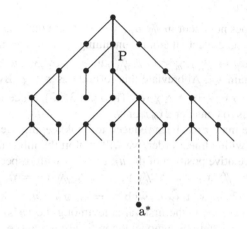

(i) P is linearly ordered since $T \models \forall xyz(x \leq y \wedge x \leq z \rightarrow y \leq z \vee z \leq y)$ and hence for any $p, q \in P \subseteq |T^*|$ we have $p \leq q$ or $q \leq p$.

(ii) Suppose P is finite with last element b, then b has a successor and hence an immediate successor in T^*, which is a predecessor of a^*. By (2) this immediate successor belongs to P. Contradiction. Hence P is infinite.

This shows that T has an infinite path. □

Quantifier Elimination Some theories have the pleasant property that they allow the reduction of formulas to a particularly simple form: one in which no quantifiers occur. Without going into a general theory of quantifier elimination, we will demonstrate the procedure in a simple case: *the theory $\mathcal{DO}$ of dense order without endpoints*, cf. Definition 3.7.3(ii); "without endpoints" is formulated as "$\forall x \exists yz(y < x \wedge x < z)$".

Let $FV(\varphi) = \{y_1, \dots, y_n\}$, where all variables actually occur in φ. By standard methods we obtain a prenex normal form φ' of φ, such that $\varphi' := Q_1 x_1 Q_2 x_2 \dots Q_m x_m \psi(x_1, \dots, x_m, y_1, \dots, y_n)$, where each Q_i is one of the quantifiers $\forall, \exists$. We will eliminate the quantifiers starting with the innermost one.

Consider the case $Q_m = \exists$. We bring ψ into disjunctive normal form $\bigvee \psi_j$, where each ψ_j is a conjunction of atoms and negations of atoms. First we observe that the negations of atoms can be eliminated in favor of atoms, since $\mathcal{DO} \vdash \neg z = z' \leftrightarrow (z < z' \vee z' < z)$ and $\mathcal{DO} \vdash \neg z < z' \leftrightarrow (z = z' \vee z' < z)$. So we may assume that the ψ_j's contain only atoms.

By plain predicate logic we can replace $\exists x_m \bigvee \psi_j$ by the equivalent formula $\bigvee \exists x_m \psi_j$.

Notation For the rest of this example we will use $\sigma \overset{*}{\leftrightarrow} \tau$ as an abbreviation for $\mathcal{DO} \vdash \sigma \leftrightarrow \tau$.

We have just seen that it suffices to consider only formulas of the form $\exists x_m \bigwedge \sigma_p$, where each σ_p is atomic. A systematic look at the conjuncts will show us what to do.

(1) If x_m does not occur in $\bigwedge \sigma_p$, we can delete the quantifier (cf. Theorem 3.5.2).
(2) Otherwise, collect all atoms containing x_m and regroup the atoms, such that we get $\bigwedge \sigma_p \overset{*}{\leftrightarrow} \bigwedge_i x_m < u_i \wedge \bigwedge_j v_j < x_m \wedge \bigwedge_k w_k = x_m \wedge \chi$, where χ does not contain x_m. Abbreviate this formula as $\tau \wedge \chi$. By predicate logic we have $\exists x_m (\tau \wedge \chi) \overset{*}{\leftrightarrow} \exists x_m \tau \wedge \chi$ (cf. Theorem 3.5.3). Since we want to eliminate $\exists x_m$, it suffices to consider $\exists x_m \tau$ only.
Now the matter has been reduced to bookkeeping. Bearing in mind that we are dealing with a linear order, we will exploit the information given by τ concerning the relative position of the u_i, v_j, w_k's with respect to x_m.

(2a) $\tau := \bigwedge x_m < u_i \wedge \bigwedge v_j < x_m \wedge \bigwedge w_k = x_m$.

Then $\exists x_m \tau \overset{*}{\leftrightarrow} \tau'$, with $\tau' := \bigwedge w_0 < u_i \wedge \bigwedge v_j < w_0 \wedge \bigwedge w_0 = w_k$ (where w_0 is the first variable among the w_k's). The equivalence follows immediately by a model theoretic argument (i.e. $\mathcal{DO} \models \exists x_m \tau \leftrightarrow \tau'$).

(2b) $\tau := \bigwedge x_m < u_i \wedge \bigwedge v_j < x_m$.

Now the properties of $\mathcal{DO}$ are essential. Observe that $\exists x_m (\bigwedge x_m < \overline{a}_i \wedge \bigwedge \overline{b}_j < x_m)$ holds in a *densely ordered* set if and only if all the a_i's lie to the right of the b_j's. So we get (by completeness) $\exists x_m \tau \overset{*}{\leftrightarrow} \bigwedge_{i,j} v_j < u_i$.

(2c) $\tau := \bigwedge x_m < u_i \wedge \bigwedge w_k = x_m$.

Then $\exists x_m \tau \overset{*}{\leftrightarrow} \bigwedge w_0 < u_i \wedge \bigwedge w_k = w_0$.

(2d) $\tau := \bigwedge v_j < x_m \wedge \bigwedge w_k = x_m$.
Cf. (2c).

(2e) $\tau := \bigwedge x_m < u_i$.
Observe that $\exists x_m \tau$ holds in all ordered sets without a left endpoint. So we have $\exists x_m \tau \overset{*}{\leftrightarrow} \top$, since we work in $\mathcal{DO}$.

(2f) $\tau := \bigwedge v_j < x_m$. Cf. (2e).

(2g) $\tau := \bigwedge w_k = x_m$.

Then $\exists x_m \tau \overset{*}{\leftrightarrow} \bigwedge w_0 = w_k$.

Remarks

 (i) The cases (2b), (2e) and (2f) make essential use of $\mathcal{DO}$.
(ii) It is often possible to introduce shortcuts, e.g. when a variable (other than x_m) occurs in two of the big conjuncts we have $\exists x_m \tau \overset{*}{\leftrightarrow} \perp$.

If the innermost quantifier is universal, we reduce it to an existential one by $\forall x_m \varphi \leftrightarrow \neg \exists x_m \neg \varphi$.
Now it is clear how to eliminate the quantifiers one by one.

Example

$$\exists xy(x < y \land \exists z(x < z \land z < y \land \forall u(u \neq z \to u < y \lor u = x)))$$
$$\overset{*}{\leftrightarrow} \exists xyz\forall u[x < y \land x < z \land z < y \land (u = z \lor u < y \lor u = x)]$$
$$\overset{*}{\leftrightarrow} \exists xyz\neg\exists u[\neg x < y \lor \neg x < z \lor \neg z < y \lor (\neg u = z \land \neg u < y \land \neg u = x)]$$
$$\overset{*}{\leftrightarrow} \exists xyz\neg\exists u[x = y \lor y < x \lor x = z \lor z < x \lor z = y \lor y < z$$
$$\lor((u < z \lor z < u) \land (u = y \lor y < u) \land (u < x \lor x < u))]$$
$$\overset{*}{\leftrightarrow} \exists xyz\neg\exists u[x = y \lor y < x \lor x = z \lor z < x \lor z = y \lor y < z$$
$$\lor(u < z \land u = y \land u < x) \lor (u < z \land u = y \land x < u)$$
$$\lor(u < z \land y < u \land u < x) \lor (u < z \land y < u \land x < u)$$
$$\lor(z < u \land u = y \land u < x) \lor (z < u \land u = y \land x < u)$$
$$\lor(z < u \land y < u \land u < x) \lor (z < u \land y < u \land x < u)].$$
$$\overset{*}{\leftrightarrow} \exists xyz\neg[x = y \lor y < x \lor x = z \lor z < x \lor z = y \lor y < z$$
$$\lor \exists u(u < z \land u = y \land u < x) \lor \exists u(u < z \land u = y \land x < u) \lor \cdots$$
$$\lor \exists u(z < u \land y < u \land x < u)].$$
$$\overset{*}{\leftrightarrow} \exists xyz\neg[x = y \lor \cdots \lor y < z \lor (y < z \land y < x) \lor (y < z \land x < y)$$
$$\lor(y < z \land y < x) \lor (y < z \land x < z) \lor (z < y \land y < x)$$
$$\lor(z < y \land x < y) \lor (z < x \land y < x) \lor \top].$$
$$\overset{*}{\leftrightarrow} \exists xyz(\neg\top).$$
$$\overset{*}{\leftrightarrow} \bot.$$

Evidently the above quantifier elimination for the theory of dense order without endpoints provides an alternative proof of its decidability. For, if φ is a sentence, then φ is equivalent to an open sentence φ'. Given the language of $\mathcal{DO}$ it is obvious that φ' is equivalent to either $\top$ or $\bot$. Hence, we have an algorithm for deciding $\mathcal{DO} \vdash \varphi$. Note that we have obtained more: $\mathcal{DO}$ is complete, since $\mathcal{DO} \vdash \varphi \leftrightarrow \bot$ or $\mathcal{DO} \vdash \varphi \leftrightarrow \top$, so $\mathcal{DO} \vdash \neg\varphi$ or $\mathcal{DO} \vdash \varphi$.

In general we cannot expect that much from quantifier elimination: e.g. the theory of algebraically closed fields admits quantifier elimination, but it is not complete (because the characteristic has not been fixed in advance); the open sentences may contain unprovable and unrefutable atoms such as $7 = 12$, $23 = 0$.

We may conclude from the existence of a quantifier elimination a certain model theoretic property, introduced by Abraham Robinson, which has turned out to be important for applications in algebra (cf. the *Handbook of Mathematical Logic*, A_4).

Definition 4.3.21 A theory T is *model complete* if for $\mathfrak{A}, \mathfrak{B} \in Mod(T)\mathfrak{A} \subseteq \mathfrak{B} \Rightarrow \mathfrak{A} \prec \mathfrak{B}$.

We can now immediately obtain the following.

Theorem 4.3.22 *If T admits quantifier elimination, then T is model complete.*

Proof Let $\mathfrak{A}$ and $\mathfrak{B}$ be models of T, such that $\mathfrak{A} \subseteq \mathfrak{B}$. We must show that $\mathfrak{A} \models \varphi(\bar{a}_1, \ldots, \bar{a}_n) \Leftrightarrow \mathfrak{B} \models \varphi(\bar{a}_1, \ldots, \bar{a}_n)$ for all $a_1, \ldots, a_n \in |\mathfrak{A}|$, where $FV(\varphi) = \{x_1, \ldots, x_n\}$.

Since T admits quantifier elimination, there is a quantifier-free $\psi(x_1, \ldots, x_n)$ such that $T \vdash \varphi \leftrightarrow \psi$.

Hence it suffices to show $\mathfrak{A} \vdash \psi(\bar{a}_1, \ldots, \bar{a}_n) \Leftrightarrow \mathfrak{B} \vdash \psi(\bar{a}_1, \ldots, \bar{a}_n)$ for a quantifier-free ψ. A simple induction establishes this equivalence. $\square$

Some theories T have a particular model that is, up to isomorphism, contained in every model of T. We call such a model a *prime model* of T.

Examples

(i) The rationals form a prime model for the theory of dense ordering without endpoints;
(ii) The field of the rationals is the prime model of the theory of fields of characteristic zero;
(iii) The standard model of arithmetic is the prime model of Peano arithmetic.

Theorem 4.3.23 *A model complete theory with a prime model is complete.*

Proof Left to the reader. $\square$

Exercises

1. Let $\mathfrak{A} = \langle A, \leq \rangle$ be a poset. Show that $Diag^+(\mathfrak{A}) \cup \{\bar{a} \neq \bar{b} \mid a \neq b, a, b \in |\mathfrak{A}|\} \cup \{\forall xy(x \leq y \vee y \leq x)\}$ has a model. (Hint: use compactness.)

   Conclude that every poset can be linearly ordered by an extension of its ordering.

2. If $f : \mathfrak{A} \cong \mathfrak{B}$ and $FV(\varphi) = \{x_1, \ldots, x_n\}$, show $\mathfrak{A} \models \varphi[\bar{a}_1, \ldots, \bar{a}_n / x_1, \ldots, x_n] \Leftrightarrow \mathfrak{B} \models \varphi[\overline{f(a_1)}, \ldots, \overline{f(a_n)}/x_1, \ldots, x_n]$.

   In particular, $\mathfrak{A} \equiv \mathfrak{B}$.

3. Let $\mathfrak{A} \subseteq \mathfrak{B}$. φ is called *universal* (*existential*) if φ is prenex with only universal (existential) quantifiers.

   (i) Show that for universal sentences φ $\mathfrak{B} \models \varphi \Rightarrow \mathfrak{A} \models \varphi$.

   (ii) Show that for existential sentences φ $\mathfrak{A} \models \varphi \Rightarrow \mathfrak{B} \models \varphi$.

   (Application: a substructure of a group is a group. This is one reason to use the similarity type $\langle -; 2, 1; 1 \rangle$ for groups, instead of $\langle -; 2; 0 \rangle$, or $\langle -; 2; 1 \rangle$, as some authors do.)

4. Let $\mathfrak{A} = \langle N, < \rangle$, $\mathfrak{B} = \langle N - \{0\}, < \rangle$.

   Show:

   (i) $\mathfrak{A} \cong \mathfrak{B}$; (ii) $\mathfrak{A} \equiv \mathfrak{B}$;

   (iii) $\mathfrak{B} \subseteq \mathfrak{A}$; (iv) not $\mathfrak{B} \prec \mathfrak{A}$.

5. (Tarski). Let $\mathfrak{A} \subseteq \mathfrak{B}$. Show $\mathfrak{A} \prec \mathfrak{B} \Leftrightarrow$ for all $\varphi \in L$ and $a_1, \ldots, a_n \in |\mathfrak{A}|$, $\mathfrak{B} \models \exists y \varphi(y, \bar{a}_1, \ldots, \bar{a}_n) \Rightarrow$ there is an element $a \in |\mathfrak{A}|$ such that $\mathfrak{B} \models \varphi(\bar{a}, \bar{a}_1, \ldots, \bar{a}_n)$, where $FV(\varphi(y, \bar{a}_1, \ldots, \bar{a}_n)) = \{y\}$. Hint: for $\Leftarrow$ show

   (i) $t^{\mathfrak{A}}(\bar{a}_1, \ldots, \bar{a}_n) = t^{\mathfrak{B}}(\bar{a}_1, \ldots, \bar{a}_n)$ for $t \in L$,

(ii) $\mathfrak{A} \models \varphi(\bar{a}_1, \ldots, \bar{a}_n) \Leftrightarrow \mathfrak{B} \models \varphi(\bar{a}_1, \ldots, \bar{a}_n)$ for $\varphi \in L$ by induction on φ (use only $\vee, \neg, \exists$).

6. Another construction of a non-standard model of arithmetic: add to the language L of arithmetic a new constant c. Show $\Gamma = Th(\mathfrak{N}) \cup \{c > \bar{n} | n \in |\mathfrak{N}|\}$ has a model $\mathfrak{M}$. Show that $\mathfrak{M} \not\cong \mathfrak{N}$. Can $\mathfrak{M}$ be countable?

7. Consider the ring Z of integers. Show that there is an $\mathfrak{A}$ such that $Z \prec \mathfrak{A}$ and $Z \not\cong \mathfrak{A}$ (a non-standard model of the integers). Show that $\mathfrak{A}$ has an "infinite prime number", p_∞.

   Let (p_∞) be the principal ideal in $\mathfrak{A}$ generated by p_∞. Show that $\mathfrak{A}/(p_\infty)$ is a field F. (Hint: look at $\forall x (\text{"}x \text{ not in } (p_\infty)\text{"} \rightarrow \exists yz(xy = 1 + zp_\infty))$, give a proper formulation and use elementary equivalence.) What is the characteristic of F? (This yields a non-standard construction of the rationals from the integers: consider the prime field.)

8. Use the non-standard model of arithmetic to show that "well-ordering" is not a first-order concept.

9. Use the non-standard model of the reals to show that "archimedean ordered field" is not a first-order concept.

10. Consider the language of identity with constants c_i $(i \in N)$, $\Gamma = \{I_1, I_2, I_3\} \cup \{c_i \neq c_j | i, j \in N, i \neq j\}$. Show that the theory of Γ is k-categorical for $k > \aleph_0$, but not $\aleph_0$-categorical.

11. Show that the condition "no finite models" in Vaught's Theorem is necessary (look at the theory of identity).

12. Let $X \subseteq |\mathfrak{A}|$. Define $X_0 = X \cup C$ where C is the set of constants of $\mathfrak{A}$, $X_{n+1} = X_n \cup \{f(a_1, \ldots, a_m) | f \text{ in } \mathfrak{A}, a_1, \ldots, a_m \in X_n\}$, $X_\omega = \bigcup \{X_n | n \in \mathbb{N}\}$.
    Show: $\mathfrak{B} = \langle X_\omega, R_1 \cap X_\omega^{r_1}, \ldots, R_n \cap X_\omega^{r_1}, f_1 | X_\omega^{a_1}, \ldots, f_m | X_\omega^{a_m}, \{c_i | i \in I\} \rangle$ is a substructure of $\mathfrak{A}$. We say that $\mathfrak{B}$ is the substructure generated by X. Show that $\mathfrak{B}$ is the smallest substructure of $\mathfrak{A}$ containing X; $\mathfrak{B}$ can also be characterized as the intersection of all substructures containing X.

13. Let $^*R$ be a non-standard model of $Th(R)$. Show that st (cf. p. 116) is a homomorphism from the ring of finite numbers onto R. What is the kernel?

14. Consider $\mathfrak{R}' = \langle R, N, <, +, \cdot, -, ^{-1}, 0, 1 \rangle$, where N is the set of natural numbers. $L(\mathfrak{R}')$ has a predicate symbol N, and we can, restricting ourselves to $+$ and $\cdot$, recover arithmetic by relativizing our formulas to N (cf. Definition 3.5.12).
    Let $\mathfrak{R}' \prec^* \mathfrak{R}' = \langle ^*R, ^*N, \ldots \rangle$. Show that $\mathfrak{N} = \langle N, <, +, \cdot, 0, 1 \rangle \prec \langle ^*N, <, +, \cdot, 0, 1 \rangle =^* \mathfrak{N}$ (hint: consider for each $\varphi \in L(\mathfrak{N})$ the relativized $\varphi^N \in L(\mathfrak{R}')$).

15. Show that any Peano structure contains $\mathfrak{N}$ as a substructure.

16. Let L be a language without identity and with at least one constant. Let $\sigma = \exists x_1 \cdots x_n \varphi(x_1, \ldots, x_n)$ and $\Sigma_\sigma = \{\varphi(t_1, \ldots, t_n) | t_i \text{ closed in } L\}$, where φ is quantifier free.
    (i) $\models \sigma \Leftrightarrow$ each $\mathfrak{A}$ is a model of at least one sentence in Σ_σ. (Hint: for each $\mathfrak{A}$, look at the substructure generated by $\emptyset$.)
    (ii) Consider Σ_σ as a set of propositions. Show that for each valuation v (in the sense of propositional logic) there is a model $\mathfrak{A}$ such that $[\![\varphi(t_1, \ldots, t_n)]\!]_v = [\![\varphi(t_1, \ldots, t_n)]\!]_{\mathfrak{A}}$, for all $\varphi(t_1, \ldots t_n) \in \Sigma_\sigma$.

(iii) Show that $\vdash \sigma \Leftrightarrow \vdash \bigvee_{i=1}^{m} \varphi(t_1^i, \ldots, t_n^i)$ for a certain m (hint: use Exercise 9, Sect. 2.5).

17. Let $\mathfrak{A}, \mathfrak{B} \in Mod(T)$ and $\mathfrak{A} \equiv \mathfrak{B}$. Show that $Diag(\mathfrak{A}) \cup Diag(\mathfrak{B}) \cup T$ is consistent (use the Compactness Theorem). Conclude that there is a model of T in which both $\mathfrak{A}$ and $\mathfrak{B}$ can be isomorphically embedded.

18. Consider the class $\mathcal{K}$ of all structures of type $\langle 1; -; 0 \rangle$ with a denumerable unary relation. Show that any $\mathfrak{A}$ and $\mathfrak{B}$ in $\mathcal{K}$ of the same cardinality $\kappa > \aleph_0$ are isomorphic. Show that $T = Th(\mathcal{K})$ is not κ-categorical for any $\kappa \geq \aleph_0$.

19. Consider a theory T of identity with axioms λ_n for all $n \in N$. In which cardinalities is T categorical? Show that T is complete and decidable. Compare the result with Exercise 10.

20. Show that the theory of dense order without endpoints is not categorical in the cardinality of the continuum.

21. Consider the structure $\mathfrak{A} = \langle \mathbb{R}, <, f \rangle$, where $<$ is the natural order, and where f is a unary function. Let L be the corresponding language. Show that there is no sentence σ in L such that $\mathfrak{A} \models \sigma \Leftrightarrow f(r) > 0$ for all $r \in R$. (Hint: consider isomorphisms $x \mapsto x + k$.)

22. Let $\mathfrak{A} = \langle A, \sim \rangle$, where $\sim$ is an equivalence relation with denumerably many equivalence classes, all of which are infinite. Show that $Th(\mathfrak{A})$ is $\aleph_0$-categorical. Axiomatize $Th(\mathfrak{A})$. Is there a finite axiomatization? Is $Th(\mathfrak{A})$ κ-categorical for $\kappa > \aleph_0$?

23. Let L be a language with one unary function symbol f. Find a sentence τ_n, which says that "f has a loop of length n", i.e. $\mathfrak{A} \models \tau_n \Leftrightarrow$ there are $a_1, \ldots, a_n \in |\mathfrak{A}|$ such that $f^{\mathfrak{A}}(a_i) = a_{i+1}(i < n)$ and $f^{\mathfrak{A}}(a_n) = a_1$. Consider a theory T with axiom set $\{\beta, \neg\tau_1, \neg\tau_2, \neg\tau_3, \ldots, \neg\tau_n, \ldots\}(n \in \omega)$, where β expresses "f is bijective".

   Show that T is κ-categorical for $\kappa > \aleph_0$. (Hint: consider the partition $\{(f^{\mathfrak{A}})^i(a) | i \in \omega\}$ in a model $\mathfrak{A}$.) Is T $\aleph_0$-categorical?

   Show that T is complete and decidable. Is T finitely axiomatizable?

24. Put $T_\forall = \{\sigma | T \vdash \sigma \text{ and } \sigma \text{ is universal}\}$. Show that $T_\forall$ axiomatizes the theory of all substructures of models of T. Note that one part follows from Exercise 3. For the converse: let $\mathfrak{A}$ be a model of $T_\forall$ and consider $Diag(\mathfrak{A}) \cup T$. Use compactness.

25. We say that a theory T is *preserved under substructures* if $\mathfrak{A} \subseteq \mathfrak{B}$ and $\mathfrak{B} \in Mod(T)$ implies $\mathfrak{A} \in Mod(T)$. (Łos–Tarski). Show that T is preserved under substructures iff T can be axiomatized by universal sentences (use Exercise 24).

26. Let $\mathfrak{A} \equiv \mathfrak{B}$, show that there exists a $\mathfrak{C}$ such that $\mathfrak{A} \prec \mathfrak{C}, \mathfrak{B} \prec \mathfrak{C}$ (up to isomorphism). Hint: assume that the set of new constants of $\hat{\mathfrak{B}}$ is disjoint with the set of new constants of $\hat{\mathfrak{A}}$. Show that $Th(\hat{\mathfrak{A}}) \cup Th(\hat{\mathfrak{B}})$ has a model.

27. Show that the ordering $<$, defined by $x < y := \exists u(y = x + Su)$ is provably transitive in Peano arithmetic, i.e. $\mathbf{PA} \vdash \forall xyz(x < y \land y < z \rightarrow x < z)$.

28. Show:
   (i) $\mathbf{PA} \vdash \forall x(0 \leq x)$ (induction on x),
   (ii) $\mathbf{PA} \vdash \forall x(x = 0 \lor \exists y(x = Sy))$(induction on x),
   (iii) $\mathbf{PA} \vdash \forall xy(x + y = y + x)$,

    (iv) **PA** $\vdash \forall y(x < y \to Sx \leq y)$ (induction on y),
    (v) **PA** $\vdash \forall xy(x < y \lor x = y \lor y < x)$ (induction on x, the case of $x = 0$ is
       simple, for the step from x to Sx use (iv)),
    (vi) **PA** $\vdash \forall y \neg \exists x(y < x \land x < Sy)$ (compare with (iv)).
29. (i) Show that the theory L_∞ of identity with "infinite universe" (cf. Sect. 4.1,
      Exercise 3 or Exercise 19 above) admits quantifier elimination.
   (ii) Show that L_∞ has a prime model.

4.4 Skolem Functions or How to Enrich Your Language

In mathematical arguments one often finds passages such as "...there is an x such that $\varphi(x)$ holds. Let a be such an element, then we see that ...". In terms of our logic, this amounts to the introduction of a constant whenever the existence of some element satisfying a certain condition has been established. The problem is: does one thus strengthen the theory in an essential way? In a precise formulation: suppose $T \vdash \exists x \varphi(x)$. Introduce a (new) constant a and replace T by $T' = T \cup \{\varphi(a)\}$. *Question*: is T' conservative over T, i.e. does $T' \vdash \psi \Rightarrow T \vdash \psi$ hold, for ψ not containing a? We have dealt with a similar problem in the context of Henkin theories (Sect. 4.1), so we can use the experience obtained there.

Theorem 4.4.1 *Let T be a theory with language L, such that $T \vdash \exists x \varphi(x)$, where $FV(\varphi) = \{x\}$, and let c be a constant not occurring in L. Then $T \cup \{\varphi(c)\}$ is conservative over T.*

Proof By Lemma 4.1.7, $T' = T \cup \{\exists x \varphi(x) \to \varphi(c)\}$ is conservative over T. If $\psi \in L$ and $T' \cup \{\varphi(c)\} \vdash \psi$, then $T' \cup \{\exists x \varphi(x)\} \vdash \psi$, or $T' \vdash \exists x \varphi(x) \to \psi$. Since T' is conservative over T we have $T \vdash \exists x \varphi(x) \to \psi$. Using $T \vdash \exists x \varphi(x)$, we get $T \vdash \psi$. (For an alternative proof see Exercise 6.)   □

The above is but a special case of a very common piece of practice; if one, in the process of proving a theorem, establishes that "for each x there is a y such that $\varphi(x, y)$", then it is convenient to introduce an auxiliary function f that picks a y for each x, such that $\varphi(x, f(x))$ holds for each x. This technique usually invokes the axiom of choice. We can put the same question in this case: if $T \vdash \forall x \exists y \varphi(x, y)$, introduce a function symbol f and replace T by $T' = T \cup \{\forall x \varphi(x, f(x))\}$. Question: is T' conservative over T? The idea of enriching the language by the introduction of extra function symbols, which take the role of choice functions, goes back to Skolem.

Definition 4.4.2 Let φ be a formula of the language L with $FV(\varphi) = \{x_1, \ldots, x_n, y\}$. Associate with φ an n-ary function symbol f_φ, called the *Skolem function (symbol)* of φ. The sentence

$$\forall x_1 \cdots x_n \left(\exists y \varphi(x_1, \ldots, x_n, y) \to \varphi(x_1, \ldots, x_n, f_\varphi(x_1, \ldots, x_n))\right)$$

is called the *Skolem axiom* for φ.

Note that the witness of Sect. 4.1 is a special case of a Skolem function (take $n = 0$): f_φ is a constant.

Definition 4.4.3 If T is a theory with language L, then $T^{sk} = T \cup \{\sigma \,|\, \sigma$ is a Skolem axiom for some formula of $L\}$ is the Skolem extension of T and its language L^{sk} extends L by including all Skolem functions for L. If $\mathfrak{A}$ is of the type of L and $\mathfrak{A}^{sk}$ an expansion of $\mathfrak{A}$ of the type of L^{sk}, such that $\mathfrak{A}^{sk} \models \sigma$ for all Skolem axioms σ of L and $|\mathfrak{A}| = |\mathfrak{A}^{sk}|$, then $\mathfrak{A}^{sk}$ is called a *Skolem expansion* of $\mathfrak{A}$.

The interpretation in $\mathfrak{A}^{sk}$ of a Skolem function symbol is called a Skolem function.

Note that a Skolem expansion contains infinitely many functions, so it is a mild extension of our notion of structure. The analogue of Lemma 4.1.7 is the following.

Theorem 4.4.4 (i) T^{sk} *is conservative over* T.
  (ii) *Each* $\mathfrak{A} \in Mod(T)$ *has a Skolem expansion* $\mathfrak{A}^{sk} \in Mod(T^{sk})$.

Proof We first show (ii). We only consider the case of formulas with $FV(\varphi) = \{x_1, \ldots, x_n, y\}$ for $n \geq 1$. The case $n = 0$ is similar, but simpler. It requires the introduction of new constants in $\mathfrak{A}$ (cf. Exercise 6). Let $\mathfrak{A} \in Mod(T)$ and $\varphi \in L$ with $FV(\varphi) = \{x_1, \ldots, x_n, y\}$. We want to find a Skolem function for φ in $\mathfrak{A}$.
  Define $V_{a_1, \ldots, a_n} = \{b \in |\mathfrak{A}| \,|\, \mathfrak{A} \models \varphi(\bar{a}_1, \ldots, \bar{a}_n, \bar{b})\}$.
  Apply AC to the set $\{V_{a_1, \ldots, a_n} \,|\, V_{a_1, \ldots, a_n} \neq \emptyset\}$: there is a choice function F such that $F(V_{a_1, \ldots, a_n}) \in V_{a_1, \ldots, a_n}$. Define a Skolem function by

$$F_\varphi(a_1, \ldots, a_n) = \begin{cases} F(V_{a_1, \ldots, a_n}) & \text{if } V_{a_1, \ldots, a_n} \neq \emptyset, \\ e & \text{else, where } e \in |\mathfrak{A}|. \end{cases}$$

Now it is a routine matter to check that indeed $\mathfrak{A}^{sk} \models \forall x_1 \ldots x_n (\exists y \varphi(x_1, \ldots, x_n, y) \to \varphi(x_1, \ldots, x_n, f_\varphi(x_1, \ldots, x_n)))$, where $F_\varphi = f_\varphi^{\mathfrak{A}^{sk}}$, and where $\mathfrak{A}^{sk}$ is the expansion of $\mathfrak{A}$ with all Skolem functions F_φ (including the "Skolem constants", i.e. witnesses). (i) Follows immediately from (ii): let $T \not\vdash \psi$ (with $\psi \in L$), then there is an $\mathfrak{A}$ such that $\mathfrak{A} \not\vdash \psi$. Since $\psi \in L$, we also have $\mathfrak{A}^{sk} \not\vdash \psi$ (cf. Sect. 4.2, Exercise 3), hence $T^{sk} \not\vdash \psi$. $\qquad\square$

Remark It is not necessary (for Theorem 4.4.4) to extend L with *all* Skolem function symbols. We may just add Skolem function symbols for some given set S of formulas of L. We then speak of the Skolem extension of T with respect to S (or with respect to φ if $S = \{\varphi\}$).

The following corollary confirms that we can introduce Skolem functions in the course of a mathematical argument, without essentially strengthening the theory.

Corollary   4.4.5 *If $T \vdash \forall x_1, \ldots, x_n \exists y \varphi(x_1, \ldots, x_n, y)$,   where $FV(\varphi) = \{x_1, \ldots, x_n, y\}$, then $T' = T \cup \{\forall x_1 \ldots x_n \varphi(x_1, \ldots, x_n, f_\varphi(x_1, \ldots, x_n))\}$ is conservative over T.*

Proof Observe that $T'' = T \cup \{\forall x_1 \ldots x_n (\exists y \varphi(x_1, \ldots, x_n, y) \to \varphi(x_1, \ldots, x_n, f(\varphi(x_1, \ldots, x_n)))\} \vdash \forall x_1 \ldots x_n \varphi(x_1, \ldots, x_n, f_\varphi(x_1, \ldots, x_n))$. So $T' \vdash \psi \Rightarrow T'' \vdash \psi$. Now apply Theorem 4.4.4.   □

The introduction of a Skolem extension of a theory T results in the "elimination" of the existential quantifier in prefixes of the form $\forall x \ldots x_n \exists y$. The iteration of this process on prenex normal forms eventually results in the elimination of all existential quantifiers.

The Skolem functions in an expanded model are by no means unique. If, however, $\mathfrak{A} \models \forall x_1 \ldots x_n \exists! y \varphi(x_1, \ldots, x_n, y)$, then the Skolem function for φ is uniquely determined; we even have $\mathfrak{A}^{sk} \models \forall x_1 \ldots x_n y (\varphi(x_1, \ldots, x_n, y) \leftrightarrow y = f_\varphi(x_1, \ldots, x_n))$. We say that φ *defines* the function F_φ in $\mathfrak{A}^{sk}$, and $\forall x_1 \ldots x_n y (\varphi(x_1, \ldots, x_n, y) \leftrightarrow y = f_\varphi(x_1, \ldots, x_n))$ is called the *definition* of F_φ in $\mathfrak{A}^{sk}$.

We may reasonably expect that with respect to Skolem functions the $\forall \exists!$-combination yields better results than the $\forall \exists$-combination. The following theorem tells us that we get substantially more than just a conservative extension result.

Theorem   4.4.6 *Let $T \vdash \forall x_1 \ldots x_n \exists! y \varphi(x_1, \ldots, x_n, y)$,   where $FV(\varphi) = \{x_1, \ldots, x_n, y\}$ and let f be an n-ary symbol not occurring in T or φ. Then $T^+ = T \cup \{\forall x_1 \ldots x_n y (\varphi(x_1, \ldots, x_n, y) \leftrightarrow y = f(x_1, \ldots, x_n))\}$ is conservative over T.*

Moreover, there is a translation $\tau \to \tau^0$ from $L^+ = L \cup \{f\}$ to L, such that

(1) $T^+ \vdash \tau \leftrightarrow \tau^0$,
(2) $T^+ \vdash \tau \Leftrightarrow T \vdash \tau^0$,
(3) $\tau = \tau^0$ *for $\tau \in L$.*

Proof

(i) We will show that f acts just like a Skolem function; in fact T^+ is equivalent to the theory T' of Corollary 4.4.5 (taking f for f_φ).
  (a) $T^+ \vdash \forall x_1 \ldots x_n \varphi(x_1, \ldots, x_n, f(x_1, \ldots, x_n))$.
     For, $T^+ \vdash \forall x_1 \ldots x_n \exists y \varphi(x_1, \ldots, x_n, y)$ and
     $T^+ \vdash \forall x_1 \ldots x_n y (\varphi(x_1, \ldots, x_n, y) \leftrightarrow y = f(x_1, \ldots, x_n))$.
     Now a simple exercise in natural deduction, involving RI_4, yields (a).
     Therefore $T' \subseteq T^+$ (in the notation of Corollary 4.4.5).
  (b) $y = f(x_1, \ldots, x_n), \forall x_1 \ldots x_n \varphi(x_1 \ldots x_n, f(x_1, \ldots, x_n)) \vdash \varphi(x_1, \ldots, x_n, y)$,
     so $T' \vdash y = f(x_1, \ldots, x_n) \to \varphi(x_1, \ldots, x_n, y)$
     and $\varphi(x_1, \ldots, x_n, y), \forall x_1 \ldots x_n \varphi(x_1, \ldots, x_n, f(x_1, \ldots, x_n))$,
     $\forall x_1 \ldots x_n \exists! y \varphi(x_1, \ldots, x_n, y) \vdash y = f(x_1, \ldots, x_n)$,
     so $T' \vdash \varphi(x_1, \ldots, x_n, y) \to y = f(x_1, \ldots, x_n)$.
     Hence $T' \vdash \forall x_1 \ldots x_n y (\varphi(x_1, \ldots, x_n, y) \leftrightarrow y = f(x_1, \ldots, x_n))$.

So $T^+ \subseteq T'$, and hence $T' = T^+$.

Now, by Corollary 4.4.5, T^+ is conservative over T.

(ii) The idea underlying the translation is to replace occurrences of $f(-)$ by a new variable and to eliminate f. Let $\tau \in L^+$ and let $f(-)$ be a term in L^+ not containing f in any of its subterms. Then $\vdash \tau(\ldots, f(-), \ldots) \leftrightarrow \exists y(y = f(-) \wedge \tau(\ldots, y, \ldots))$, where y does not occur in τ, and $T^+ \vdash \tau(\ldots, f(-), \ldots) \leftrightarrow \exists y(\varphi(-, y) \wedge \tau(\ldots, y, \ldots))$. The right-hand side contains one occurrence of f less than τ. Iteration of the procedure leads to the required f-free formula τ^0. The reader can provide the details of a precise inductive definition of τ^0; note that one need only consider atomic τ (the translation extends trivially to all formulas). Hint: define something like "f-depth" of terms and atoms. From the above description of τ^0 it immediately follows that $T^+ \vdash \tau \leftrightarrow \tau^0$. Now (2) follows from (i) and (1). Finally (3) is evident. $\square$

As a special case we get the *explicit definition* of a function.

Corollary 4.4.7 *Let* $FV(t) = \{x_1, \ldots, x_n\}$ *and* $f \notin L$. *Then* $T^+ = T \cup \{\forall x_1 \ldots x_n(t = f(x_1, \ldots, x_n))\}$ *is conservative over* T.

Proof We have $\forall x_1 \ldots x_n \exists! y(y = t)$, so the definition of f, as in Theorem 4.4.6, becomes $\forall x_1 \ldots x_n y(y = t \leftrightarrow y = f(x_1, \ldots, x_n))$, which, by the predicate and identity rules, is equivalent to $\forall x_1 \ldots x_n(t = f(x_1, \ldots, x_n))$. $\square$

We call $f(x_1, \ldots, x_n) = t$ the *explicit definition* of f. One can also add new predicate symbols to a language in order to replace formulas by atoms.

Theorem 4.4.8 *Let* $FV(\varphi) = \{x_1, \ldots, x_n\}$ *and let* Q *be a predicate symbol not in* L. *Then*

(i) $T^+ = T \cup \{\forall x_1 \ldots x_n(\varphi \leftrightarrow Q(x_1, \ldots, x_n))\}$ *is conservative over* T.
(ii) *There is a translation* $\tau \to \tau^0$ *into* L *such that*
    (1) $T^+ \vdash \tau \leftrightarrow \tau^0$,
    (2) $T^+ \vdash \tau \Leftrightarrow T \vdash \tau^0$,
    (3) $\tau = \tau^0$ *for* $\tau \in L$.

Proof Similar to, but simpler than, the above. We indicate the steps; the details are left to the reader.

(a) Let $\mathfrak{A}$ be of the type of L. Expand $\mathfrak{A}$ to $\mathfrak{A}^+$ by adding a relation $Q^+ = \{\langle a_1, \ldots, a_n \rangle | \mathfrak{A} \models \varphi(\overline{a}_1, \ldots, \overline{a}_n)\}$.
(b) Show $\mathfrak{A} \models T \Leftrightarrow \mathfrak{A}^+ \models T^+$ and conclude (i).
(c) Imitate the translation of Theorem 4.4.6. $\square$

We call the extensions shown in Theorem 4.4.6, Corollary 4.4.7 and Theorem 4.4.8, *extensions by definition*. The sentences

$$\forall x_1 \ldots x_n y(\varphi \leftrightarrow y = f(x_1, \ldots, x_n)),$$

$\forall x_1 \ldots x_n (f(x_1, \ldots, x_n) = t)$,
$\forall x_1 \ldots x_n (\varphi \leftrightarrow Q(x_1, \ldots, x_n))$,
are called the *defining axioms* for f and Q respectively.

Extension by definition belongs to the daily practice of mathematics (and science in general). If a certain notion, definable in a given language, plays an important role in our considerations, then it is convenient to have a short, handy notation for it.

Think of "x is a prime number", "x is equal to y or less than y", "z is the maximum of x and y", etc.

Examples 1. *Characteristic functions*

Consider a theory T with (at least) two constants c_0, c_1, such that $T \vdash c_0 \neq c_1$. Let $FV(\varphi) = \{x_1, \ldots, x_n\}$, then $T \vdash \forall x_1 \ldots x_n \exists! y ((\varphi \wedge y = c_1) \vee (\neg \varphi \wedge y = c_0))$. (Show this directly or use the Completeness Theorem.)

The defining axiom for the characteristic function K_φ is $\forall x_1 \ldots x_n y [(\varphi \wedge y = c_1) \vee (\neg \varphi \wedge y = c_0)) \leftrightarrow y = K_\varphi (x_1, \ldots, x_n))$.

2. *Definition by (primitive) recursion.*

In arithmetic one often introduces functions by recursion, e.g. $x!, x^y$. The study of these and similar functions belongs to recursion theory; here we only note that we can conservatively add symbols and axioms for them. *Fact* (Gödel, Davis, Matijasevich): each recursive function is definable in **PA**, in the sense that there is a formula φ of **PA** such that

(i) $\mathbf{PA} \vdash \forall x_1 \ldots x_n \exists! y \varphi(x_1, \ldots, x_n, y)$ and
(ii) for $k_1, \ldots, k_n, m \in N f(k_1, \ldots, k_n) = m \Rightarrow \mathbf{PA} \vdash \varphi(\overline{k}_1, \ldots, \overline{k}_n, \overline{m})$.

For details see [Smoryński 1991, Davis 1958].

Before ending this chapter, let us briefly return to the topic of Skolem functions and Skolem expansions. As we remarked before, the introduction of Skolem functions allows us to dispense with certain existential quantifiers in formulas. We will exploit this idea to rewrite formulas as universal formulas (in an extended language!).

First we transform the formula φ into prenex normal form φ'. Let us suppose that $\varphi' = \forall x_1 \ldots x_n \exists y \psi(x_1, \ldots, x_n, y, z_1, \ldots, z_k)$, where $z_1, \ldots, z_k$ are all the free variables in φ. Now consider
$$T^* = T \cup \{\forall x_1 \ldots x_n z_1 \ldots z_k (\exists y \psi(x_1, \ldots, x_n, y, z_1, \ldots, z_k) \rightarrow \psi(x_1, \ldots, x_n,$$
$f(x_1, \ldots, x_n, z_1, \ldots, z_k), z_1, \ldots, z_k))\}$.

By Theorem 4.4.4 T^* is conservative over T, and it is a simple exercise in logic to show that
$$T^* \vdash \forall x_1 \ldots x_n \exists y \psi(-, y, -) \leftrightarrow \forall x_1 \ldots x_n \psi(-, f(\ldots), -).$$
We now repeat the process and eliminate the next existential quantifier in the prefix of ψ; in finitely many steps we obtain a formula φ^s in prenex normal form without existential quantifiers, which, in a suitable conservative extension of T obtained by a series of Skolem expansions, is equivalent to φ.

Warning The *Skolem form* φ^s differs in kind from other normal forms, in the sense that it is not *logically equivalent* to φ.

Theorem 4.4.4 shows that the adding of Skolem axioms to a theory is conservative, so we can safely operate with Skolem forms. The Skolem form φ^s has the property that is satisfiable if and only if φ is so (cf. Exercise 4). Therefore it is sometimes called the Skolem form for satisfiability. There is a dual Skolem form φ_s (cf. Exercise 5), which is valid if and only if φ is so. φ_s is called the Skolem form for validity.

Example $\forall x_1 \exists y_1 \exists y_2 \forall x_2 \exists y_3 \forall x_3 \forall x_4 \exists y_4 \; \varphi(x_1, x_2, x_3, x_4, y_1, y_2, y_3, y_4, z_1, z_2)$.
Step 1. Eliminate y_1:
$\quad \forall x_1 \exists y_2 \forall x_2 \exists y_3 \forall x_3 \forall x_4 \exists y_4 \; \varphi(x_1, x_2, x_3, x_4, f(x_1, z_1, z_2), y_2, y_3, y_4, z_1, z_2)$.
Step 2. Eliminate y_2:
$\quad \forall x_1 x_2 \exists y_3 \forall x_3 x_4 \exists y_4 \; \varphi(\ldots, f(x_1, z_1, z_2), g(x_1, z_1, z_2), y_3, y_4, z_1, z_2)$.
Step 3. Eliminate y_3:
$\quad \forall x_1 x_2 x_3 x_4 \exists y_4 \; \varphi(\ldots, f(x_1, z_1, z_2), g(x_1, z_1, z_2), h(x_1, x_2, z_1, z_2), y_4, z_1, z_2)$.
Step 4. Eliminate y_4:
$\quad \forall x_1 x_2 x_3 x_4 \quad \varphi(\ldots, f(x_1, z_1, z_2), g(x_1, z_1, z_2), h(x_1, x_2, z_1, z_2), k(x_1, x_2, x_3, x_4, z_1, z_2), z_1, z_2)$.

In Skolem expansions we have functions available which pick elements for us. We can exploit this phenomenon to obtain elementary extensions.

Theorem 4.4.9 *Consider $\mathfrak{A}$ and $\mathfrak{B}$ of the same type. If $\mathfrak{B}^{\text{sk}}$ is a Skolem expansion of $\mathfrak{B}$ and $\mathfrak{A}^* \subseteq \mathfrak{B}^{\text{sk}}$, where $\mathfrak{A}^*$ is some expansion of $\mathfrak{A}$, then $\mathfrak{A} \prec \mathfrak{B}$.*

Proof We use Exercise 5 of Sect. 4.3. Let $a_1, \ldots, a_n \in |\mathfrak{A}|$, $\mathfrak{B} \models \exists y \varphi(y, \overline{a}_1, \ldots, \overline{a}_n) \Leftrightarrow \mathfrak{B}^{\text{sk}} \models \varphi(f_\varphi(\overline{a}_1, \ldots, \overline{a}_n), \overline{a}_1, \ldots, \overline{a}_n)$, where f_φ is the Skolem function for φ. Since $\mathfrak{A}^* \subseteq \mathfrak{B}^{\text{sk}}$, $f_\varphi^{\mathfrak{A}^*}(a_1, \ldots, a_n) = f_\varphi^{\mathfrak{B}^{\text{sk}}}(a_1, \ldots, a_n)$ and so $b = (f_\varphi(\overline{a}_1, \ldots, \overline{a}_n))^{\mathfrak{B}^{\text{sk}}} = (f_\varphi(\overline{a}_1, \ldots, \overline{a}_n))^{\mathfrak{A}^*} \in |\mathfrak{A}|$.
   Hence $\mathfrak{B}^{\text{sk}} \models \varphi(\overline{b}, \overline{a}_1, \ldots, \overline{a}_n)$. This shows $\mathfrak{A} \prec \mathfrak{B}$.   □

Definition 4.4.10 Let $X \subseteq |\mathfrak{A}|$. The Skolem hull $\mathfrak{S}_X$ of X is the substructure of $\mathfrak{A}$ which is the reduct of the structure generated by X in the Skolem expansion $\mathfrak{A}^{\text{sk}}$ of $\mathfrak{A}$ (cf. Exercise 12, Sect. 4.3).

In other words $\mathfrak{S}_X$ is the smallest substructure of $\mathfrak{A}$, containing X, which is closed under all Skolem functions (including the constants).

Corollary 4.4.11 *For all $X \subseteq |\mathfrak{A}| \; \mathfrak{S}_X \prec \mathfrak{A}$.*

We now immediately get the strengthening of the Downward Skolem–Löwenheim Theorem formulated in Theorem 4.3.12, by observing that the cardinality of a substructure generated by X is the maximum of the cardinalities of X and of the language. This also holds in the present case, where infinitely many Skolem functions are added to the language.

Exercises

1. Consider the example concerning the characteristic function.
   (i) Show $T^+ \vdash \forall x_1 \ldots x_n (\varphi \leftrightarrow K_\varphi(x_1, \ldots, x_n) = c_1)$.
   (ii) Translate $K_\varphi(x_1, \ldots, x_n) = K_\varphi(y_1, \ldots, y_n)$.
   (iii) Show $T^+ \vdash \forall x_1 \ldots x_n y_1, \ldots, y_n (K_\varphi(x_1, \ldots, x_n) = K_\varphi(y_1, \ldots, y_n)) \leftrightarrow \forall x_1 \ldots x_n \varphi(x_1, \ldots, x_n) \vee \forall x_1 \ldots x_n \neg \varphi(x_1, \ldots, x_n)$.
2. Determine the Skolem forms of
   (a) $\forall y \exists x (2x^2 + yx - 1 = 0)$,
   (b) $\forall \varepsilon \exists \delta (\varepsilon > 0 \rightarrow (\delta > 0 \wedge \forall x (|x - \bar{a}| < \delta \rightarrow |f(x) - f(\bar{a})| < \varepsilon)))$,
   (c) $\forall x \exists y (x = f(y))$,
   (d) $\forall x y (x < y \rightarrow \exists u (u < x) \wedge \exists v (y < v) \wedge \exists w (x < v \wedge w < y))$,
   (e) $\forall x \exists y (x = y^2 \vee x = -y^2)$.
3. Let σ^s be the Skolem form of σ. Consider only sentences.
   (i) Show that $\Gamma \cup \{\sigma^s\}$ is conservative over $\Gamma \cup \{\sigma\}$.
   (ii) Put $\Gamma^s = \{\sigma^s | \sigma \in \Gamma\}$. Show that for finite Γ, Γ^s is conservative over Γ.
   (iii) Show that Γ^s is conservative over Γ for arbitrary Γ.
4. A formula φ with $FV(\varphi) = \{x_1, \ldots, x_n\}$ is called *satisfiable* if there is an $\mathfrak{A}$ and $a_1, \ldots, a_n \in |\mathfrak{A}|$ such that $\mathfrak{A} \models \varphi(\bar{a}, \ldots, \bar{a}_n)$. Show that φ is satisfiable iff φ^s is satisfiable.
5. Let σ be a sentence in prenex normal form. We define the *dual Skolem form σ_s* of σ as follows: let $\sigma = (Q_1 x_1) \ldots (Q_n x_n) \tau$, where τ is quantifier free and the Q_i are quantifiers. Consider $\sigma' = (\overline{Q}_1 x_1) \ldots (\overline{Q}_n x_n) \neg \tau$, where $\overline{Q}_i = \forall, \exists$ iff $Q_i = \exists, \forall$. Suppose $(\sigma')^s = (\overline{Q}_{i_1} x_{i_1}) \ldots (\overline{Q}_{i_k} x_{i_k}) \neg \tau'$; then $\sigma_s = (Q_{i_1} x_{i_1}) \ldots (Q_{i_k} x_{i_k}) \tau'$.

   In words: eliminate from σ the universal quantifiers and their variables just as the existential ones in the case of the Skolem form. We end up with an existential sentence.

   Example $(\forall x \exists y \forall z \varphi(x, y, z))_s = \exists y \varphi(c, y, f(y))$.

   We suppose that L has at least one constant symbol.
   (a) Show that for all (prenex) sentences σ, $\models \sigma$ iff $\models \sigma_s$. (Hint: look at Exercise 4.) Hence the name: "Skolem form for *validity*".
   (b) Prove *Herbrand's theorem*:

   $$\vdash \sigma \; \leftrightarrow \; \vdash \bigvee_{i=1}^{m} \sigma_s'(t_1^i, \ldots, t_n^i)$$

   for some m, where σ_s' is obtained from σ_s by deleting the quantifiers. The $t_j^i (i \leq m, j \leq n)$ are certain closed terms in the dual Skolem expansion of L. Hint: look at $\neg(\neg\sigma)^s$. Use Exercise 16, Sect. 4.3.
6. Let $T \vdash \exists x \varphi(x)$, with $FV(\varphi) = \{x\}$. Show that any model $\mathfrak{A}$ of T can be expanded to a model $\mathfrak{A}^*$ of T with an extra constant c such that $\mathfrak{A}^* \models \varphi(c)$. Use this for an alternative proof of Theorem 4.4.1.

7. Consider I_∞ the theory of identity "with infinite universe" with axioms $\lambda_n (n \in N)$ and I'_∞ with extra constants $c_i (i \in N)$ and axioms $c_i \neq c_j$ for $i \neq j$, $i, j \in N$. Show that I'_∞ is conservative over I_∞.

4.5 Ultraproducts

So far we have exploited the strength of the interaction between logic and its interpretations; the completeness theorem is the key to that part of logic, it provides us with the compactness theorem, yields non-standard models, characterizes theories, etc. In the present section we will look at the model theoretic side from a different point of view.

One might wonder how far one can get without the formal tools of logic; more specifically, how far will the model theoretic approach get us? In this section we will introduce a tool for model construction that is very useful for applications in mathematics, and that allows us to forget for a minute the derivations side of logic. The notion of ultraproduct was introduced in the 1950s by Łos.

There are many techniques that construct new structures out of old ones. Here is a simple example: let $\mathfrak{A}$ and $\mathfrak{B}$ be groups; then the Cartesian product $\mathfrak{A} \times \mathfrak{B}$ consists of pairs (a, b) with $a \in A$, $b \in B$. Multiplication is defined coordinate-wise: $(a_1, b_1) \cdot (a_2, b_2) = a_1 \cdot a_2, b_1 \cdot b_2)$ and the identity is the pair of identities. The result is again a group. Note, however, that the Cartesian product of two fields is not a field.

We will generalize this notion of product so that the properties of the product can be kept under control. This requires a little bit of set theory.

Definition 4.5.1 Let I be a non-empty set. $\mathcal{F} \subseteq \mathcal{P}(I)$ is a *filter* if

 (i) $A, B \in \mathcal{F} \Rightarrow A \cap B \in \mathcal{F}$
 (ii) $A \in \mathcal{F}, A \subseteq B \Rightarrow B \in \mathcal{F}$
 (iii) $\emptyset \notin \mathcal{F}$ (the filter is *proper*)

We assume $\mathcal{F} \neq \emptyset$. If $\mathcal{F}$ is maximal, then $\mathcal{F}$ is called an *ultrafilter*. $\mathcal{F}$ is called a *free filter* if $\cap \mathcal{F} = \emptyset$.

Example 4.5.2 The following sets are filters:

1. $\mathcal{F} = \{A \mid A \supseteq A_0\}$, for $A_0 \neq \emptyset$. If $A_0 = \{a_0\}$, then clearly $\mathcal{F}$ is an ultrafilter; this filter is called a *principal ultrafilter*.
2. $\mathcal{F} = \{A \mid I - A$ is finite$\}$, for infinite I. The A's in this filter are called *cofinite* sets.
3. $\mathcal{F} = \{A \subseteq [0, 1] \mid \mu(A) = 1\}$. Here μ is the Lebesgue measure.

The following provides us with lots of filters.

Definition 4.5.3 $G \subseteq \mathcal{P}(I)$ has the finite intersection property (*fip*) if

$$A_1, \ldots, A_n \in G \Rightarrow A_1 \cap \cdots \cap A_n \neq \emptyset.$$

Lemma 4.5.4 *If G has the finite intersection property, then G is contained in a filter.*

Proof Define $\mathcal{F} = \{A \mid A \supseteq B_1 \cap \cdots \cap B_k$ for $B_1, \ldots, B_k \in G$ and $k \in \mathbb{N}\}$.

(i) If $A, A' \in \mathcal{F}$, then $A \supseteq B_1 \cap \cdots \cap B_k$ and $A' \supseteq B_1' \cap \cdots \cap B_l'$ for certain $B_1, \ldots, B_k, B_1', \ldots, B_l' \in G$. So $A \cap A' \supseteq B_1 \cap \cdots \cap B_k \cap B_1' \cap \cdots \cap B_l'$, which implies $A \cap A' \in \mathcal{F}$.
(ii) $A \in \mathcal{F}, A \subseteq A' \Rightarrow A' \in \mathcal{F}$. Trivial.
(iii) $\emptyset \notin \mathcal{F}$. Trivial. $\qquad\square$

Exercise 4.5.5 Show that this $\mathcal{F}$ is the smallest filter containing G. We say that G *generates* $\mathcal{F}$.

Lemma 4.5.6 *Let $\mathcal{F} \subseteq \mathcal{P}(I)$ be a filter. The following are equivalent:*

(i) *$\mathcal{F}$ is an ultrafilter.*
(ii) *$A \in \mathcal{F}$ or $A^c \in \mathcal{F}$ for all $A \subseteq X$.*
(iii) *$A \cup B \in \mathcal{F} \Rightarrow A \in \mathcal{F}$ or $B \in \mathcal{F}$.*

Proof (i) $\Rightarrow$ (ii) Assume that A, $A^c \notin \mathcal{F}$. Claim: $\{A\} \cup \mathcal{F}$ has the finite intersection property. Suppose $\{A\} \cup \mathcal{F}$ does not have the finite intersection property. Consider $A_1 \cap \cdots \cap A_n$ where $A_i \in \mathcal{F}$. If $(A_1 \cap \cdots \cap A_n) \cap A = \emptyset$, then $A^c \supseteq A_1 \cap \cdots \cap A_n$. But then $A^c \in \mathcal{F}$—contradiction. Let $\mathcal{G} \subseteq \mathcal{P}(X)$ be generated by $\mathcal{F} \cup \{A\}$, then $\mathcal{G} \supseteq \mathcal{F}$ and $\mathcal{G} \neq \mathcal{F}$; this contradicts the maximality of $\mathcal{F}$. Hence $A \in \mathcal{F}$ or $A^c \in \mathcal{F}$.

(ii) $\Rightarrow$ (iii) Suppose $A \cup B \in \mathcal{F}$. Now suppose A, $B \notin \mathcal{F}$. Then by (ii), $A^c, B^c \in \mathcal{F}$. Because $\mathcal{F}$ is a filter, also $A^c \cap B^c \in \mathcal{F}$, which is equivalent to $(A \cup B)^c \in \mathcal{F}$. But now it follows that $(A \cup B) \cap (A \cup B)^c = \emptyset \in \mathcal{F}$. Contradiction.

(iii) $\Rightarrow$ (i) Assume that $\mathcal{F}$ is not maximal. Then $\mathcal{F}$ is a proper subset of a filter $\mathcal{F}'$. Therefore there is a set A in $\mathcal{F}'$ which is not in $\mathcal{F}$.

Now $A \cup A^c = I \in \mathcal{F}$, therefore, by (ii), $A \in \mathcal{F}$ or $A^c \in \mathcal{F}$. In fact $A^c \in \mathcal{F}$, hence $A^c \in \mathcal{F}'$. We know that $A \in \mathcal{F}'$, hence $\emptyset \in \mathcal{F}'$. Contradiction, so $\mathcal{F}$ is an ultrafilter. $\square$

Exercise 4.5.7 If $\mathcal{F}$ is a free filter, then $\mathcal{F}$ does not contain finite sets.

Corollary 4.5.8

(i) *If $\mathcal{F}$ is a free ultrafilter, then $\mathcal{F}$ contains all cofinite sets.*
(ii) *If I is finite and $\mathcal{F} \subseteq \mathcal{P}(I)$ is an ultrafilter, then $\mathcal{F}$ is not free (it is principal).*

Theorem 4.5.9 (AC) *Each filter is contained in an ultrafilter.*

Proof We will use Zorn's lemma. Suppose we have a filter $\mathcal{F} \subseteq \mathcal{P}(X)$. Let Z be the set of all filters containing $\mathcal{F}$ partially ordered by $\subseteq$. Let $\mathcal{K}$ be a chain in Z. Define $\mathcal{F}^* = \bigcup \mathcal{K}$. Claim: $\mathcal{F}^*$ is a filter.

(i) $A, B \in \mathcal{F}^* \Rightarrow A \in \mathcal{F}_1, B \in \mathcal{F}_2$ for certain $\mathcal{F}_1, \mathcal{F}_2 \in \mathcal{K}$. Say $\mathcal{F}_1 \subseteq \mathcal{F}_2$, then
$A, B \in \mathcal{F}_2 \Rightarrow A \cap B \in \mathcal{F}_2 \Rightarrow A \cap B \in \mathcal{F}^*$.

(ii) Suppose $A \in \mathcal{F}^*$ and $B \supseteq A$. This means that $A \in \mathcal{F}$ for some $\mathcal{F} \in \mathcal{K}$. Because $\mathcal{F}$ is a filter it follows that $B \in \mathcal{F}$, which implies that $B \in \mathcal{F}^*$.

(iii) $\emptyset \notin \mathcal{F}^*$. Trivial.

Note that $\mathcal{F} \in \mathcal{K} \Rightarrow \mathcal{F} \subseteq \mathcal{F}^*$. We now may apply Zorn's lemma: there is a maximal $\mathcal{F}_m \in Z$. This is the required ultrafilter.   □

Corollary 4.5.10 *There is a free ultrafilter on every infinite set I.*

Proof The intersection of all cofinite sets of I is empty (why?). So take the $\mathcal{F}$ generated by the cofinite sets and extend it to an ultrafilter.   □

Exercise 4.5.11

(1) Let $\mathcal{F}$ be a free ultrafilter. Show that $A \in \mathcal{F}$ and B is finite $\Rightarrow A - B \in \mathcal{F}$.
(2) Show that if $G \subseteq \mathcal{P}(\mathbb{N})$ and G is countable $\Rightarrow$ G does not generate a free ultrafilter.
(3) Let $\mathcal{F}$ be a filter (an ultrafilter) on I and $A \in \mathcal{F}$, then $\mathcal{F} \cap \mathcal{P}(A)$ is a filter (an ultrafilter) on A.
(4) The set of all filters is closed under arbitrary intersection and unions of chains.

Our next step is the general definition of Cartesian products of structures with the same similarity type. Consider an indexed set $\{\mathfrak{A}_i \mid i \in I\}$; for the universe of the Cartesian product of the set we simply take the Cartesian product of the universes and we define the relations and operations coordinate-wise. The language L is fixed.

Definition 4.5.12 (Cartesian Product of Structures)

1. $\prod_{i \in I} A_i = \{f : I \to \bigcup_{i \in I} A_i \mid f(i) \in A_i\}$. For convenience, we put $\prod_{i \in I} A_i = A$.
2. $R^A(f_1, \ldots, f_n) \Leftrightarrow \forall i \in I(R^{A_i}(f_1(i), \ldots, f_n(i)))$.
3. $F^A(f_1, \ldots, f_n) = \lambda i. F^{A_i}(f_1(i), \ldots, f_n(i))$.
4. $c^A = \lambda i. c^{A_i}$.
5. $\prod_{i \in I} \mathfrak{A}_i = \langle A, R_1^A, \ldots, R_n^A, F_1^A, \ldots, F_n^A, \{c_j^A \mid j \in J\}\rangle$.

We will denote this product structure by $\mathfrak{A}$ when no confusion arises.

Example 4.5.13 $\prod_{i \in \mathbb{N}} \mathfrak{A}_i$ where $\mathfrak{A}_i = \langle \mathbb{N}, +, \cdot, \{0, 1\}\rangle$ Since this is the denumerable product of the natural numbers, we are considering all functions from $\mathbb{N}$ to $\mathbb{N}$. One can visualize those in the lattice in the first quadrant of the plane.

From now on we will practice a harmless bit of abuse of language by using $\mathfrak{A}$ both for the structure and for its universe; the reader will have no difficulty discerning the two meanings.

Lemma 4.5.14 *Let $\mathcal{F}$ be a filter on I; put $f_1 \sim f_2 \Leftrightarrow \{i \in I \mid f_1(i) = f_2(i)\} \in \mathcal{F}$, where $f_1, f_2 \in \mathfrak{A}$. Then $\sim$ is an equivalence relation on $\mathfrak{A}$.*

Proof

(i) $f_1 \sim f_1$. Note that $\{i \in I \mid f_1(i) = f_1(i)\} = I \in \mathcal{F}$.

(ii) $f_1 \sim f_2 \Rightarrow f_2 \sim f_1$, by definition.

(iii) *Transitivity*. Assume $f_1 \sim f_2$ and $f_2 \sim f_3$.

Put $A_1 = \{i \in I \mid f_1(i) = f_2(i)\}$ and $A_2 = \{i \in I \mid f_2(i) = f_3(i)\}$, then $A_1, A_2 \in (\mathcal{F})$, and hence $A_1 \cap A_2 \in \mathcal{F}$. Now $A_1 \cap A_2 \subseteq \{i \mid f_1(i) = f_3(i)\}$. So $\{i \in I \mid f_1(i) = f_3(i)\} \in \mathcal{F}$, and hence $f_1 \sim f_3$. $\qquad\qquad\qquad\qquad\qquad\qquad\qquad\qquad\qquad\qquad\qquad\qquad\quad\square$

Notation The proper notation for the equivalence relation with respect to $\mathcal{F}$ would be $\sim_{\mathcal{F}}$. When no confusion arises, we will stick to $\sim$. The equivalence class of f under $\sim_{\mathcal{F}}$ will be denoted by $f/\mathcal{F}$.

The equivalence classes of the elements of the Cartesian product $\mathfrak{A}$ will serve as the elements of a new structure. For this purpose we have to define the relations and operations.

Lemma 4.5.15

1. *If $f_1 \sim g_1, \ldots, f_n \sim g_n$ then $\{i \in I \mid R^{\mathfrak{A}_i}(f_1(i), \ldots, f_n(i))\} \in \mathcal{F} \Leftrightarrow \{i \in I \mid R^{\mathfrak{A}_i}(g_1(i), \ldots, g_n(i))\} \in \mathcal{F}$*

2. *If $f_1 \sim g_1, \ldots, f_n \sim g_n$, then $F^A(f_1(i), \ldots, f_n(i)) \sim F^A(g_1(i), \ldots, g_n(i))$*

Proof (a) $\Rightarrow$: Put $A_1 = \{i \in I \mid f_1(i) = g_1(i)\}, \ldots, A_n = \{i \in I \mid f_n(i) = g_n(i)\}$. Let $A_1, \ldots, A_n \in \mathcal{F}$ and $B = \{i \in I \mid R^{\mathfrak{A}_i}(f_1(i), \ldots, f_n(i))\} \in \mathcal{F}$ be given. If $i \in A_1 \cap \cdots \cap A_n \cap B$, then $f_1(i) = g_1(i), \ldots, f_n(i) = g_n(i)$ and $R^{\mathfrak{A}_i}(f_1(i), \ldots, f_n(i))$ so $R^{\mathfrak{A}_i}(g_1(i), \ldots, g_n(i))$. So $A_1 \cap \cdots \cap A_n \cap B \subseteq \{i \in I \mid R^{\mathfrak{A}_i}(g_1(i), \ldots, g_n(i))\}$. And since $A_1 \cap \cdots \cap A_n \cap B \in \mathcal{F}$, it follows that $\{i \in I \mid R^{\mathfrak{A}_i}(g_1(i), \ldots, g_n(i))\} \in \mathcal{F}$.

$\Leftarrow$: Similar.

(b) Let $A_1 \cap \cdots \cap A_n \in \mathcal{F}$, as in (a). Consider the set $C = \{i \in I \mid \mathcal{F}^{\mathfrak{A}_i}(f_1(i), \ldots, f_n(i)) = \mathcal{F}^{\mathfrak{A}_i}(g_1(i), \ldots, g_n(i))\}$. Suppose $i \in A_1 \cap \cdots \cap A_n$, then $f_1(i) = g_1(i), \ldots, f_n(i) = g_n(i)$ and therefore $\mathcal{F}^{\mathfrak{A}_i}(f_1(i), \ldots, f_n(i)) = \mathcal{F}^{\mathfrak{A}_i}(g_1(i), \ldots, g_n(i))\}$. This tells us that $A_1 \cap \cdots \cap A_n \subseteq C$, and since $A_1 \cap \cdots \cap A_n \in \mathcal{F}$, it follows that $C \in \mathcal{F}$. $\square$

Now we can define a product structure modulo a filter.

Definition 4.5.16 Let $\mathcal{F}$ be a filter on I, then

$$\mathfrak{A}/\mathcal{F} = \prod_{i \in I} \mathfrak{A}_i / \mathcal{F} = \left\langle \prod_{i \in I} A_i / \mathcal{F}, \tilde{R}_1, \ldots, \tilde{R}_n, \tilde{F}_1, \ldots, \tilde{F}_m, \{\tilde{c}_j \mid j \in J\} \right\rangle$$

where $\prod_{i \in I} A_i / \mathcal{F} = \{f/\mathcal{F} \mid f \in \prod_{i \in I} A_i\}$
$\tilde{R}_k(f_1/\mathcal{F}, \dots, f_n/\mathcal{F}) = \{i \in I \mid R_k^{\mathfrak{A}_i}(f_1(i), \dots, f_n(i))\} \in \mathcal{F}$
$\tilde{F}_k(f_1/\mathcal{F}, \dots, f_n/\mathcal{F}) = F_k^{\mathfrak{A}}(f_1, \dots, f_n)/\mathcal{F}$
$\tilde{c}_j = c_j^{\mathfrak{A}}/\mathcal{F}.$

Observe that $\tilde{R}_k$ and $\tilde{\mathcal{F}}_k$ are well defined by Lemma 4.5.15.

Notation $\prod_F \mathfrak{A}_i := \prod_{i \in I} \mathfrak{A}_i / \mathcal{F}.$
    $\prod_F \mathfrak{A}_i$ is called the *reduced product* of the $\mathfrak{A}_i$'s. $\prod_F \mathfrak{A}_i$ is an *ultraproduct* if $\mathcal{F}$ is an ultrafilter.

Our next task is to interpret terms of the language in the reduced product.

Lemma 4.5.17 $[\![t(\overline{f_1/\mathcal{F}}, \dots, \overline{f_n/\mathcal{F}})]\!]_{\prod_F \mathfrak{A}_i} = \lambda i.[\![t(f_1(i), \dots, f_n(i))]\!]_{\prod_F \mathfrak{A}_i}/\mathcal{F}.$

Proof Induction on t.

- $t = c$:
  $[\![c]\!]_{\prod_F \mathfrak{A}_i} = \lambda i.c(i)/\mathcal{F} = \lambda i.[\![c]\!]_{\mathfrak{A}_i}/\mathcal{F}.$
- $t = \overline{f}$:
  $[\![\overline{f}]\!]_{\prod_F \mathfrak{A}_i} = f/\mathcal{F} = \lambda i.f(i)/\mathcal{F} = \lambda i.[\![f]\!]_{\mathfrak{A}_i}/\mathcal{F}.$
- $t = G(t_1, \dots, t_k)$:
  $[\![t(\overline{f_1/\mathcal{F}}, \dots, \overline{f_n/\mathcal{F}})]\!]_{\prod_F \mathfrak{A}_i}$
  $= [\![G(t_1(\overline{f_1/\mathcal{F}}, \dots, \overline{f_n/\mathcal{F}}), \dots, t_k(\overline{f_1/\mathcal{F}}, \dots, \overline{f_n/\mathcal{F}}))]\!]_{\prod_F \mathfrak{A}_i}$
  $= \tilde{G}([\![t_1(\overline{f_1/\mathcal{F}}, \dots, \overline{f_n/\mathcal{F}})]\!]_{\prod_F \mathfrak{A}_i}, \dots, [\![t_k(\overline{f_1/\mathcal{F}}, \dots, \overline{f_n/\mathcal{F}})]\!]_{\prod_F \mathfrak{A}_i})$
  $\overset{\text{IH}}{=} \tilde{G}(\lambda i.[\![t_1(f_1(i), \dots, f_n(i))]\!]_{\mathfrak{A}_i}/\mathcal{F}, \dots, \lambda i.[\![t_k(f_1(i), \dots, f_n(i))]\!]_{\mathfrak{A}_i}/\mathcal{F})$
  $\overset{\text{def}}{=} \lambda i.G^{\mathfrak{A}_i}([\![t_1(f_1(i), \dots, f_n(i))]\!]_{\mathfrak{A}_i}, \dots, [\![t_k(f_1(i), \dots, f_n(i))]\!]_{\mathfrak{A}_i})/\mathcal{F}$
  $= \lambda i.[\![G(t_1(f_1(i), \dots, f_n(i)), \dots, t_k(f_1(i), \dots, f_n(i)))]\!]_{\mathfrak{A}_i}/\mathcal{F}$
  $= \lambda i.[\![t(t_1(f_1(i), \dots, f_n(i)), \dots, t_k(f_1(i), \dots, f_n(i)))]\!]_{\mathfrak{A}_i}/\mathcal{F}.$ □

Lemma 4.5.18

$$\prod_F \mathfrak{A}_i \models t = s \quad \Leftrightarrow \quad \{i \in I \mid [\![t]\!]_{\mathfrak{A}_i} = [\![s]\!]_{\mathfrak{A}_i}\} \in \mathcal{F}.$$

Proof $\prod_F \mathfrak{A}_i \models t = s \Leftrightarrow [\![t]\!]_{\prod_F \mathfrak{A}_i} = [\![s]\!]_{\prod_F \mathfrak{A}_i} \Leftrightarrow \lambda i.[\![t]\!]_{\mathfrak{A}_i}/\mathcal{F} = \lambda i.[\![s]\!]_{\mathfrak{A}_i}/\mathcal{F} \Leftrightarrow \{i \in I \mid [\![t]\!]_{\mathfrak{A}_i} = [\![s]\!]_{\mathfrak{A}_i}\} \in \mathcal{F}.$ □

Theorem 4.5.19 (Fundamental Theorem of Ultraproducts, Łos) *Let $\mathcal{F}$ be an ultrafilter, then*

$$\prod_F \mathfrak{A}_i \models \varphi(\overline{f_1/\mathcal{F}}, \dots, \overline{f_n/\mathcal{F}}) \quad \Leftrightarrow \quad \{i \in I \mid \mathfrak{A}_i \models \varphi(\overline{f_1(i)}, \dots, \overline{f_n(i)})\} \in \mathcal{F}.$$

Proof Induction on φ.

1. Observe that in general φ has parameters in $\prod_F \mathfrak{A}_i$. φ is a sentence in the extended language $L(\prod_F \mathfrak{A}_i)$.
2. Note that we can restrict ourselves to formulas without terms, i.e. the following is proved for closed $\varphi(t_1, \ldots, t_n)$, see Corollary 3.5.9:

$$\mathfrak{B} \models \varphi(t_1, \ldots, t_n) \quad \Leftrightarrow \quad \mathfrak{B} \models \varphi(\overline{[\![t_1]\!]_{\mathfrak{B}}}, \ldots, \overline{[\![t_n]\!]_{\mathfrak{B}}})$$

- $\varphi = (t = s)$. See Lemma 4.5.18.
- $\varphi = R(\overline{g_1/\mathcal{F}}, \ldots, \overline{g_m/\mathcal{F}})$. $\prod_F \mathfrak{A}_i \models R(\overline{g_1/\mathcal{F}}, \ldots, \overline{g_m/\mathcal{F}}) \Leftrightarrow$
  $\{i \in I \mid R^{\mathfrak{A}_i}(g_1(i), \ldots, g_m(i))\} \in \mathcal{F} \Leftrightarrow \{i \in I \mid \mathfrak{A}_i \models R(\overline{g_1(i)}, \ldots, \overline{g_m(i)})\} \in \mathcal{F}$.
- $\varphi = \varphi_1 \wedge \varphi_2$. $\prod_F \mathfrak{A}_i \models \varphi_1 \wedge \varphi_2 \overset{\text{def}}{\Longleftrightarrow} \prod_F \mathfrak{A}_i \models \varphi_1$ and $\prod_F \mathfrak{A}_i \models \varphi_2 \overset{\text{IH}}{\Longleftrightarrow}$
  $\{i \in I \mid \mathfrak{A}_i \models \varphi_1\} \in \mathcal{F}$ and $\{i \in I \mid \mathfrak{A}_i \models \varphi_2\} \in \mathcal{F}$.
  Put $X_1 = \{i \in I \mid \mathfrak{A}_i \models \varphi_1\}$ and $X_2 = \{i \in I \mid \mathfrak{A}_i \models \varphi_2\}$.
  Now $X_1 \cap X_2 \in \mathcal{F}$ and also $X_1 \cap X_2 \subseteq \{i \in I \mid \mathfrak{A}_i \models \varphi_1 \wedge \varphi_2\} \in \mathcal{F}$, hence
  $\{i \in I \mid \mathfrak{A}_i \models \varphi_1 \wedge \varphi_2\} \in \mathcal{F}$. The converse is obvious.
- $\varphi = \neg \psi$. $\prod_F \mathfrak{A}_i \models \neg \psi(\overline{f_1/\mathcal{F}}, \ldots, \overline{f_n/\mathcal{F}}) \Leftrightarrow$
  $\prod_F \mathfrak{A}_i \not\models \psi(\overline{f_1/\mathcal{F}}, \ldots, \overline{f_n/\mathcal{F}}) \overset{\text{IH}}{\Longleftrightarrow} \{i \in I \mid \mathfrak{A}_i \models \psi(\overline{f_1(i)}, \ldots, \overline{f_n(i)})\} \notin \mathcal{F}$.
  Since $\mathcal{F}$ is an ultrafilter this last statement is equivalent to
  $\{i \in I \mid \mathfrak{A}_i \not\models \psi(\overline{f_1(i)}, \ldots, \overline{f_n(i)})\} \in \mathcal{F}$ and hence to
  $\{i \in I \mid \mathfrak{A}_i \models \neg\psi(\overline{f_1(i)}, \ldots, \overline{f_n(i)})\} \in \mathcal{F}$.
- $\varphi = \forall x \psi(x)$. $\prod_F \mathfrak{A}_i \models \forall x \psi(x, \overline{f_1/\mathcal{F}}, \ldots, \overline{f_n/\mathcal{F}}) \Leftrightarrow$
  $\prod_F \mathfrak{A}_i \models \psi(\overline{g/\mathcal{F}}, \overline{f_1/\mathcal{F}}, \ldots, \overline{f_n/\mathcal{F}})$ for all $g \in \prod_F \mathfrak{A}_i \overset{\text{IH}}{\Longleftrightarrow}$
  $\{i \in I \mid \mathfrak{A}_i \models \psi(\overline{g(i)}, \overline{f_1(i)}, \ldots, \overline{f_n(i)})\} \in \mathcal{F}$ for all $g \in \prod_{i \in I} A_i$.
  Note that for any $a \in \mathfrak{A}_i$ we can find a $g \in \prod_{i \in I} \mathfrak{A}_i$ such that $g(i) = a$: take an arbitrary g' and define

$$g(i) = \begin{cases} g'(i) & \text{if } i \neq j \\ a & \text{if } = j. \end{cases}$$

So we get for all $a \in |\mathfrak{A}_i|$:

$$X = \{i \in I \mid \mathfrak{A}_i \models \psi(\overline{a}, \overline{f_1(i)}, \ldots, \overline{f_n(i)})\} \in \mathcal{F}. \tag{1}$$

Assume now that $\{i \in I \mid \mathfrak{A}_i \models \forall x \psi(x, \overline{f_1(i)}, \ldots, \overline{f_n(i)})\} \notin \mathcal{F}$. Then
$\{i \in I \mid \mathfrak{A}_i \not\models \forall x \psi(x, \overline{f_1(i)}, \ldots, \overline{f_n(i)})\} \in \mathcal{F}$, and hence

$$Y = \{i \in I \mid \mathfrak{A}_i \models \exists x \neg \psi(x, \overline{f_1(i)}, \ldots, \overline{f_n(i)})\} \in \mathcal{F}. \tag{2}$$

Since $X \cap Y \in \mathcal{F}$, we have $X \cap Y \neq \emptyset$. So pick an $i \in X \cap Y$, then by (2) $\mathfrak{A}_i \models \psi(\overline{b}, \overline{f_1(i)}, \ldots, \overline{f_n(i)})$ for all $b \in A_i$, and by (1) $\mathfrak{A}_i \models \exists x \neg \psi(x, \overline{f_1(i)}, \ldots, \overline{f_n(i)})$, that is, there is a $b \in A_i$ such that $\mathfrak{A}_i \models \neg\psi(\overline{b}, \overline{f_1(i)}, \ldots, \overline{f_n(i)})$. Contradiction.
Hence $\{i \in I \mid \mathfrak{A}_i \models \forall x \psi(x, \overline{f_1(i)}, \ldots, \overline{f_n(i)})\} \in \mathcal{F}$.
The converse is left to the reader. $\qquad\square$

Definition 4.5.20 $\mathfrak{A}_{\mathcal{F}}^I = \prod_F \mathfrak{A}_i$ where $\mathfrak{A}_i = \mathfrak{A}$ for all i. This is called the *ultra-power* of $\mathfrak{A}$.

Corollary 4.5.21 $\mathfrak{A} \prec \mathfrak{A}_F^I$.

Proof Consider the embedding $\lambda i.a/\mathcal{F}$. This gives you all the constant functions modulo $\mathcal{F}$. Now $\mathfrak{A}^I_F \models \varphi(\hat{a}_1/\mathcal{F}, \ldots, \hat{a}_k/\mathcal{F}) \Leftrightarrow \{i \in I \mid \mathfrak{A} \models \varphi(a_1, \ldots, a_k)\} = I \in \mathcal{F}$, where $\hat{a}_j = \lambda i \cdot a_j$ are constant functions with values a_j. We say that $\mathfrak{A}$ is elementarily embedded in $\prec \mathfrak{A}^I_F$, and the function $\lambda i.a/\mathcal{F}$ is an elementary embedding. □

We continue Example 4.5.13. To avoid notational confusion, we denote the index set, i.e. $\mathbb{N}$, by I. The (horizontal) rows are the standard numbers, embedded in the product (modulo $\mathcal{F}$). Consider $d(i) = i$ (the diagonal). Claim: $d > \hat{n}/\mathcal{F}$ for all standard n. $\mathfrak{A}^I_F \models d > \hat{n}(i)/\mathcal{F} \Leftrightarrow \{i \in \mathbb{N} \mid d(i) > \hat{n}(i)\} \in \mathcal{F}\{i \in I \mid d(i) > \hat{n}(i)\} = \{i \in I \mid i > n\}$ is cofinite, so it is in $\mathcal{F}$.

Let us define an infinite prime number: $f(i) = p_i$ (the ith prime). We show that $f/\mathcal{F}$ is a prime:

$$\mathbb{N}^I_{\mathcal{F}} \models \forall xy(xy = \overline{f/\mathcal{F}} \rightarrow x = 1 \vee y = 1) \quad \Leftrightarrow$$

$$\{i \in \mid \mathbb{N} \models \forall xy(xy = p_i \rightarrow x = 1 \vee y = 1)\} \in \mathcal{F} \quad \Leftrightarrow$$

$$\{i \in I \mid \mathbb{N} \models \forall xy(xy = \overline{f(i)} \rightarrow x = 1 \vee y = 1)\} \in \mathcal{F}$$

Since p_i is a prime number, this set is I (i.e. $\mathbb{N}$), which, being cofinite, belongs to $\mathcal{F}$. This shows that $f/\mathcal{F}$ is a prime in $\mathbb{N}^I_{\mathcal{F}}$.

Theorem 4.5.22 (Ultrafilter Compactness Theorem) *Let $\mathcal{K} = Mod(\Gamma)$. If for every finite $\Delta \subseteq \Gamma$ there is an $\mathfrak{A}_\Delta \in K$ with $\mathfrak{A}_\Delta \models \Delta$, then there is an ultraproduct $\mathfrak{B}$ of $\mathfrak{A}_\Delta$'s such that $\mathfrak{B} \models \Gamma$.*

Proof We may assume that Γ is infinite. Let $\mathfrak{A}_\Delta \models \Delta$, and let I be the set of finite subsets of Γ.

Define for a $\varphi \in \Gamma$: $S_\varphi = \{\Delta \in I \mid \varphi \in \Delta\}$. The family of S_φ's has the finite intersection property: $\{\varphi_1, \ldots, \varphi_k\} \subset S_{\varphi_1} \cap \cdots \cap S_{\varphi_k}$. By Lemma 4.5.4 and Theorem 4.5.9 there is an ultrafilter $\mathcal{F}$ containing the S_φ's.

If $\Delta \in S_\varphi$ then $\varphi \in \Delta$ so $\mathfrak{A}_\Delta \models \varphi$, and therefore $S_\varphi \subseteq \{\Delta \mid \mathfrak{A}_\Delta \models \varphi\}$. So $\{\Delta \mid \mathfrak{A}_\Delta \models \varphi\} \in \mathcal{F}$.

Now apply the fundamental theorem: $\{\Delta \mid \mathfrak{A}_\Delta \models \varphi\} \in \mathcal{F} \Leftrightarrow \prod_{\mathcal{F}} \mathfrak{A}_\Delta \models \varphi$. Hence $\prod_{\mathcal{F}} \mathfrak{A}_\Delta \models \Gamma$. □

This yields a purely model theoretic proof of the Compactness Theorem.

Corollary 4.5.23 (The Compactness Theorem) *If $Mod(\Delta) \neq \emptyset$ for all finite $\Delta \subseteq \Gamma$ then $Mod(\Gamma) \neq \emptyset$.*

This proof doesn't use any logic at all. That is indeed most satisfying, as with the Compactness Theorem one gets enough logic back. One can get, so to speak, the advantages of the Compactness Theorem, with all its virtues, by completely algebraic means.

The following is a mild variation of the ultrafilter compactness theorem.

Theorem 4.5.24 *Every $\mathfrak{A}$ is embedded in an ultraproduct of its finitely generated substructures.*

Proof For a given structure $\mathfrak{A}$ we consider $\Gamma = Diag(\mathfrak{A})$, with the language containing constants for all elements of $\mathfrak{A}$. Let Δ be a finite subset of Γ; the new constants of Δ are $\overline{a_1}, \ldots, \overline{a_n}$. The substructure $\mathfrak{A}_\Delta$ of $\mathfrak{A}$ is generated by $a_1, \ldots, a_n$. As Δ is a subset of the diagram, $\mathfrak{A}_\Delta \models \Delta$.

Let $\mathcal{F}$ be an ultrafilter containing the finite subsets Δ of Γ, then $\{\Delta \mid \mathfrak{A}_\Delta \models \Delta\} \in \mathcal{F}$, and hence $\varPi_{\mathcal{F}} A_\Delta \models \Gamma$, so $\mathfrak{A} \hookrightarrow \varPi_{\mathcal{F}} \mathfrak{A}_\Delta$. $\qquad\square$

Exercise 4.5.25 If $\mathcal{F} = \{X \subseteq I \mid i_0 \in X\}$ (i.e. $\mathcal{F}$ is a principal ultrafilter) then $\prod_F \mathfrak{A}_i \cong \mathfrak{A}_{i_0}$.

Definition 4.5.26

(i) $\mathcal{K}$ is a *basic elementary class* if $\mathcal{K} = Mod(\varphi)$, for some sentence φ (i.e. $\mathcal{K}$ is finitely axiomatizable).
(ii) $\mathcal{K}$ is an *elementary class* if $\mathcal{K} = Mod(\Gamma)$, for some set of sentences Γ (i.e. $\mathcal{K}$ is axiomatizable).
(iii) $\mathcal{K}$ is *closed under elementary equivalence* if $\mathfrak{A} \in \mathcal{K}, \mathfrak{A} \equiv \mathfrak{B} \Rightarrow \mathfrak{B} \in \mathcal{K}$.
(vi) $\mathcal{K}$ is *closed under ultraproducts* if $\mathfrak{A}_i \in \mathcal{K}$, $\mathcal{F}$ is an ultrafilter $\Rightarrow \prod_{\mathcal{F}} \mathfrak{A}_i \in \mathcal{K}$.

Axiomatizability and finite axiomatizability are by nature syntactic notions; as it turns out, they also have strictly model theoretic characterizations.

Theorem 4.5.27

(i) $\mathcal{K}$ *is an elementary class* $\Leftrightarrow$ $\mathcal{K}$ *is closed under elementary equivalence and ultraproducts.*
(ii) $\mathcal{K}$ *is a basic elementary class* $\Leftrightarrow$ $\mathcal{K}$ *and* $\mathcal{K}^c$ *are closed under elementary equivalence and ultraproducts.*

Proof (i) ($\Rightarrow$) $\mathcal{K} = Mod(\Gamma)$ for some Γ.
$\mathcal{K}$ *is closed under elementary equivalence.* Choose an arbitrary $\mathfrak{A} \in \mathcal{K}$. Let $\mathfrak{B} \equiv \mathfrak{A}$. This means $\mathfrak{A} \models \varphi \Leftrightarrow \mathfrak{B} \models \varphi$, for all sentences φ. Because $\mathfrak{A} \models \varphi$ for all $\varphi \in \Gamma$, it now follows that $\mathfrak{B} \models \varphi$ for all $\varphi \in \Gamma$, so $\mathfrak{B} \in Mod(\Gamma) = \mathcal{K}$.
$\mathcal{K}$ *is closed under ultraproducts.* Let $\mathfrak{A}_i \in K$ for $i \in I$ and let $\mathcal{F}$ be an ultrafilter. Define $\mathfrak{A} = \prod_{\mathcal{F}} \mathfrak{A}_i$. We know $\mathfrak{A} \models \varphi \Leftrightarrow \{i \in I \mid \mathfrak{A}_i \models \varphi\} \in \mathcal{F}$. If $\varphi \in \Gamma$, then $\{i \in I \mid A_i \models \varphi\} = I \in \mathcal{F}$. So $\mathfrak{A} \models \varphi$ for all $\varphi \in \Gamma$, which means that $\mathfrak{A} \in Mod(\Gamma) = \mathcal{K}$.

($\Leftarrow$): We have to find an axiom set for $\mathcal{K}$, i.e. a set Γ such that $\Gamma \models \varphi \Leftrightarrow \varphi \in \mathcal{K}$. Clearly, $Th(\mathcal{K})$ is a plausible candidate, so let us try $\Gamma = Th(\mathcal{K})$. Since $\mathcal{K} \subseteq Mod(Th(\mathcal{K}))$, we only have to show "$\supseteq$".. So let $\mathfrak{A} \in Mod(\Gamma)$. Consider $Th(\hat{\mathfrak{A}})$ (i.e. the theory of $\mathfrak{A}$ in the extended language $L(\mathfrak{A})$).

Claim: Any finite $\Delta \subseteq Th(\hat{\mathfrak{A}})$ has a model in $\mathcal{K}$.

Pick such a $\Delta = \{\varphi_1 \ldots \varphi_n\}$ and put $\sigma = \varphi_1 \wedge \cdots \wedge \varphi_n$. Let $\overline{a_1}, \ldots, \overline{a_k}$ be the new constants occurring in σ; we define $\sigma^* = \sigma[z_1, \ldots, z_k / \overline{a_1}, \ldots, \overline{a_k}]$, where $z_1, \ldots, z_k$ are fresh variables. Now suppose that $\exists \vec{z} \sigma^*$ holds in no $\mathfrak{B}$ in $\mathcal{K}$; then $\neg \exists \vec{z} \sigma^* \in \Gamma$, hence $\mathfrak{A} \models \neg \exists \vec{z} \sigma^*$, i.e. $\mathfrak{A} \models \forall \neg \vec{z} \sigma^*$, which contradicts $\hat{\mathfrak{A}} \models \Gamma$. Therefore there is for each Δ an $\mathfrak{A}_\Delta \in \mathcal{K}$ with $\mathfrak{A}_\Delta \models \Delta$. By Theorem 4.5.24, $\prod_F \mathfrak{A}_\Delta \in Mod(Th(\mathfrak{A}))$

for a suitable utrafilter $\mathcal{F}$. Therefore $\mathfrak{A} \overset{\prec}{\hookrightarrow} \prod_F \mathfrak{A}_\Delta$ (by Lemma 4.3.9), and thus $\mathfrak{A} \equiv \prod_F \mathfrak{A}_\Delta$. $\mathcal{K}$ is closed under ultrafilters and $\equiv$, hence $\mathfrak{A} \in \mathcal{K}$.

(ii) By (i) and Lemma 4.2.10. $\square$

Application 4.5.28

(i) The class of well-ordered sets is not elementary. $\mathbb{N}$ is well-ordered, but $\mathfrak{A} = \mathbb{N}_{\mathcal{F}}^{\mathbb{N}}$ (for some free ultrafilter $\mathcal{F}$) has infinite descending sequences. Provide an example.

(ii) The class of all trees is not elementary. $\mathcal{F}$act: $\mathbb{N}$ is a tree. Consider the structure $\mathfrak{A}$ under (i). There is an infinite element d. The top cannot be reached in a finite number of immediate predecessor steps from d.

(iii) The class of all fields of positive characteristic is not elementary. Consider $\mathbb{F}_{p_i} = \mathbb{Z}/(p_i)$ (i.e. the prime field of characteristic p_i). Take a free ultrafilter $\mathcal{F}$ on $\mathbb{N}$. $\Pi_{\mathcal{F}} \mathbb{F}_{p_i} = \mathfrak{A}$ is a field.

Since $\{i \mid \mathbb{F}_{p_j} \models p_i \neq 0\} = \{i\}^c$ is cofinite, we have $\mathfrak{A} \models p_i \neq 0$. Hence $\Pi_{\mathcal{F}} \mathbb{F}_{p_i}$ has characteristic 0.

(vi) The class of archimedean ordered fields is not elementary. Consider $\mathfrak{A} = \mathbb{Q}_{\mathcal{F}}^{\mathbb{N}}$, where $\mathcal{F}$ is a free ultrafilter over $\mathbb{N}$. $\mathfrak{A}$ has infinite elements, for example $d = \lambda i.i/\mathcal{F}$. $\mathfrak{A} \models d > r$ for all $r \in \mathbb{Q}$. So there is no standard $\mathbf{n}$ such that $\mathbf{n} > d$. In other words the series $1 + 1, 1 + 1 + 1, 1 + 1 + 1 + 1, \ldots$ will remain below d.

Example 4.5.29 Let $\mathfrak{A} = \mathbb{Z}_{\mathcal{F}}^{\mathbb{N}}$, where $\mathcal{F}$ is a free ultrafilter over $\mathbb{N}$.
$\mathfrak{A}$ has infinite numbers (non-standard numbers) including infinite primes: put $f(i) = p_i$. Now $\mathfrak{A} \models$ "$f/\mathcal{F}$ is a prime", that is $\mathfrak{A} \models \forall xy(xy = \overline{f/\mathcal{F}} \to x = 1 \lor y = 1$). This infinite prime—let us call it p_∞—generates an ideal I. We claim that I is a maximal ideal. Fortunately there is a way to formulate this in first-order logic. Recall that the maximality of the prime ideal (p) for an ordinary prime is expressed by "for all n there is an m such that $mn \equiv 1 \mod p$". This is formalized by the formula $\sigma \colon \forall x \exists yz(xy = 1 + zp)$. As σ holds for all standard primes, it also holds for p_∞ by the fundamental theorem. Therefore $\mathfrak{A}/(p_\infty)$ is a field. Since it cannot have a positive characteristic, it has characteristic 0. A field of characteristic 0 contains as its prime field the rationals, so we can recover the rationals from $\mathfrak{A}/(p_\infty)$. Hence we have in a roundabout way constructed the rationals directly from the finite prime fields (and hence from $\mathbb{N}$) by model theoretic means.

Corollary 4.5.30 (a) *A group G can be ordered $\Leftrightarrow$ all of its finitely generated subgroups can be ordered.*

(b) *Every torsion-free abelian group can be ordered.*

Proof

(a) Extend the language of group theory with $<$. The theory of abelian groups has the axiom set Γ_1, Γ_2 consists of the ordering axioms $+ \forall xyz(x < y \to x + z < y + z)$. Apply Lemma 4.5.24 to the extended language/similarity type.

(b) Let G_Δ be generated by $\{a_1, \ldots, a_n\}$. By the fundamental theorem of abelian groups $G_\Delta \cong \mathbb{Z}^n$ for some n. $\mathbb{Z}^n$ can be ordered lexicographically and thus so can G_Δ. Now apply (a).   □

Theorem 4.5.31 *If* $\mathfrak{A} \equiv \mathfrak{B}$ *then there exists a* $\mathfrak{C}$ *such that* $\mathfrak{A} \prec \mathfrak{C}$ *and* $\mathfrak{B} \prec \mathfrak{C}$.

We will not prove this theorem here.

Theorem 4.5.32 *If* $\mathfrak{A} \equiv \mathfrak{B}$ *then* $\mathfrak{A} \overset{\prec}{\to} \mathfrak{B}^I_{\mathcal{F}}$, *for some* I, *and some ultrafilter* $\mathcal{F}$.

Proof ($\Leftarrow$): Trivial.

($\Rightarrow$): Let Δ's be a finite subset of $Th(\hat{\mathfrak{A}})$ (in $L(\mathfrak{A})$). Say $\Delta = \{\varphi_1, \ldots, \varphi_n\}$, where $\varphi_1, \ldots, \varphi_n$ contain new constants $\overline{a_1}, \ldots, \overline{a_k}$. Define $\varphi_\Delta = \varphi_1 \wedge \cdots \wedge \varphi_n$ and $\hat{\varphi}_\Delta = \varphi_\Delta[z_1, \ldots, z_k/a_1, \ldots, a_k]$, where $z_1, \ldots, z_k$ are fresh variables. $\mathfrak{A} \models \exists z_1, \ldots, z_k \varphi_\Delta$ and $\mathfrak{A} \equiv \mathfrak{B}$, hence $\mathfrak{B} \models \exists z_1, \ldots, z_k \hat{\varphi}_\Delta$. Now by Lemma 4.3.9 $\mathfrak{B} \models \hat{\varphi}_\Delta(b_1, \ldots, b_k)$ for certain $b_1, \ldots, b_k \in |\mathfrak{B}|$; note that we had to expand $\mathfrak{B}$ in order to accommodate the new constants. By Theorem 4.5.22 $\hat{\mathfrak{B}}^I_F \models Th(\hat{\mathfrak{A}})$. So by Lemma 4.3.9 $\mathfrak{A} \overset{\prec}{\to} \prod_{\mathcal{F}} \mathfrak{B}_\Delta$ in $L(\mathfrak{A})$. By taking reducts we get the desired result: $\mathfrak{B}_\Delta : \mathfrak{A} \overset{\prec}{\to} \mathfrak{B}^I_{\mathcal{F}}$ in L. □

Exercise 4.5.33 (1) $\forall i \in I (\mathfrak{A}_i \hookrightarrow \mathfrak{B}_i) \Rightarrow \prod_{\mathcal{F}} \mathfrak{A}_i \hookrightarrow \prod_{\mathcal{F}} \mathfrak{B}_i$.

(2) $\forall i \in I (\mathfrak{A}_i \overset{\prec}{\to} \mathfrak{B}_i) \Rightarrow \prod_{\mathcal{F}} \mathfrak{A}_i \overset{\prec}{\to} \prod_{\mathcal{F}} \mathfrak{B}_i$.

Theorem 4.5.34 *Let* $\mathcal{K}$ *be an elementary class.* $\mathfrak{A}$ *can be embedded into an element of* $\mathcal{K}$ $\Leftrightarrow$ *all finitely generated substructures of* $\mathfrak{A}$ *can be embedded into elements of* $\mathcal{K}$.

Proof ($\Rightarrow$): trivial.

($\Leftarrow$): For each finitely generated $\mathfrak{A}_i$ of $\mathfrak{A}$ there is a $\mathfrak{B}_i \in \mathcal{K}$ so that $\mathfrak{A}_i \hookrightarrow \mathfrak{B}_i \in \mathcal{K}$. So by Theorem 4.5.24 $\mathfrak{A} \hookrightarrow \prod_{\mathcal{F}} \mathfrak{A}_i \hookrightarrow \prod_{\mathcal{F}} \mathfrak{B}_i \in \mathcal{K}$. □

Application 4.5.35 Recall that a Boolean algebra is atomic iff

$$\forall x (x > 0 \to \exists y (0 < y \le x \wedge \forall z (0 < z \le y \to z = y))).$$

Facts: (1) An atomic Boolean algebra is isomorphic to a subset of the powerset of the atoms.

(2) A finitely generated Boolean algebra is atomic (see P.R. Halmos, Lectures on Boolean Algebras).

By Theorem 4.5.24 any Boolean algebra $\mathfrak{A}$ is embedded into an ultraproduct of its finitely generated subalgebras (which are atomic). The atomic Boolean algebra's are elementary, so $\mathfrak{A}$ is embedded in an atomic Boolean algebra. Hence each $\mathfrak{A}$ is isomorphic to a sub-Boolean algebra of a $\mathcal{P}(X)$ (Stone representation theorem).

Application 4.5.36 (Algebraic closures) Given a field $\mathfrak{A}$ we will exhibit a field $\mathfrak{B}$ extending $\mathfrak{A}$, in which all polynomials $p(x)$ of positive degree have zeros.

Fact: For each finite set of non-constant polynomials $p_1(x), \ldots, p_k(x) \in \mathfrak{A}[x]$ there is an extension $\mathfrak{A}'$ of $\mathfrak{A}$ such that $\mathfrak{A}' \models \exists x p_1(x) = 0 \wedge \cdots \exists x p_k(x) = 0$. Define $\Gamma = \{\exists x p(x) = 0 \mid p(x) \text{ a polynomial of positive degree over } \mathfrak{A}\} \cup Diag(\mathfrak{A})$, $I = \{\Delta \subseteq \Gamma \mid \Delta \text{ finite }\}$, $S_p = \{\Delta \mid \exists x p(x) = 0 \in \Delta\}$. The S_p's have the finite intersection property: $\{\exists x p_1(x) = 0, \ldots, \exists x p_k(x) = 0\} \in S_{p_1} \cap \cdots \cap S_{p_k}$ (so $\mathfrak{A}_0 \models Diag(\mathfrak{A})$). For each Δ there is an $\mathfrak{A}_\Delta$ such that $\mathfrak{A}_\Delta \models \Delta$. Let $\mathcal{F}$ be an ultrafilter on I extending the S_p's. $\Pi_\mathcal{F} \mathfrak{A}_\Delta \models \exists_x p(x) = 0 \Leftrightarrow \{\Delta \mid \mathfrak{A}_\Delta \models \exists_x p(x) = 0\} \in \mathcal{F}$. $\Pi_\mathcal{F} \mathfrak{A}_\wedge \models Diag(\mathfrak{A})$ implies $\mathfrak{A} \hookrightarrow \Pi_\mathcal{F} \mathfrak{A}_\Delta$. We have found one particular extension in which all the polynomials have zeros. Then we consider the smallest extension of $\mathfrak{A}$ inside $\Pi_\mathcal{F} \mathfrak{A}_\Delta$; this, on algebraic grounds, is the algebraic closure.

For more on ultraproducts, see: P.C. Eklof, *Ultraproducts for Algebraists* in Handbook of Mathematical Logic, Elsevier, Amsterdam, 1977; H. Schoutens, *The Use of Ultraproducts in Commutative Algebra*, Springer, 2010; J.L. Bell and A.B. Slomson, *Models and Ultraproducts: An Introduction* (Dover Books on Mathematics), 2006; C.C. Chang and H.J. Keisler, *Model Theory*. Elsevier, Amsterdam 1990 (3rd ed.).

Chapter 5
Second-Order Logic

In first-order predicate logic the variables range over *elements* of a structure, in particular the quantifiers are interpreted in the familiar way as "for all elements a of $|\mathfrak{A}|$ ..." and "there exists an element a of $|\mathfrak{A}|$ ...". We will now allow a second kind of variable ranging over subsets of the universe and its Cartesian products, i.e. relations over the universe.

The introduction of these second-order variables is not the result of an unbridled pursuit of generality; one is often forced to take all subsets of a structure into consideration. Examples are "each bounded non-empty set of reals has a supremum", "each non-empty set of natural numbers has a smallest element", "each ideal is contained in a maximal ideal". Already the introduction of the reals on the basis of the rationals requires quantification over sets of rationals, as we know from the theory of Dedekind cuts.

Instead of allowing variables for (and quantification over) sets, one can also allow variables for functions. However, since we can reduce functions to sets (or relations), we will restrict ourselves here to second-order logic with set variables.

When dealing with second-order arithmetic we can restrict our attention to variables ranging over subsets of N, since there is a coding of finite sequences of numbers to numbers, e.g. via Gödel's β-function, or via prime factorization. In general we will, however, allow for variables for relations.

The introduction of the syntax of second-order logic is so similar to that of first-order logic that we will leave most of the details to the reader.

The *alphabet* consists of symbols for

(i) individual variables: $x_0, x_1, x_2, \ldots$,
(ii) individual constants: $c_0, c_1, c_2, \ldots$,

and for each $n \geq 0$,

(iii) n-ary set (predicate) variables: $X_0^n, X_1^n, X_2^n, \ldots$,
(iv) n-ary set (predicate) constants: $\bot, P_0^n, P_1^n, P_2^n, \ldots$,
 (v) connectives : $\wedge, \rightarrow, \vee, \neg, \leftrightarrow, \exists, \forall$.
     Finally we have the usual auxiliary symbols: (,) , , .

D. van Dalen, *Logic and Structure*, Universitext, DOI 10.1007/978-1-4471-4558-5_5,
© Springer-Verlag London 2013

Remark There are denumerably many variables of each kind. The number of constants may be arbitrarily large.

Formulas are inductively defined by:

(i) $X_i^0, P_i^0, \perp \in FORM$,
(ii) for $n > 0$ $X^n(t_1, \ldots, t_n) \in FORM$, $P^n(t_1, \ldots, t_n) \in FORM$,
(iii) *FORM* is closed under the propositional connectives,
(iv) *FORM* is closed under first- and second-order quantification.

Notation We will often write $\langle x_1, \ldots, x_n \rangle \in X^n$ for $X^n(x_1, \ldots, x_n)$ and we will usually drop the superscript in X^n.

The semantics of second-order logic is defined in the same manner as in the case of first-order logic.

Definition 5.1 A *second-order structure* is a sequence $\mathfrak{A} = \langle A, A^*, c^*, R^* \rangle$, where $A^* = \langle A_n | n \in N \rangle, c^* = \{c_i | i \in N\} \subseteq A$, $R^* = \langle R_i^n | i, n \in N \rangle$, and $A_n \subseteq \mathcal{P}(A^n), R_i^n \in A_n$.

In words: a second-order structure consists of a universe A of individuals and second-order universes of n-ary relations ($n \geq 0$), individual constants and set (relation) constants, belonging to the various universes.

In case each A_n contains *all* n-ary relations (i.e. $A_n = \mathcal{P}(A^n)$), we call $\mathfrak{A}$ *full*.

Since we have listed $\perp$ as a 0-ary predicate constant, we must accommodate it in the structure $\mathfrak{A}$.

In accordance with the customary definitions of set theory, we write $0 = \emptyset, 1 = \{0\}$ and $2 = \{0, 1\}$. Also we take $A^0 = 1$, and hence $A_0 \subseteq \mathcal{P}(A^0) = \mathcal{P}(1) = 2$. By convention we assign 0 to $\perp$. Since we also want a distinct 0-ary predicate (proposition)$\top := \neg \perp$, we put $1 \in A_0$. So, in fact, $A_0 = \mathcal{P}(A^0) = 2$.

Now, in order to define *validity* in $\mathfrak{A}$, we mimic the procedure of first-order logic. Given a structure $\mathfrak{A}$, we introduce an extended language $L(\mathfrak{A})$ with names $\overline{S}$ for all elements S of A and $A_n(n \in N)$. The constants R_i^n are interpretations of the corresponding constant symbols P_i^n.

We define $\mathfrak{A} \models \varphi$, φ is *true* or *valid* in $\mathfrak{A}$, for closed φ.

Definition 5.2

(i) $\mathfrak{A} \models \overline{S}$ if $S = 1$,
(ii) $\mathfrak{A} \models \overline{S}^n(\overline{s}_1, \ldots, \overline{s}_n)$ if $\langle s_1, \ldots, s_n \rangle \in S^n$,
(iii) the propositional connectives are interpreted as usual (cf. Definition 2.2.1, Lemma 3.4.5),
(iv) $\mathfrak{A} \models \forall x \varphi(x)$ if $\mathfrak{A} \models \varphi(\overline{s})$ for all $s \in A$,
      $\mathfrak{A} \models \exists x \varphi(x)$ if $\mathfrak{A} \models \varphi(\overline{s})$ for some $s \in A$,
(v) $\mathfrak{A} \models \forall X^n \varphi(X^n)$ if $\mathfrak{A} \models \varphi(S^n)$ for all $S^n \in A_n$,
      $\mathfrak{A} \models \exists X^n \varphi(X^n)$ if $\mathfrak{A} \models \varphi(\overline{S}^n)$ for some $S^n \in A_n$.

If $\mathfrak{A} \models \varphi$ we say that φ is *true*, or valid, in $\mathfrak{A}$.

As in first-order logic we have a natural deduction system, which consists of the usual rules for first-order logic, plus extra rules for second-order quantifiers,

$$\frac{\varphi}{\forall X^n \varphi} \forall^2 I \qquad \frac{\forall X^n \varphi}{\varphi^*} \forall^2 E$$

$$[\varphi]$$

$$\frac{\varphi^*}{\exists X^n \varphi} \exists^2 I \qquad \qquad \frac{\exists X^n \varphi \quad \overset{\vdots}{\psi}}{\psi} \exists^2 E$$

where the conditions on $\forall^2 I$ and $\exists^2 E$ are the usual ones, and φ^* is obtained from φ by replacing each occurrence of $X^n(t_1, \ldots, t_n)$ by $\sigma(t_1, \ldots, t_n)$ for a certain formula σ, such that no free variables among the t_i become bound after the substitution.

Note that $\exists^2 I$ gives us the traditional *comprehension schema*:

$$\exists X^n \forall x_1 \ldots x_n [\varphi(x_1, \ldots, x_n) \leftrightarrow X^n(x_1, \ldots, x_n)],$$

where X^n may not occur free in φ.

Proof

$$\frac{\forall x_1 \ldots x_n (\varphi(x_1, \ldots, x_n) \leftrightarrow \varphi(x_1, \ldots, x_n))}{\exists X^n \forall x_1 \ldots x_n (\varphi(x_1, \ldots, x_n) \leftrightarrow X^n(x_1, \ldots, x_n))} \exists^2 I.$$

Since the top line is derivable, we have a proof of the desired principle. Conversely, $\exists^2 I$ follows from the comprehension principle, given the ordinary rules of logic. The proof is sketched here ($\vec{x}$ and $\vec{t}$ stand for sequences of variables or terms; assume that X^n does not occur in σ).

$$\frac{\dfrac{[\forall \vec{x}(\sigma(\vec{x}) \leftrightarrow X^n(\vec{x}))]}{\sigma(\vec{t}) \leftrightarrow X^n(\vec{t}) \qquad \varphi(\ldots, \sigma(\vec{t}), \ldots)}{\exists X^n \forall \vec{x}(\sigma(\vec{x}) \leftrightarrow X^n(\vec{x})) \qquad \dfrac{\dfrac{\varphi(\ldots, X^n(\vec{t}), \ldots)}{\exists X^n \varphi(\ldots, X^n(\vec{t}), \ldots)} *}{}}{\exists X^n \varphi(\ldots, X^n(\vec{t}), \ldots)}} \dagger$$

$\square$

In † a number of steps are involved, i.e. those necessary for the substitution theorem. In * we have applied a harmless $\exists$-introduction, in the sense that we went from an instance involving a variable to an existence statement, exactly as in first-order logic. This seems to beg the question, as we want to justify $\exists^2$-introduction. However, on the basis of the ordinary quantifier rules we have justified something much stronger than * on the assumption of the comprehension schema, namely the introduction of the existential quantifier, given a *formula* σ and not merely a variable or a constant.

Since we can define $\forall^2$ from $\exists^2$ a similar argument works for $\forall^2 E$.

The extra strength of the second-order quantifier rules lies in $\forall^2 I$ and $\exists^2 E$. We can make this precise by considering second-order logic as a special kind of first-order logic (i.e. "flattening" second-order logic). The basic idea is to introduce special predicates to express the relation between a predicate and its arguments.

So let us consider a first-order logic with a sequence of predicates Ap_0, Ap_1, Ap_2, Ap_3, ..., such that each Ap_n is $(n + 1)$-ary. We think of $Ap_n(x, y_1, \ldots, y_n)$ as $x^n(y_1, \ldots, y_n)$.

For $n = 0$ we get $Ap_0(x)$ as a first-order version of X^0, but that is in accordance with our intentions. X^0 is a proposition (i.e. something that can be assigned a truth value), and so is $Ap_0(x)$. We now have a logic in which all variables are first order, so we can apply all the results from the preceding chapters.

For the sake of a natural simulation of second-order logic we add unary predicates $V, U_0, U_1, U_2, \ldots$, to be thought of as "is an element", "is a o-ary predicate (i.e. proposition)" "is a 1-ary predicate", etc.

We now have to indicate axioms of our first-order system that embody the characteristic properties of second-order logic.

(i) $\forall xyz(U_i(x) \wedge U_j(y) \wedge V(z) \rightarrow x \neq y \wedge y \neq z \wedge z \neq x)$ for all $i \neq j$.
   (i.e. the U_i's are pairwise disjoint, and disjoint from V).
(ii) $\forall x y_1 \ldots y_n(Ap_n(x, y_1, \ldots, y_n) \rightarrow U_n(x) \wedge \bigwedge_i V(y_i))$ for $n \geq 1$.
   (i.e. if $x, y_1, \ldots, y_n$ are in the relation Ap_n, then think of x as a predicate, and the y_i's as elements).
(iii) $U_0(C_0), V(C_{2i+1})$, for $i \geq 0$, and $U_n(C_{3i \cdot 5^n})$, for $i, n \geq 0$.
   (i.e. certain constants are designated as "elements" and "predicates").
(iv) $\forall z_1 \ldots z_m \exists x[U_n(x) \wedge \forall y_1 \ldots y_n(\bigwedge V(y_i) \rightarrow (\varphi^* \leftrightarrow Ap_n(x, y_1, \ldots, y_n)))]$,
   where $x \notin FV(\varphi^*)$, see below. (The first-order version of the comprehension schema. We assume that $FV(\varphi) \subseteq \{z_1, \ldots, z_n, y_1, \ldots, y_n\}$.
(v) $\neg Ap_0(C_0)$. So there is a 0-ary predicate for "falsity".)

We claim that the first-order theory given by the above axioms represents second-order logic in the following precise way: we can translate second-order logic in the language of the above theory such that derivability is faithfully preserved.

The translation is obtained by assigning suitable symbols to the various symbols of the alphabet of second-order logic and defining an inductive procedure for converting composite strings of symbols. We put

$$(x_i)^* := x_{2i+1},$$

$$(c_i)^* := c_{2i+1}, \quad \text{for } i \geq 0,$$

$$(X_i^n)^* := x_{3i \cdot 5^n},$$

$$(P_i^n)^* := c_{3i \cdot 5^n}, \quad \text{for } i \geq 0, n \geq 0',$$

$$(X_i^0)^* := Ap_0(x_{3i}), \quad \text{for } i \geq 0,$$

$$\left(P_i^0\right)^* := Ap_0(c_{3i}), \quad \text{for } i \geq 0,$$
$$(\bot)^* := Ap_0(c_0).$$

Furthermore:

$$(\varphi \Box \psi)^* := \varphi^* \Box \psi^* \quad \text{for binary connectives } \Box \text{ and,}$$
$$(\neg \varphi)^* := \neg \varphi^* \quad \text{and,}$$
$$\left(\forall x_i \varphi(x_i)\right)^* := \forall x_i^* \left(V\left(x_i^*\right) \to \varphi^*\left(x_i^*\right)\right),$$
$$\left(\exists x_i \varphi(x_i)\right)^* := \exists x_i^* \left(V\left(x_i^*\right) \wedge \varphi^*\left(x_i^*\right)\right),$$
$$\left(\forall X_i^n \varphi\left(X_i^n\right)\right)^* := \forall \left(X_i^n\right)^* \left(U_n\left(\left(X_i^n\right)^*\right) \to \varphi^*\left(\left(X_i^n\right)^*\right)\right),$$
$$\left(\exists X_i^n \varphi\left(X_i^n\right)\right)^* := \exists \left(X_i^n\right)^* \left(U_n\left(\left(X_i^n\right)^*\right) \wedge \varphi^*\left(\left(X_i^n\right)^*\right)\right).$$

It is a tedious but routine job to show that $\vdash_2 \varphi \Leftrightarrow \vdash_1 \varphi^*$, where 2 and 1 refer to derivability in the respective second-order and first-order systems.

Note that the above translation could be used as an excuse for not doing second-order logic at all, were it not for the fact that the first-order version is not nearly as natural as the second-order one. Moreover, it obscures a number of interesting and fundamental features; e.g. validity in all principal models (see below) makes sense for the second-order version, whereas it is rather an extraneous matter for the first-order version.

Definition 5.3 A second-order structure $\mathfrak{A}$ is called a *model* of second-order logic if the comprehension schema is valid in $\mathfrak{A}$.

If $\mathfrak{A}$ is full (i.e. $A_n = \mathcal{P}(A^n)$ for all n), then we call $\mathfrak{A}$ a *principal* (or *standard*) model.

From the notion of model we get two distinct notions of "second-order validity": (i) true in all models, (ii) true in all principal models.

Recall that $\mathfrak{A} \models \varphi$ was defined for arbitrary second-order structures; we will use $\models \varphi$ for "true in all models".

By the standard induction on derivations we get $\vdash_2 \varphi \Rightarrow \models \varphi$.

Using the above translation into first-order logic we also get $\models \varphi \Rightarrow \vdash_2 \varphi$. Combining these results we get the following theorem.

Theorem 5.4 (Completeness Theorem) $\vdash_2 \varphi \Leftrightarrow \models \varphi$.

Obviously, we also have $\models \varphi \Rightarrow \varphi$ is true in all principal models. The converse, however, is not the case. We can make this plausible by the following argument:

(i) We can define the notion of a unary function in second-order logic, and hence the notions "bijective" and "surjective". Using these notions we can formulate a sentence σ, which states "the universe (of individuals) is finite" (any injection of the universe into itself is a surjection).

(ii) Consider $\Gamma = \{\sigma\} \cup \{\lambda_n | n \in N\}$. Γ is consistent, because each finite subset $\{\sigma, \lambda_{n_1}, \ldots, \lambda_{n_k}\}$ is consistent, since it has a second-order model, namely the principal model over a universe with n elements, where $n = \max\{n_1, \ldots, n_k\}$.

So, by the Completeness Theorem above Γ has a second-order model. Suppose now that Γ has a principal model $\mathfrak{A}$. Then $|\mathfrak{A}|$ is actually Dedekind finite, and (assuming the axiom of choice) *finite*. Say $\mathfrak{A}$ has n_0 elements, then $\mathfrak{A} \not\models \lambda_{n_0+1}$. Contradiction.

We see that Γ has no principal model. Hence the Completeness Theorem fails for validity w.r.t. principal models (and likewise compactness). To find a sentence that holds in all principal models but fails in some model a more refined argument is required.

A peculiar feature of second-order logic is the definability of all the usual connectives in terms of $\forall$ and $\rightarrow$.

Theorem 5.5

(a) $\vdash_2 \bot \leftrightarrow \forall X^0.X^0$,

(b) $\vdash_2 \varphi \wedge \psi \leftrightarrow \forall X^0((\varphi \rightarrow (\psi \rightarrow X^0)) \rightarrow X^0)$,

(c) $\vdash_2 \varphi \vee \psi \leftrightarrow \forall X^0((\varphi \rightarrow X^0) \wedge (\psi \rightarrow X^0) \rightarrow X^0)$,

(d) $\vdash_2 \exists x \varphi \leftrightarrow \forall X^0(\forall x(\varphi \rightarrow X^0) \rightarrow X^0)$,

(e) $\vdash_2 \exists X^n \varphi \leftrightarrow \forall X^0(\forall X^n((\varphi \rightarrow X^0) \rightarrow X^0)$.

Proof (a) is obvious.

(b)

$$
\cfrac{
\cfrac{
\cfrac{[\varphi \wedge \psi]}{\varphi} \qquad \cfrac{[\varphi \rightarrow (\psi \rightarrow X^0)] \qquad [\varphi \wedge \psi]}{\psi \rightarrow X^0 \qquad \qquad \psi}
}{
\cfrac{X^0}{\cfrac{(\varphi \rightarrow (\psi \rightarrow X^0)) \rightarrow X^0}{\cfrac{\forall X^0((\varphi \rightarrow (\psi \rightarrow X^0)) \rightarrow X^0)}{\varphi \wedge \psi \rightarrow \forall X^0((\varphi \rightarrow (\psi \rightarrow X^0)) \rightarrow X^0)}}}
}
}{}
$$

Conversely,

$$
\cfrac{\cfrac{\cfrac{[\varphi] \quad [\psi]}{\varphi \wedge \psi}}{\cfrac{\psi \rightarrow (\varphi \wedge \psi)}{\varphi \rightarrow (\psi \rightarrow (\varphi \wedge \psi))}} \qquad \cfrac{\cfrac{\forall X^0((\varphi \rightarrow (\psi \rightarrow X^0)) \rightarrow X^0)}{\varphi \rightarrow (\psi \rightarrow (\varphi \wedge \psi)) \rightarrow \varphi \wedge \psi}}{}}{\varphi \wedge \psi} \forall^2 E
$$

(d)

$$\cfrac{\cfrac{\exists x\varphi(x) \qquad \cfrac{[\varphi(x)] \qquad \cfrac{[\forall x(\varphi(x) \to X)]}{\varphi(x) \to X}}{X}}{\forall x(\varphi(x) \to X) \to X}}{\cfrac{\forall x(\varphi(x) \to X) \to X}{\forall X(\forall x(\varphi(x) \to X) \to X)}} .$$

Conversely,

$$\cfrac{\cfrac{\cfrac{\cfrac{[\varphi(x)]}{\exists x\varphi(x)}}{\varphi(x) \to \exists x\varphi(x)}}{\forall x(\varphi(x) \to \exists x\varphi(x))} \qquad \cfrac{\forall X(\forall x(\varphi(x) \to X) \to X)}{\forall x(\varphi(x) \to \exists x\varphi(x)) \to \exists x\varphi(x)}}{\exists x\varphi(x)} .$$

(c) and (e) are left to the reader. □

In second-order logic we also have natural means to define identity for individuals. The underlying idea, going back to Leibniz, is that equals have exactly the same properties.

Definition 5.6 (Leibniz Identity) $x = y := \forall X(X(x) \leftrightarrow X(y))$.

This defined identity has the desired properties, i.e. it satisfies $I_1, \ldots, I_4$.

Theorem 5.7
(i) $\vdash_2 x = x$.
(ii) $\vdash_2 x = y \to y = x$.
(iii) $\vdash_2 x = y \wedge y = z \to x = z$.
(iv) $\vdash_2 x = y \to (\varphi(x) \to \varphi(y))$.

Proof Obvious. □

In case the logic already has an identity relation for individuals, say $\dot{=}$, we can show the following.

Theorem 5.8 $\vdash_2 x \dot{=} y \leftrightarrow x = y$.

Proof $\to$ is obvious, by I_4. $\leftarrow$ is obtained as follows:

$$\cfrac{x \dot{=} x \qquad \cfrac{\forall X(X(x) \leftrightarrow X(y))}{x \dot{=} x \leftrightarrow x \dot{=} y}}{x \dot{=} y} .$$

In $\forall^2 E$ we have substituted $z = x$ for $X(z)$. □

We can also use second-order logic to extend Peano arithmetic to second-order arithmetic.

We consider a second-order logic with (first-order) identity and one binary predicate constant S, which represents, intuitively, the successor relation. The following special axioms are added:

1. $\exists!x\forall y\neg S(y,x)$.
2. $\forall x\exists!yS(x,y)$.
3. $\forall xyz(S(x,z)\wedge S(y,z)\rightarrow x=y)$.

For convenience we extend the language with numerals and the successor function. This extension is conservative anyway, under the following axioms:

(i) $\forall y\neg S(y,\bar{0})$,
(ii) $S(\bar{n},\overline{n+1})$,
(iii) $y=x^{+}\leftrightarrow S(x,y)$.

We now write down the *induction axiom* (N.B., not a schema, as in first-order arithmetic, but a proper axiom!).

4. $\forall X(X(0)\wedge\forall x(X(x)\rightarrow X(x^{+}))\rightarrow\forall xX(x))$.

The extension from first-order to second-order arithmetic is *not* conservative. It is, however, beyond our modest means to prove this fact.

One can also use the idea behind the induction axiom to give a (inductive) definition of the class of natural numbers in a second-order logic with axioms (1), (2), (3): N is the smallest class containing 0 and closed under the successor operation.

Let $v(x):=\forall X[(X(0)\wedge\forall y(X(y)\rightarrow X(y^{+}))\rightarrow X(x)]$.

Then, by the comprehension axiom $\exists Y\forall x(v(x)\leftrightarrow Y(x))$.

As yet we cannot assert the existence of a unique Y satisfying $\forall x(v(x)\leftrightarrow Y(x))$, since we have not yet introduced identity for second-order terms.

Therefore, let us add identity relations for the various second-order terms, plus their obvious axioms.

Now we can formulate the axiom of extensionality.

Axiom of Extensionality

$$\forall\vec{x}\big(X^{n}(\vec{x})\leftrightarrow Y^{n}(\vec{x})\big)\leftrightarrow X^{n}=Y^{n}.$$

So, finally, with the help of the axiom of extensionality, we can assert $\exists!Y\forall x(v(x)\leftrightarrow Y(x))$. Thus we can conservatively add a unary predicate constant N with axiom $\forall x(v(x)\leftrightarrow N(x))$.

The axiom of extensionality is on the one hand rather basic—it allows definition by abstraction ("the set of all x, such that …")—and on the other hand rather harmless—we can always turn a second-order model without extensionality into one with extensionality by taking a quotient with respect to the equivalence relation induced by $=$.

Exercises

1. Show that the restriction on X^n in the comprehension schema cannot be dropped (consider $\neg X(x)$).

2. Show $\Gamma \vdash_2 \varphi \Leftrightarrow \Gamma^* \vdash_1 \varphi^*$ (where $\Gamma^* = \{\psi^* | \psi \in \Gamma\}$).

   Hint: use induction on the derivation, with the comprehension schema and simplified $\forall$-, $\exists$-rules. For the quantifier rules it is convenient to consider an intermediate step consisting of a replacement of the free variable by a fresh constant of the proper kind.

3. Prove (c) and (e) of Theorem 5.5.

4. Prove Theorem 5.7.

5. Give a formula $\varphi(X^2)$, which states that X^2 is a function.

6. Give a formula $\varphi(X^2)$ which states that X^2 is a linear order.

7. Give a sentence σ which states that the individuals can be linearly ordered without having a last element (σ can serve as an infinity axiom).

8. Given second-order arithmetic with the successor function, give axioms for addition as a ternary relation.

9. Let a second-order logic with a binary predicate constant $<$ be given with extra axioms that make $<$ a dense linear ordering without endpoints. We write $x < y$ for $<(x, y)$. X is a *Dedekind cut* if $\exists x\, X(x) \wedge \exists x \neg X(x) \wedge \forall x (X(x) \wedge y < x \rightarrow X(y))$. Define a partial ordering on the Dedekind cuts by putting $X \leq X' :=$ $\forall x (X(x) \rightarrow X'(x))$. Show that this partial order is total.

10. Consider the first-order version of second-order logic (involving the predicates Ap_n, U_n, V) with the axiom of extensionality. Any model $\mathfrak{A}$ of this first-order theory can be "embedded" in the principal second-order model over $L_{\mathfrak{A}} = \{a \in |\mathfrak{A}| | \mathfrak{A} \models V(\overline{a})\}$, as follows.

    Define for any $r \in U_n$ $f(r) = \{\langle a_1, \ldots, a_n \rangle | \mathfrak{A} \models Ap_n(\overline{r}, \overline{a_1}, \ldots, \overline{a_n})\}$.

    Show that f establishes an "isomorphic" embedding of $\mathfrak{A}$ into the corresponding principal model. Hence principal models can be viewed as unique maximal models of second-order logic.

11. Formulate the axiom of choice—for each number x there is a set $X \ldots$—in second-order arithmetic.

12. Show that in Definition 5.6 a single implication suffices.

Chapter 6
Intuitionistic Logic

6.1 Constructive Reasoning

In the preceding chapters, we have been guided by the following, seemingly harmless extrapolation from our experience with finite sets: infinite universes can be surveyed in their totality. In particular can we in a global manner determine whether $\mathfrak{A} \models \exists x \varphi(x)$ holds, or not? To adapt Hermann Weyl's phrasing: we are used to thinking of infinite sets not merely as defined by a property, but as sets whose elements are so to speak spread out in front of us, so that we can run through them just as an officer in the police department goes through his file. This view of the mathematical universe is an attractive but rather unrealistic idealization. If one takes our limitations in the face of infinite totalities seriously, then one has to read a statement like "there is a prime number greater than $10^{10^{10}}$" in a stricter way than "it is impossible that the set of primes is exhausted before $10^{10^{10}}$". For we cannot inspect the set of natural numbers at a glance and detect a prime. We have to *exhibit* a prime p greater than $10^{10^{10}}$.

Similarly, one might be convinced that a certain problem (e.g. the determination of the saddle point of a zero-sum game) has a solution on the basis of an abstract theorem (such as Brouwer's fixed point theorem). Nonetheless one cannot always exhibit a solution. What one needs is a *constructive* method (proof) that determines the solution.

Let us give one more example to illustrate the restrictions of abstract methods. Consider the problem: "Are there two irrational numbers a and b such that a^b is rational?" We apply the following smart reasoning: suppose $\sqrt{2}^{\sqrt{2}}$ is rational, then we have solved the problem. Should $\sqrt{2}^{\sqrt{2}}$ be irrational then $(\sqrt{2}^{\sqrt{2}})^{\sqrt{2}}$ is rational. In both cases there is a solution, so the answer to the problem is yes. However, should somebody ask us to produce such a pair a, b, then we have to engage in some serious number theory in order to come up with the right choice between the numbers mentioned above.

Evidently, statements can be read in a non-constructive way, as we did in the preceding chapters, and in a constructive way. In this chapter we will briefly sketch the

D. van Dalen, *Logic and Structure*, Universitext, DOI 10.1007/978-1-4471-4558-5_6,
© Springer-Verlag London 2013

logic one uses in constructive reasoning. In mathematics the practice of constructive procedures and reasoning has been advocated by a number of people, but the founding fathers of constructive mathematics clearly are L. Kronecker and L.E.J. Brouwer. The latter presented a complete program for the rebuilding of mathematics on a constructive basis. Brouwer's mathematics (and the accompanying logic) is called *intuitionistic*, and in this context the traditional non-constructive mathematics (and logic) is called *classical*.

There are a number of philosophical issues connected with intuitionism, for which we refer the reader to the literature, cf. *Dummett, Troelstra–van Dalen*.

Since we can no longer base our interpretations of logic on the fiction that the mathematical universe is a predetermined totality which can be surveyed as a whole, we have to provide a heuristic interpretation of the logical connectives in intuitionistic logic. We will base our heuristics on the proof interpretation put forward by A. Heyting. A similar semantics was proposed by A. Kolmogorov; the proof interpretation is called the Brouwer–Heyting–Kolmogorov (BHK) interpretation.

The point of departure is that a statement φ is considered to be true (or to hold) if we have a proof for it. By a proof we mean a mathematical construction that establishes φ, not a deduction in some formal system. For example, a proof of "$2+3 = 5$" consists of the successive constructions of 2, 3 and 5, followed by a construction that adds 2 and 3, followed by a construction that compares the outcome of this addition and 5.

The primitive notion is here "a proves φ", where we understand by a proof a (for our purpose unspecified) construction. We will now indicate how proofs of composite statements depend on proofs of their parts.

($\wedge$) a proves $\varphi \wedge \psi := a$ is a pair $\langle b, c \rangle$ such that b proves φ and c proves ψ.

($\vee$) a proves $\varphi \vee \psi := a$ is a pair $\langle b, c \rangle$ such that b is a natural number and if $b = 0$ then c proves φ, if $b \neq 0$ then c proves ψ.

($\rightarrow$) a proves $\varphi \rightarrow \psi := a$ is a construction that converts any proof p of φ into a proof $a(p)$ of ψ.

($\bot$) no a proves $\bot$.

  In order to deal with the quantifiers we assume that some domain D of objects is given.

($\forall$) a proves $\forall x \varphi(x) := a$ is a construction such that for each $b \in D$ $a(b)$ proves $\varphi(\bar{b})$.

($\exists$) a proves $\exists x \varphi(x) := a$ is a pair (b, c) such that $b \in D$ and c proves $\varphi(\bar{b})$.

The above explanation of the connectives serves as a means of giving the reader a feeling for what is and what is not correct in intuitionistic logic. It is generally considered to be the intended intuitionistic meaning of the connectives.

Examples

1. $\varphi \wedge \psi \rightarrow \varphi$ is true, for let $\langle a, b \rangle$ be a proof of $\varphi \wedge \psi$, then the construction c with $c(a, b) = a$ converts a proof of $\varphi \wedge \psi$ into a proof of φ. So c proves $(\varphi \wedge \psi \rightarrow \varphi)$.

2. $(\varphi \wedge \psi \rightarrow \sigma) \rightarrow (\varphi \rightarrow (\psi \rightarrow \sigma))$. Let a prove $\varphi \wedge \psi \rightarrow \sigma$, i.e. a converts each proof $\langle b, c \rangle$ of $\varphi \wedge \psi$ into a proof $a(b, c)$ of σ. Now the required proof p of

$\varphi \to (\psi \to \sigma)$ is a construction that converts each proof b of φ into a $p(b)$ of $\psi \to \sigma$. So $p(b)$ is a construction that converts a proof c of ψ into a proof $(p(b))(c)$ of σ. Recall that we had a proof $a(b, c)$ of σ, so put $(p(b))(c) = a(b, c)$; let q be given by $q(c) = a(b, c)$, then p is defined by $p(b) = q$. Clearly, the above contains the description of a construction that converts a into a proof p of $\varphi \to (\psi \to \sigma)$. (For those familiar with the λ-notation: $p = \lambda b.\lambda c.a(b, c)$, so $\lambda a.\lambda b.\lambda c.a(b, c)$ is the proof we are looking for.)

3. $\neg \exists x \varphi(x) \to \forall x \neg \varphi(x)$.

   We will now argue a bit more informally. Suppose we have a construction a that reduces a proof of $\exists x \varphi(x)$ to a proof of $\bot$. We want a construction p that produces for each $d \in D$ a proof of $\varphi(\overline{d}) \to \bot$, i.e. a construction that converts a proof of $\varphi(\overline{d})$ into a proof of $\bot$. So let b be a proof of $\varphi(\overline{d})$, then $\langle d, b \rangle$ is a proof of $\exists x \varphi(x)$, and $a(d, b)$ is a proof of $\bot$. Hence p with $(p(d))(b) = a(d, b)$ is a proof of $\forall x \neg \varphi(x)$. This provides us with a construction that converts a into p.

The reader may try to justify some statements for himself, but he should not worry if the details turn out to be too complicated. A convenient handling of these problems requires a bit more machinery than we have at hand (e.g. λ-notation). Note, by the way, that the whole procedure is not unproblematic since we assume a number of closure properties of the class of constructions.

Now that we have given a rough heuristics of the meaning of the logical connectives in intuitionistic logic, let us move on to a formalization. As it happens, the system of natural deduction is almost right. The only rule that lacks constructive content is that of reductio ad absurdum (RAA). As we have seen (p. 36), an application of *RAA* yields $\vdash \neg\neg\varphi \to \varphi$, but for $\neg\neg\varphi \to \varphi$ to hold informally we need a construction that transforms a proof of $\neg\neg\varphi$ into a proof of φ. Now a proves $\neg\neg\varphi$ if a transforms each proof b of $\neg\varphi$ into a proof of $\bot$, i.e. there cannot be a proof b of $\neg\varphi$. b itself should be a construction that transforms each proof c of φ into a proof of $\bot$. So we know that there cannot be a construction that turns a proof of φ into a proof of $\bot$, but that is a long way from the required proof of φ! (See Example 1.)

6.2 Intuitionistic Propositional and Predicate Logic

We adopt all the rules of natural deduction for the connectives $\vee, \wedge, \to, \bot, \exists, \forall$ with the exception of the rule *RAA*. In order to cover both propositional and predicate logic in one sweep we allow in the alphabet (cf. Sect. 3.3, p. 55) 0-ary predicate symbols, usually called proposition symbols.

Strictly speaking we deal with a derivability notion different from the one introduced earlier (cf. p. 34), since *RAA* is dropped; therefore we should use a distinct notation, e.g. $\vdash_i$. However, we will continue to use $\vdash$ when no confusion arises.

We can now adopt all results of the preceding parts that did not make use of *RAA*. The following list may be helpful.

Lemma 6.2.1

(1) $\vdash \varphi \wedge \psi \leftrightarrow \psi \wedge \varphi$ $(p.\,31)$
(2) $\vdash \varphi \vee \psi \leftrightarrow \psi \vee \varphi$
(3) $\vdash (\varphi \wedge \psi) \wedge \sigma \leftrightarrow \varphi \wedge (\psi \wedge \sigma)$
(4) $\vdash (\varphi \vee \psi) \vee \sigma \leftrightarrow \varphi \vee (\psi \vee \sigma)$
(5) $\vdash \varphi \vee (\psi \wedge \sigma) \leftrightarrow (\varphi \vee \psi) \wedge (\varphi \vee \sigma)$
(6) $\vdash \varphi \wedge (\psi \vee \sigma) \leftrightarrow (\varphi \wedge \psi) \vee (\varphi \wedge \sigma)$
(7) $\vdash \varphi \rightarrow \neg\neg\varphi$ $(p.\,31)$
(8) $\vdash (\varphi \rightarrow (\psi \rightarrow \sigma)) \leftrightarrow (\varphi \wedge \psi \rightarrow \sigma)$ $(p.\,31)$
(9) $\vdash \varphi \rightarrow (\psi \rightarrow \varphi)$ $(p.\,35)$
(10) $\vdash \varphi \rightarrow (\neg\varphi \rightarrow \psi)$ $(p.\,35)$
(11) $\vdash \neg(\varphi \vee \psi) \leftrightarrow \neg\varphi \wedge \neg\psi$
(12) $\vdash \neg\varphi \vee \neg\psi \rightarrow \neg(\varphi \wedge \psi)$
(13) $\vdash (\neg\varphi \vee \psi) \rightarrow (\varphi \rightarrow \psi)$
(14) $\vdash (\varphi \rightarrow \psi) \rightarrow (\neg\psi \rightarrow \neg\varphi)$ $(p.\,35)$
(15) $\vdash (\varphi \rightarrow \psi) \rightarrow ((\psi \rightarrow \sigma) \rightarrow (\varphi \rightarrow \sigma))$ $(p.\,35)$
(16) $\vdash \perp \leftrightarrow (\varphi \wedge \neg\varphi)$ $(p.\,35)$
(17) $\vdash \exists x(\varphi(x) \vee \psi(x)) \leftrightarrow \exists x\varphi(x) \vee \exists x\psi(x)$
(18) $\vdash \forall x(\varphi(x) \wedge \psi(x)) \leftrightarrow \forall x\varphi(x) \wedge \forall x\psi(x)$
(19) $\vdash \neg\exists x\varphi(x) \leftrightarrow \forall x\neg\varphi(x)$
(20) $\vdash \exists x\neg\varphi(x) \rightarrow \neg\forall x\varphi(x)$
(21) $\vdash \forall x(\varphi \rightarrow \psi(x)) \leftrightarrow (\varphi \rightarrow \forall x\psi(x))$
(22) $\vdash \exists x(\varphi \rightarrow \psi(x)) \rightarrow (\varphi \rightarrow \exists x\psi(x))$
(23) $\vdash (\varphi \vee \forall x\psi(x)) \rightarrow \forall x(\varphi \vee \psi(x))$
(24) $\vdash (\varphi \wedge \exists x\psi(x)) \leftrightarrow \exists x(\varphi \wedge \psi(x))$
(25) $\vdash \exists x(\varphi(x) \rightarrow \psi) \rightarrow (\forall x\varphi(x) \rightarrow \psi)$
(26) $\vdash \forall x(\varphi(x) \rightarrow \psi) \leftrightarrow (\exists x\varphi(x) \rightarrow \psi).$

(Observe that (19) and (20) are special cases of (26) and (25).)

All of those theorems can be proved by means of straightforward application of the rules. Some well-known theorems are conspicuously absent, and in some cases there is only an implication one way; we will show later that these implications cannot, in general, be reversed.

From a constructive point of view *RAA* is used to derive strong conclusions from weak premises. For example, in $\neg(\varphi \wedge \psi) \vdash \neg\varphi \vee \neg\psi$ the premise is weak (something has no proof) and the conclusion is strong, it asks for an effective decision. One cannot expect to get such results in intuitionistic logic. Instead there is a collection of weak results, usually involving negations and double negations.

Lemma 6.2.2

(1) $\vdash \neg\varphi \leftrightarrow \neg\neg\neg\varphi$
(2) $\vdash (\varphi \wedge \neg\psi) \rightarrow \neg(\varphi \rightarrow \psi)$
(3) $\vdash (\varphi \rightarrow \psi) \rightarrow (\neg\neg\varphi \rightarrow \neg\neg\psi)$
(4) $\vdash \neg\neg(\varphi \rightarrow \psi) \leftrightarrow (\neg\neg\varphi \rightarrow \neg\neg\psi)$

(5) $\vdash \neg\neg(\varphi \wedge \psi) \leftrightarrow (\neg\neg\varphi \wedge \neg\neg\psi)$
(6) $\vdash \neg\neg\forall x\varphi(x) \rightarrow \forall x\neg\neg\varphi(x)$.

In order to abbreviate derivations we will use the notation $\frac{\Gamma}{\varphi}$ in a derivation when there is a derivation for $\Gamma \vdash \varphi$ (Γ has 0, 1 or 2 elements).

Proof (1) $\neg\varphi \rightarrow \neg\neg\neg\varphi$ follows from Lemma 6.2.1 (7). For the converse we again use Lemma 6.2.1(7):

$$
\cfrac{\cfrac{\cfrac{[\varphi]^1 \quad \varphi \rightarrow \neg\neg\varphi}{\neg\neg\varphi} \quad [\neg\neg\neg\varphi]^2}{\cfrac{\bot}{\neg\varphi}\,1}}{\neg\neg\neg\varphi \rightarrow \neg\varphi}\,2
\qquad
\cfrac{\cfrac{\cfrac{\cfrac{\overline{[\varphi \wedge \neg\psi]^2}}{\varphi} \quad [\varphi \rightarrow \psi]^1}{\psi} \quad \cfrac{[\varphi \wedge \neg\psi]^2}{\neg\psi}}{\cfrac{\bot}{\neg(\varphi \rightarrow \psi)}\,1}}{(\varphi \wedge \neg\psi) \rightarrow \neg(\varphi \rightarrow \psi)}\,2
$$

$$
\cfrac{\cfrac{\cfrac{[\neg\neg\varphi]^3 \quad \cfrac{\cfrac{[\varphi]^1 \quad [\varphi \rightarrow \psi]^4}{\psi} \quad [\neg\psi]^2}{\cfrac{\bot}{\neg\varphi}\,1}}{\cfrac{\bot}{\neg\neg\psi}\,2}}{\cfrac{\neg\neg\varphi \rightarrow \neg\neg\psi}{3}}}{(\varphi \rightarrow \psi) \rightarrow (\neg\neg\varphi \rightarrow \neg\neg\psi)}\,4
$$

We prove (3) also by using (14) and (15) from Lemma 6.2.1.
(4) Apply the intuitionistic half of the contraposition (Lemma 6.2.1 (14)) to (2):

$$
\cfrac{\cfrac{\cfrac{\cfrac{[\neg\neg(\varphi \rightarrow \psi)]^4}{\neg(\varphi \wedge \neg\psi)} \quad \cfrac{[\varphi]^1 \quad [\neg\psi]^2}{\varphi \wedge \neg\psi}}{\cfrac{\bot}{\neg\varphi}\,1} \quad [\neg\neg\varphi]^3}{\cfrac{\bot}{\neg\neg\psi}\,2}}{\cfrac{\neg\neg\varphi \rightarrow \neg\neg\psi}{3}}}{\neg\neg(\varphi \rightarrow \psi) \rightarrow (\neg\neg\varphi \rightarrow \neg\neg\psi)}\,4
$$

For the converse we apply some facts from Lemma 6.2.1:

$$
\cfrac{\cfrac{\cfrac{\cfrac{[\neg(\varphi \to \psi)]^1}{\neg(\neg\varphi \vee \psi)}}{\neg\neg\varphi \wedge \neg\psi}}{\neg\neg\varphi} \qquad [\neg\neg\varphi \to \neg\neg\psi]^2}{\cfrac{\cfrac{\neg\neg\psi \qquad \cfrac{\cfrac{\cfrac{[\neg(\varphi \to \psi)]^1}{\neg(\neg\varphi \vee \psi)}}{\neg\neg\varphi \wedge \neg\psi}}{\neg\psi}}{\cfrac{\bot}{\neg\neg(\varphi \to \psi)}\,1}}{(\neg\neg\varphi \to \neg\neg\psi) \to \neg\neg(\varphi \to \psi)}\,2}
$$

(5) $\to$: Apply (3) to $\varphi \wedge \psi \to \varphi$ and $\varphi \wedge \psi \to \psi$. The derivation of the converse is given below.

$$
\cfrac{\cfrac{\cfrac{[\neg(\varphi \wedge \psi)]^3 \qquad \cfrac{[\varphi]^1 \quad [\psi]^2}{\varphi \wedge \psi}}{\cfrac{\bot}{\neg\varphi}\,1} \qquad [\neg\neg\varphi \wedge \neg\neg\psi]^4}{\cfrac{\cfrac{\bot}{\neg\psi}\,2 \qquad \cfrac{[\neg\neg\varphi \wedge \neg\neg\psi]^4}{\neg\neg\psi}}{\cfrac{\bot}{\neg\neg(\varphi \wedge \psi)}\,3}}}{(\neg\neg\varphi \wedge \neg\neg\psi) \to \neg\neg(\varphi \wedge \psi)}\,4
$$

(6) $\quad \vdash \exists x \neg\varphi(x) \to \neg\forall x \varphi(x),$ Lemma 6.2.1 (20)

so $\quad \neg\neg\forall x \varphi(x) \to \neg\exists x \neg\varphi(x),$ Lemma 6.2.1 (14)

hence $\neg\neg\forall x \varphi(x) \to \forall x \neg\neg\varphi(x),$ Lemma 6.2.1 (19).

Most of the straightforward meta-theorems of propositional and predicate logic carry over to intuitionistic logic. The following theorems can be proved by a tedious but routine induction.   □

Theorem 6.2.3 (Substitution Theorem for Derivations) *If $\mathcal{D}$ is a derivation and $\$$ a propositional atom, then $\mathcal{D}[\varphi/\$]$ is a derivation if the free variables of φ do not occur bound in $\mathcal{D}$.*

Theorem 6.2.4 (Substitution Theorem for Derivability) *If $\Gamma \vdash \sigma$ and is a propositional atom, then $\Gamma[\varphi/\$] \vdash \sigma[\varphi/\$]$, where the free variables of φ do not occur bound in σ or Γ.*

Theorem 6.2.5 (Substitution Theorem for Equivalence)

$$\Gamma \vdash (\varphi_1 \leftrightarrow \varphi_2) \rightarrow \big(\psi[\varphi_1/\$] \leftrightarrow \psi[\varphi_2/\$]\big),$$

$$\Gamma \vdash \varphi_1 \leftrightarrow \varphi_2 \Rightarrow \Gamma \vdash b\psi[\varphi_1/\$] \leftrightarrow \psi[\varphi_2/\$],$$

where is an atomic proposition, the free variables of φ_1 and φ_2 do not occur bound in Γ or ψ and the bound variables of ψ do not occur free in Γ.

The proofs of the above theorems are left to the reader. Theorems of this kind are always suffering from unaesthetic variable conditions. In practical applications one always renames bound variables or considers only closed hypotheses, so that there is not much to worry about. For precise formulations cf. Chap. 7.

The reader will have observed from the heuristics that $\vee$ and $\exists$ carry most of the burden of constructiveness. We will demonstrate this once more in an informal argument.

There is an effective procedure to compute the decimal expansion of $\pi (3.1415927\ldots)$. Let us consider the statement $\varphi_n :=$ in the decimal expansion of π there is a sequence of n consecutive sevens.

Clearly $\varphi_{100} \rightarrow \varphi_{99}$ holds, but there is no evidence whatsoever for $\neg\varphi_{100} \vee \varphi_{99}$.

The fact that $\wedge, \rightarrow, \forall, \bot$ do not ask for the kind of decisions that $\vee$ and $\exists$ require, is more or less confirmed by the following.

Theorem 6.2.6 *If φ does not contain $\vee$ or $\exists$ and all atoms but $\bot$ in φ are negated, then $\vdash \varphi \leftrightarrow \neg\neg\varphi$.*

Proof Induction on φ.

We leave the proof to the reader. (Hint: apply Lemma 6.2.2.)   □

By definition intuitionistic predicate (propositional) logic is a subsystem of the corresponding classical systems. Gödel and Gentzen have shown, however, that by interpreting the classical disjunction and existence quantifier in a weak sense, we can embed classical logic into intuitionistic logic. For this purpose we introduce a suitable translation.

Definition 6.2.7 The mapping $^\circ : FORM \rightarrow FORM$ is defined by

(i) $\bot^\circ := \bot$ and $\varphi^\circ := \neg\neg\varphi$ for atomic φ distinct from $\bot$,
(ii) $(\varphi \wedge \psi)^\circ := \varphi^\circ \wedge \psi^\circ$,
(iii) $(\varphi \vee \psi)^\circ := \neg(\neg\varphi^\circ \wedge \neg\psi^\circ)$,
(iv) $(\varphi \rightarrow \psi)^\circ := \varphi^\circ \rightarrow \psi^\circ$,
(v) $(\forall x\varphi(x))^\circ := \forall x\varphi^\circ(x)$,
(vi) $(\exists x\varphi(x))^\circ := \neg\forall x\neg\varphi^\circ(x)$.

This mapping is called the Gödel translation.
We define $\Gamma^\circ = \{\varphi^\circ | \varphi \in \Gamma\}$. The relation between classical derivability ($\vdash_c$) and intuitionistic derivability ($\vdash_i$) is given by the following.

Theorem 6.2.8 $\Gamma \vdash_c \varphi \Leftrightarrow \Gamma^\circ \vdash_i \varphi^\circ$.

Proof It follows from the preceding chapters that $\vdash_c \varphi \leftrightarrow \varphi^\circ$, therefore $\Leftarrow$ is an immediate consequence of $\Gamma \vdash_i \varphi \Rightarrow \Gamma \vdash_c \varphi$.

For $\Rightarrow$, we use induction on the derivation $\mathcal{D}$ of φ from Γ.

1. $\varphi \in \Gamma$, then also $\varphi^\circ \in \Gamma^\circ$ and hence $\Gamma^\circ \vdash_i \varphi^\circ$.
2. The last rule of $\mathcal{D}$ is a propositional introduction or elimination rule. We consider two cases:

$$
\begin{array}{ll}
\rightarrow I \quad \begin{array}{c} [\varphi] \\ \mathcal{D} \\ \dfrac{\psi}{\varphi \rightarrow \psi} \end{array} &
\begin{array}{l}
\text{Induction hypothesis } \Gamma^\circ, \varphi^\circ \vdash_i \psi^\circ. \\
\text{By } \rightarrow I \; \Gamma^\circ \vdash_i \varphi^\circ \rightarrow \psi^\circ, \text{ and so by definition} \\
\Gamma^\circ \vdash_i (\varphi \rightarrow \psi)^\circ.
\end{array}
\end{array}
$$

$$
\vee E \quad
\begin{array}{ccc}
 & [\varphi] & [\psi] \\
\mathcal{D} & \mathcal{D}_1 & \mathcal{D}_2 \\
\varphi \vee \psi & \sigma & \sigma \\
\hline
\end{array}
\qquad
\begin{array}{l}
\text{Induction hypothesis :} \Gamma^\circ \vdash_i (\varphi \vee \psi)^\circ, \\
\Gamma^\circ, \varphi^\circ \vdash_i \sigma^\circ \; \Gamma^\circ, \psi^\circ \vdash_i \sigma^\circ \\
\text{(where } \Gamma \text{ contains all uncancelled} \\
\text{hypotheses involved).}
\end{array}
$$

$$
\sigma
$$

$\Gamma^\circ \vdash_i \neg(\neg\varphi^\circ \wedge \neg\psi^\circ), \Gamma^\circ \vdash_i \varphi^\circ \rightarrow \sigma^\circ, \Gamma^\circ \vdash_i \psi^\circ \rightarrow \sigma^\circ$.

The result follows from the derivation below:

$$
\begin{array}{c}
\dfrac{\dfrac{[\varphi^\circ]^1 \quad \varphi^\circ \rightarrow \sigma^\circ}{\sigma^\circ} \quad [\neg\sigma^\circ]^3}{\dfrac{\bot}{\neg\varphi^\circ}1} \qquad
\dfrac{\dfrac{[\psi^\circ]^2 \quad \psi^\circ \rightarrow \sigma^\circ}{\sigma^\circ} \quad [\neg\sigma^\circ]^3}{\dfrac{\bot}{\neg\psi^\circ}2} \\[6pt]
\neg(\neg\varphi^\circ \wedge \neg\psi^\circ) \qquad\qquad \neg\varphi^\circ \wedge \neg\psi^\circ \\
\hline
\dfrac{\bot}{\neg\neg\sigma^\circ}3 \\
\hline\hline
\sigma^\circ
\end{array}
$$

The remaining rules are left to the reader.

3. The last rule of $\mathcal{D}$ is the falsum rule. This case is obvious.
4. The last rule of $\mathcal{D}$ is a quantifier introduction or elimination rule. Let us consider two cases:

$$
\forall I \quad
\begin{array}{c}
\mathcal{D} \\
\dfrac{\varphi(x)}{\forall x \varphi(x)}
\end{array}
\qquad
\begin{array}{l}
\text{Induction hypothesis: } \Gamma^\circ \vdash_i \varphi(x)^\circ \\
\text{By } \forall I \; \Gamma^\circ \vdash_i \forall x \varphi(x)^\circ, \text{ so } \Gamma^\circ \vdash_i (\forall x \varphi(x))^\circ.
\end{array}
$$

$$\exists E: \quad \begin{array}{ccc} & [\varphi(x)] & \\ \mathcal{D} & \mathcal{D}_1 & \\ \dfrac{\exists x \varphi(x) \qquad \sigma}{\sigma} & & \end{array}$$

Induction hypothesis: $\Gamma^\circ \vdash_i (\exists x \varphi(x))^\circ$,
$\Gamma^\circ, \varphi(x)^\circ \vdash_i \sigma^\circ$.
So $\Gamma^\circ \vdash_i (\neg \forall x \neg \varphi(x))^\circ$ and
$\Gamma^\circ \vdash_i \forall x (\varphi(x)^\circ \to \sigma^\circ)$.

$$\cfrac{\neg \forall x \neg \varphi(x)^\circ \qquad \cfrac{\cfrac{\cfrac{\cfrac{\cfrac{\forall x (\varphi(x)^\circ \to \sigma^\circ)}{[\varphi(x)^\circ]^1 \qquad \varphi(x)^\circ \to \sigma^\circ}}{\sigma^\circ \qquad [\neg \sigma^\circ]^2}}{\cfrac{\bot}{\neg \varphi(x)^\circ} \, 1}}{\forall x \neg \varphi(x)^\circ}}{\cfrac{\bot}{\neg \neg \sigma^\circ} \, 2}}{\sigma^\circ}$$

We now get $\Gamma^\circ \vdash_i \sigma^\circ$.

5. The last rule of $\mathcal{D}$ is *RAA*.

$$\begin{array}{c} [\neg \varphi] \\ \mathcal{D} \\ \dfrac{\bot}{\varphi} \end{array}.$$

Induction hypothesis $\Gamma^\circ, (\neg \varphi)^\circ \vdash_i \bot$.
so $\Gamma^\circ \vdash_i \neg\neg \varphi^\circ$, and hence by Theorem 6.2.6 $\Gamma^\circ \vdash_i \varphi^\circ$ □

Let us call formulas in which all atoms occur negated, and those which contain only the connectives $\wedge, \to, \forall, \bot$, *negative*.

The special role of $\vee$ and $\exists$ is underlined by the following.

Corollary 6.2.9 *Classical predicate (propositional) logic is conservative over intuitionistic predicate (propositional) logic with respect to negative formulas, i.e.* $\vdash_c \varphi \Leftrightarrow \vdash_i \varphi$ *for negative* φ.

Proof φ°, for negative φ, is obtained by replacing each atom p by $\neg \neg p$. Since all atoms occur negated we have $\vdash_i \varphi^\circ \leftrightarrow \varphi$ (apply Lemma 6.2.2(1) and Theorem 6.2.6). The result now follows from Theorem 6.2.8. □

In some particular theories (e.g. arithmetic) the atoms are *decidable*, i.e. $\Gamma \vdash \varphi \vee \neg \varphi$ for atomic φ. For such theories one may simplify the Gödel translation by putting $\varphi^\circ := \varphi$ for atomic φ.

Observe that Corollary 6.2.9 tells us that intuitionistic logic is consistent iff classical logic is so (a not very surprising result!).

For propositional logic we have a somewhat stronger result than Theorem 6.2.8.

Theorem 6.2.10 (Glivenko's Theorem) $\vdash_c \varphi \Leftrightarrow \vdash_i \neg\neg\varphi$.

Proof Show by induction on φ that $\vdash_i \varphi^\circ \leftrightarrow \neg\neg\varphi$ (use Lemma 6.2.2), and apply Theorem 6.2.8. $\qquad\qquad\square$

6.3 Kripke Semantics

There are a number of (more or less formalized) semantics for intuitionistic logic that allow for a completeness theorem. We will concentrate here on the semantics introduced by Kripke since it is convenient for applications and it is fairly simple.

Heuristic Motivation   Think of an idealized mathematician (in this context traditionally called the *creative subject*), who extends both his knowledge and his universe of objects in the course of time. At each moment k he has a stock Σ_k of sentences, which he, by some means, has recognized as true and a stock A_k of objects which he has constructed (or created). Since at every moment k the idealized mathematician has various choices for his future activities (he may even stop altogether), the stages of his activity must be thought of as being *partially ordered*, and not necessarily linearly ordered. How will the idealized mathematician interpret the logical connectives? Evidently the interpretation of a composite statement must depend on the interpretation of its parts; e.g. the idealized mathematician has established φ or (and) ψ at stage k if he has established φ at stage k or (and) ψ at stage k. The implication is more cumbersome, since $\varphi \to \psi$ may be known at stage k without φ or ψ being known. Clearly, the idealized mathematician knows $\varphi \to \psi$ at stage k if he knows that if at any future stage (including k) φ is established, also ψ is established. Similarly $\forall x \varphi(x)$ is established at stage k if at any future stage (including k) for all objects a that exist at that stage $\varphi(\overline{a})$ is established.

Evidently in the case of the universal quantifier we must take the future into account since *for all elements* means more than just "for all elements that we have constructed so far"! Existence, on the other hand, is not relegated to the future. The idealized mathematician knows at stage k that $\exists x \varphi(x)$ if he has constructed an object a such that at stage k he has established $\varphi(\overline{a})$. Of course, there are many observations that could be made, for example that it is reasonable to add "in principle" to a number of clauses. This takes care of large numbers, choice sequences, etc. Think of $\forall x y \exists z (z = x^y)$; does the idealized mathematician really construct 10^{10} as a succession of units? For this and similar questions the reader is referred to the literature.

We will now formalize the above sketched semantics.

For a first introduction it is convenient to consider a language without function symbols. Later it will be simple to extend the language.

We consider models for some language L.

Definition 6.3.1 A *Kripke model* is a quadruple $\mathcal{K} = \langle K, \Sigma, C, D \rangle$, where K is a (non-empty) partially ordered set, C a function defined on the constants of L, D a set-valued function on K, Σ a function on K such that

- $C(c) \in D(k)$ for all $k \in K$,
- $D(k) \neq \emptyset$ for all $k \in K$,
- $\Sigma(k) \subseteq At_k$ for all $k \in K$,

where At_k is the set of all atomic sentences of L with constants for the elements of $D(k)$. D and Σ satisfy the following conditions:

(i) $k \leq l \Rightarrow D(k) \subseteq D(l)$,
(ii) $\bot \notin \Sigma(k)$, for all k,
(iii) $k \leq l \Rightarrow \Sigma(k) \subseteq \Sigma(l)$.

$D(k)$ is called the *domain* of $\mathcal{K}$ at k, the elements of K are called *nodes* of $\mathcal{K}$. Instead of "φ has auxiliary constants for elements of $D(k)$" we say for short "φ has parameters in $D(k)$".

Σ assigns to each node the "basic facts" that hold at k, the conditions (i), (ii), (iii) merely state that the collection of available objects does not decrease in time, that a falsity is never established and that a basic fact that once has been established remains true in later stages. The constants are interpreted by the same elements in all domains (they are *rigid designators*).

Note that D and Σ together determine at each node k a classical structure $\mathfrak{A}(k)$ (in the sense of Definition 3.2.1). The universe of $\mathfrak{A}(k)$ is $D(k)$ and the relations of $\mathfrak{A}(k)$ are given by $\Sigma(k)$ as the positive diagram: $\langle \vec{a} \rangle \in R^{\mathfrak{A}(k)}$ iff $R(\vec{a}) \in \Sigma(k)$. The conditions (i) and (iii) above tell us that the universes are increasing:

$$k \leq l \quad \Rightarrow \quad |\mathfrak{A}(k)| \subseteq |\mathfrak{A}(l)|$$

and that the relations are increasing:

$$k \leq l \quad \Rightarrow \quad R^{\mathfrak{A}(k)} \subseteq R^{\mathfrak{A}(l)}.$$

Furthermore $c^{\mathfrak{A}(k)} = c^{\mathfrak{A}(l)}$ for all k and l.

In $\Sigma(k)$ there are also propositions, something we did not allow in classical predicate logic. Here it is convenient for treating propositional and predicate logic simultaneously.

The function Σ tells us which atoms are "true" in k. We now extend Σ to all sentences.

Lemma 6.3.2 Σ *has a unique extension to a function on K (also denoted by Σ) such that $\Sigma(k) \subseteq Sent_k$, the set of all sentences with parameters in $D(k)$, satisfying:*

(i) $\varphi \vee \psi \in \Sigma(k) \Leftrightarrow \varphi \in \Sigma(k)$ *or* $\psi \in \Sigma(k)$,
(ii) $\varphi \wedge \psi \in \Sigma(k) \Leftrightarrow \varphi \in \Sigma(k)$ *and* $\psi \in \Sigma(k)$,
(iii) $\varphi \rightarrow \psi \in \Sigma(k) \Leftrightarrow$ *for all* $l \geq k$ $(\varphi \in \Sigma(l) \Rightarrow \psi \in \Sigma(l))$,
(iv) $\exists x \varphi(x) \in \Sigma(k) \Leftrightarrow$ *there is an* $a \in D(k)$ *such that* $\varphi(\bar{a}) \in \Sigma(k)$,
(v) $\forall x \varphi(x) \in \Sigma(k) \Leftrightarrow$ *for all* $l \geq k$ *and for all* $a \in D(l)$ $\varphi(\bar{a}) \in \Sigma(l)$.

Proof Immediate. We simply define $\varphi \in \Sigma(k)$ for all $k \in K$ simultaneously by induction on φ. $\qquad\square$

Notation We write $k \Vdash \varphi$ for $\varphi \in \Sigma(k)$, pronouncing it as "k forces φ".
Exercise for the reader: reformulate (i)–(v) above in terms of forcing.

Corollary 6.3.3 (i) $k \Vdash \neg\varphi \Leftrightarrow$ *for all* $l \geq k$ $l \nVdash \varphi$.
  (ii) $k \Vdash \neg\neg\varphi \Leftrightarrow$ *for all* $l \geq k$ *there exists a* $p \geq l$ *such that* $(p \Vdash \varphi)$.

Proof $k \Vdash \neg\varphi \Leftrightarrow k \Vdash \varphi \to \bot \Leftrightarrow$ for all $l \geq k(l \Vdash \varphi \Rightarrow l \Vdash \bot) \Leftrightarrow$ for all $l \geq k$ $l \nVdash \varphi$.
$k \Vdash \neg\neg\varphi \Leftrightarrow$ for all $l \geq k$ $l \nVdash \neg\varphi \Leftrightarrow$ for all $l \geq k$ not (for all $p \geq l$ $p \nVdash \varphi$) $\Leftrightarrow$ for all
$l \geq k$ there is a $p \geq l$ such that $p \Vdash \varphi$. $\square$

The monotonicity of Σ for atoms is carried over to arbitrary formulas.

Lemma 6.3.4 (Monotonicity of $\Vdash$) $k \leq l$, $k \Vdash \varphi \Rightarrow l \Vdash \varphi$.

Proof Induction on φ.

Atomic φ: The lemma holds by Definition 6.3.1.
$\varphi = \varphi_1 \wedge \varphi_2$: Let $k \Vdash \varphi_1 \wedge \varphi_2$ and $k \leq l$, then $k \Vdash \varphi_1 \wedge \varphi_2 \Leftrightarrow k \Vdash \varphi_1$ and $k \Vdash \varphi_2 \Rightarrow$
  (ind. hyp.) $l \Vdash \varphi_1$ and $l \Vdash \varphi_2 \Leftrightarrow l \Vdash \varphi_1 \wedge \varphi_2$.
$\varphi = \varphi_1 \vee \varphi_2$: Mimic the conjunction case.
$\varphi = \varphi_1 \to \varphi_2$ Let $k \Vdash \varphi_1 \to \varphi_2$, $l \geq k$. Suppose $p \geq l$ and $p \Vdash \varphi_1$ then, since $p \geq k$,
  $p \Vdash \varphi_2$. Hence $l \Vdash \varphi_1 \to \varphi_2$.
$\varphi = \exists x \varphi_1(x)$: Immediate.
$\varphi = \forall x \varphi_1(x)$: Let $k \Vdash \forall x \varphi_1(x)$ and $l \geq k$. Suppose $p \geq l$ and $a \in D(p)$, then, since
  $p \geq k$, $p \Vdash \varphi_1(\overline{a})$. Hence $l \Vdash \forall x \varphi_1(x)$. $\square$

We will now present some examples, which refute classically true formulas. It
suffices to indicate which atoms are forced at each node. We will simplify the presentation by drawing the partially ordered set and indicating the atoms forced at
each node. For propositional logic no domain function is required (equivalently,
a constant one, say $D(k) = \{0\}$), so we simplify the presentation accordingly.

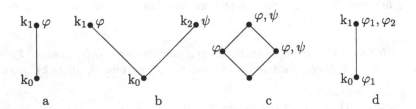

(a) In the bottom node no atoms are known, in the second one only φ, to be precise
    $k_0 \nVdash \varphi$, $k_1 \Vdash \varphi$. By Corollary 6.3.3 $k_0 \Vdash \neg\neg\varphi$, so $k_0 \nVdash \neg\neg\varphi \to \varphi$. Note, however,
    that $k_0 \nVdash \neg\varphi$, since $k_1 \Vdash \varphi$. So $k_0 \nVdash \varphi \vee \neg\varphi$.
(b) $k_i \nVdash \varphi \wedge \psi$ ($i = 0, 1, 2$), so $k_0 \Vdash \neg(\varphi \wedge \psi)$. By definition, $k_0 \Vdash \neg\varphi \vee \neg\psi \Leftrightarrow$
    $k_0 \Vdash \neg\varphi$ or $k_0 \Vdash \neg\psi$. The first is false, since $k_1 \Vdash \varphi$, and the latter is false, since
    $k_2 \Vdash \psi$. Hence $k_0 \nVdash \neg(\varphi \wedge \psi) \to \neg\varphi \vee \neg\psi$.

(c) The bottom node forces $\psi \to \varphi$, but it does not force $\neg\psi \vee \varphi$ (why?). So it does not force $(\psi \to \varphi) \to (\neg\psi \vee \varphi)$.

(d) In the bottom node the following implications are forced: $\varphi_2 \to \varphi_1, \varphi_3 \to \varphi_2, \varphi_3 \to \varphi_1$, but none of the converse implications is forced, hence $k_0 \nVdash (\varphi_1 \leftrightarrow \varphi_2) \vee (\varphi_2 \leftrightarrow \varphi_3) \vee (\varphi_3 \leftrightarrow \varphi_1)$.

We will analyse the last example a bit further. Consider a Kripke model with two nodes as in d, with some assignment Σ of atoms. We will show that for four arbitrary propositions $\sigma_1, \sigma_2, \sigma_3, \sigma_4 k_0 \Vdash \bigvee_{1 \leq i < j \leq 4} \sigma_i \leftrightarrow \sigma_j$, i.e. from any four propositions at least two are equivalent.

There are a number of cases. (1) At least two of $\sigma_1, \sigma_2, \sigma_3, \sigma_4$ are forced in k_0. Then we are done. (2) Just one σ_i is forced in k_0. Then of the remaining propositions, either two are forced in k_1, or two of them are not forced in k_1. In both cases there are σ_j and $\sigma_{j'}$, such that $k_0 \Vdash \sigma_j \leftrightarrow \sigma_{j'}$. (3) No σ_i is forced in k_0. Then we may repeat the argument under (2).

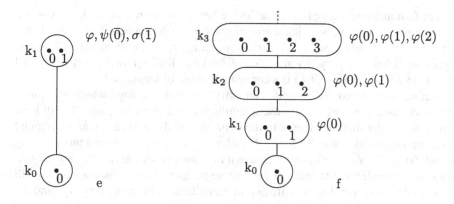

(e) (i) $k_0 \Vdash \varphi \to \exists x \sigma(x)$, for the only node that forces φ is k_1, and indeed $k_1 \Vdash \sigma(1)$, so $k_1 \Vdash \exists x \sigma(x)$.
Now suppose $k_0 \Vdash \exists x(\varphi \to \sigma(x))$, then, since $D(k_0) = \{0\}, k_0 \Vdash \varphi \to \sigma(0)$. But $k_1 \Vdash \varphi$ and $k_1 \nVdash \sigma(0)$.
Contradiction. Hence $k_0 \nVdash (\varphi \to \exists x \sigma(x)) \to \exists x(\varphi \to \sigma(x))$.

Remark $(\varphi \to \exists x \sigma(x)) \to \exists x(\varphi \to \sigma(x))$ is called the *independence of premise principle*. It is not surprising that it fails in some Kripke models, for $\varphi \to \exists x \sigma(x)$ tells us that the required element a for $\sigma(\bar{a})$ may depend on the proof of φ (in our heuristic interpretation); while in $\exists x(\varphi \to \sigma(x))$, the element a must be found independently of φ. So the right-hand side is stronger.

(ii) $k_0 \Vdash \neg\forall x \psi(x) \Leftrightarrow k_i \nVdash \forall x \psi(x)(i = 0, 1)$. $k_1 \nVdash \psi(\bar{1})$, so we have shown $k_0 \Vdash \neg\forall x \psi(x)$. $k_0 \Vdash \exists x \neg\psi(x) \Leftrightarrow k_0 \Vdash \neg\psi(\bar{0})$. However, $k_1 \Vdash \psi(\bar{0})$, so $k_0 \nVdash \exists x \neg\psi(x)$. Hence $k_0 \nVdash \neg\forall x \psi(x) \to \exists x \neg\psi(x)$.

(iii) A similar argument shows $k_0 \nVdash (\forall x \psi(x) \to \tau) \to \exists x(\psi(x) \to \tau)$, where τ is not forced in k_1.

(f) $D(k_i) = \{0, \ldots, i\}$, $\Sigma(k_i) = \{\varphi(0), \ldots, \varphi(i - 1)\}$, $k_0 \Vdash \forall x \neg\neg\varphi(x) \Leftrightarrow$ for all
i $k_i \Vdash \neg\neg\varphi(j)$, $j \le i$. The latter is true since for all $p > i$ $k_p \Vdash \varphi(j)$, $j \le i$.
Now $k_0 \Vdash \neg\neg\forall x \varphi(x) \Leftrightarrow$ for all i there is a $j \ge i$ such that $k_j \Vdash \forall x \varphi(x)$. But no
k_j forces $\forall x \varphi(x)$. So $k_0 \nVdash \forall x \neg\neg\varphi(x) \to \neg\neg\forall x \varphi(x)$.

Remark We have seen that $\neg\neg\forall x \varphi(x) \to \forall x \neg\neg\varphi(x)$ is derivable and it is easily
seen that it holds in all Kripke models, but the converse fails in some models. The
schema $\forall x \neg\neg\varphi(x) \to \neg\neg\forall x \varphi(x)$ is called the *double negation shift* (DNS).

The next thing to do is to show that Kripke semantics is sound for intuitionistic
logic.
We define a few more notions for sentences:

 (i) $\mathcal{K} \Vdash \varphi$ if $k \Vdash \varphi$ for all $k \in K$.
(ii) $\Vdash \varphi$ if $\mathcal{K} \Vdash \varphi$ for all $\mathcal{K}$.

For formulas containing free variables we have to be more careful. Let φ contain
free variables, then we say that $k \Vdash \varphi$ iff $k \Vdash Cl(\varphi)$ (the universal closure). For a
set Γ and a formula φ with free variables $x_{i_0}, x_{i_1}, x_{i_2}, \ldots$ (which we will denote
by $\vec{x}$), we define $\Gamma \Vdash \varphi$ by: for all $\mathcal{K}, k \in K$ and for all $(\vec{a} \in D(k))$ $[k \Vdash \psi(\vec{a})$ for all
$\psi \in \Gamma \Rightarrow k \Vdash \varphi(\vec{a})]$. ($\vec{a} \in D(k)$ is a convenient abuse of language.)
Before we proceed we introduce an extra abuse of language which will prove
extremely useful: we will freely use quantifiers in our meta-language. It will have
struck the reader that the clauses in the definition of the Kripke semantics abound
with expressions like "for all $l \ge k$", "for all $a \in D(k)$". It saves quite a bit of writing
to use "$\forall l \ge k$", "$\forall a \in D(k)$" instead, and it increases systematic readability to boot.
By now the reader is well used to the routine phrases of our semantics, so he will
have no difficulty avoiding a confusion of quantifiers in the meta-language and the
object language.
By way of example we will reformulate the preceding definition:

$$\Gamma \Vdash \varphi := (\forall \mathcal{K})(\forall k \in K)(\forall \vec{a} \in D(k))\big[\forall \psi \in \Gamma (k \Vdash \psi(\vec{a})) \Rightarrow k \Vdash \varphi(\vec{a})\big].$$

There is a useful reformulation of this "semantic consequence" notion.

Lemma 6.3.5 *Let Γ be finite, then $\Gamma \Vdash \varphi \Leftrightarrow \Vdash Cl(\bigwedge \Gamma \to \varphi)$ (where $Cl(X)$ is the
universal closure of X).*

Proof Left to the reader.   □

Theorem 6.3.6 (Soundness Theorem) $\Gamma \vdash \varphi \Rightarrow \Gamma \Vdash \varphi$.

Proof Use induction on the derivation $\mathcal{D}$ of φ from Γ. We will abbreviate "$k \Vdash \psi(\vec{a})$
for all $\psi \in \Gamma$" by "$k \Vdash \Gamma(\vec{a})$". The model $\mathcal{K}$ is fixed in the proof.

(1) $\mathcal{D}$ consists of just φ, then obviously $k \Vdash \Gamma(\vec{a}) \Rightarrow k \Vdash \varphi(\vec{a})$ for all k and $(\vec{a}) \in$
     $D(k)$.

(2) $\mathcal{D}$ ends with an application of a derivation rule.

$(\wedge I)$ Induction hypothesis: $\forall k \forall \vec{a} \in D(k)(k \Vdash \Gamma(\vec{a}) \Rightarrow k \Vdash \varphi_i(\vec{a}))$, for $i = 1, 2$. Now choose a $k \in K$ and $\vec{a} \in D(k)$ such that $k \Vdash \Gamma(\vec{a})$, then $k \Vdash \varphi_1(\vec{a})$ and $k \Vdash \varphi_2(\vec{a})$, so $k \Vdash (\varphi_1 \wedge \varphi_2)(\vec{a})$.

Note that the choice of $\vec{a}$ did not really play a role in this proof. To simplify the presentation we will suppress reference to $\vec{a}$, when it does not play a role.

$(\wedge E)$ Immediate.

$(\vee I)$ Immediate.

$(\vee E)$ Induction hypothesis: $\forall k(k \Vdash \Gamma \Rightarrow k \Vdash \varphi \vee \psi)$, $\forall k(k \Vdash \Gamma, \varphi \Rightarrow k \Vdash \sigma)$, $\forall k(k \Vdash \Gamma, \psi \Rightarrow k \Vdash \sigma)$. Now let $k \Vdash \Gamma$, then by the ind. hyp. $k \Vdash \varphi \vee \psi$, so $k \Vdash \varphi$ or $k \Vdash \psi$. In the first case $k \Vdash \Gamma, \varphi$, so $k \Vdash \sigma$. In the second case $k \Vdash \Gamma, \psi$, so $k \Vdash \sigma$. In both cases $k \Vdash \sigma$, so we are done.

$(\rightarrow I)$ Induction hypothesis: $(\forall k)(\forall \vec{a} \in D(k))(k \Vdash \Gamma(\vec{a}), \varphi(\vec{a}) \Rightarrow k \Vdash \psi(\vec{a}))$. Now let $k \Vdash \Gamma(\vec{a})$ for some $\vec{a} \in D(k)$. We want to show $k \Vdash (\varphi \rightarrow \psi)(\vec{a})$, so let $l \geq k$ and $l \Vdash \varphi(\vec{a})$. By monotonicity $l \Vdash \Gamma(\vec{a})$, and $\vec{a} \in D(l)$, so the ind. hyp. tells us that $l \Vdash \psi(\vec{a})$. Hence $\forall l \geq k(l \Vdash \varphi(\vec{a}) \Rightarrow l \Vdash \psi(\vec{a}))$, so $k \Vdash (\varphi \rightarrow \psi)(\vec{a})$.

$(\rightarrow E)$ Immediate.

$(\bot)$ Induction hypothesis: $\forall k(k \Vdash \Gamma \Rightarrow k \Vdash \bot)$. Since, evidently, no k can force Γ, $\forall k(k \Vdash \Gamma \Rightarrow k \Vdash \varphi)$ is correct.

$(\forall I)$ The free variables in Γ are $\vec{x}$, and z does not occur in the sequence $\vec{x}$. Induction hypothesis: $(\forall k)(\forall \vec{a}, b \in D(k))(k \Vdash \Gamma(\vec{a}) \Rightarrow k \Vdash \varphi(\vec{a}, b))$. Now let $k \Vdash \Gamma(\vec{a})$ for some $\vec{a} \in D(k)$, we must show $k \Vdash \forall z \varphi(\vec{a}, z)$. So let $l \geq k$ and $b \in D(l)$. By monotonicity $l \Vdash \Gamma(\vec{a})$ and $\vec{a} \in D(l)$, so by the ind. hyp. $l \Vdash \varphi(\vec{a}, b)$. This shows $(\forall l \geq k)(\forall b \in D(l))(l \Vdash \varphi(\vec{a}, b))$, and hence $k \Vdash \forall z \varphi(\vec{a}, z)$.

$(\forall E)$ Immediate.

$(\exists I)$ Immediate.

$(\exists E)$ Induction hypothesis: $(\forall k)(\forall \vec{a} \in D(k))(k \Vdash \Gamma(\vec{a}) \Rightarrow k \Vdash \exists z \varphi(\vec{a}, z))$ and $(\forall k)(\forall \vec{a}, b \in D(k))(k \Vdash \varphi(\vec{a}, b), k \Vdash \Gamma(\vec{a}) \Rightarrow k \Vdash \sigma(\vec{a}))$. Here the variables in Γ and σ are $\vec{x}$, and z does not occur in the sequence $\vec{x}$. Now let $k \Vdash \Gamma(\vec{a})$, for some $\vec{a} \in D(k)$, then $k \Vdash \exists z \varphi(\vec{a}, z)$. So let $k \Vdash \varphi(\vec{a}, b)$ for some $b \in D(k)$. By the induction hypothesis $k \Vdash \sigma(\vec{a})$. $\qquad\square$

For the Completeness Theorem we need some notions and a few lemmas.

Definition 6.3.7 A set of sentences Γ is a *prime theory* with respect to a language L if

(i) Γ is closed under $\vdash$,
(ii) $\varphi \vee \psi \in \Gamma \Rightarrow \varphi \in \Gamma$ or $\psi \in \Gamma$,
(iii) $\exists x \varphi(x) \in \Gamma \Rightarrow \varphi(c) \in \Gamma$ for some constant c in L.

The following is an analogue of the Henkin construction combined with a maximal consistent extension.

Lemma 6.3.8 *Let Γ and φ be closed, then if $\Gamma \nvdash \varphi$, there is a prime theory Γ' in a language L', extending Γ such that $\Gamma' \nvdash \varphi$.*

Proof In general one has to extend the language L of Γ by a suitable set of "witnessing" constants. So we extend the language L of Γ by a denumerable set of constants to a new language L'. The required theory Γ' is obtained by series of extensions $\Gamma_0 \subseteq \Gamma_1 \subseteq \Gamma_2 \ldots$. We put $\Gamma_0 := \Gamma$.

Let Γ_k be given such that $\Gamma_k \nvdash \varphi$ and Γ_k contains only finitely many new constants. We consider two cases.

k is even. Look for the first existential sentence $\exists x \psi(x)$ in L' that has not yet been treated, such that $\Gamma_k \vdash \exists x \psi(x)$. Let c be the first new constant not in Γ_k. Now put $\Gamma_{k+1} := \Gamma_k \cup \{\psi(c)\}$.

k is odd. Look for the first disjunctive sentence $\psi_1 \vee \psi_2$ with $\Gamma_k \vdash \psi_1 \vee \psi_2$ that has not yet been treated. Note that not both $\Gamma_k, \psi_1 \vdash \varphi$ and $\Gamma_k, \psi_2 \vdash \varphi$ for then by $\vee E$ $\Gamma_k \vdash \varphi$.

Now we put:

$$\Gamma_{k+1} := \begin{cases} \Gamma_k \cup \{\psi_1\} & \text{if } \Gamma_k, \psi_1 \nvdash \varphi \\ \Gamma_k \cup \{\psi_2\} & \text{otherwise.} \end{cases}$$

Finally:

$$\Gamma' := \bigcup_{k \geq 0} \Gamma_k.$$

There are a few things to be shown:

1. $\Gamma' \nvdash \varphi$. We first show $\Gamma_i \nvdash \varphi$ by induction on i. For $i = 0$, $\Gamma_0 \nvdash \varphi$ holds by assumption. The induction step is obvious for i odd. For i even we suppose $\Gamma_{i+1} \vdash \varphi$. Then $\Gamma_i, \psi(c) \vdash \varphi$. Since $\Gamma_i \vdash \exists x \psi(x)$, we get $\Gamma_i \vdash \varphi$ by $\exists E$, which contradicts the induction hypothesis. Hence $\Gamma_{i+1} \nvdash \varphi$, and therefore by complete induction $\Gamma_i \nvdash \varphi$ for all i.

   Now, if $\Gamma' \vdash \varphi$ then $\Gamma_i \vdash \varphi$ for some i. Contradiction.

2. Γ' is a prime theory.

   (a) Let $\psi_1 \vee \psi_2 \in \Gamma'$ and let k be the least number such that $\Gamma_k \vdash \psi_1 \vee \psi_2$. Clearly $\psi_1 \vee \psi_2$ has not been treated before stage k, and $\Gamma_h \vdash \psi_1 \vee \psi_2$ for $h \geq k$. Eventually $\psi_1 \vee \psi_2$ has to be treated at some stage $h \geq k$, so then $\psi_1 \in \Gamma_{h+1}$ or $\psi_2 \in \Gamma_{h+1}$, and hence $\psi_1 \in \Gamma'$ or $\psi_2 \in \Gamma'$.

   (b) Let $\exists x \psi(x) \in \Gamma'$, and let k be the least number such that $\Gamma_k \vdash \exists x \psi(x)$. For some $h \geq k$ $\exists x \psi(x)$ is treated, and hence $\psi(c) \in \Gamma_{h+1} \subseteq \Gamma'$ for some c.

   (c) Γ' is closed under $\vdash$. If $\Gamma' \vdash \psi$, then $\Gamma' \vdash \psi \vee \psi$, and hence by (a) $\psi \in \Gamma'$.

Conclusion: Γ' is a prime theory containing Γ, such that $\Gamma' \nvdash \varphi$. □

The next step is to construct for closed Γ and φ with $\Gamma \nvdash \varphi$, a Kripke model, with $\mathcal{K} \Vdash \Gamma$ and $k \nVdash \varphi$ for some $k \in K$.

Lemma 6.3.9 (Model Existence Lemma) *If $\Gamma \nvdash \varphi$ then there is a Kripke model $\mathcal{K}$ with a bottom node k_0 such that $k_0 \Vdash \Gamma$ and $k_0 \nVdash \varphi$.*

Proof We first extend Γ to a suitable prime theory Γ' such that $\Gamma' \nvdash \varphi$. Γ' has the language L' with the set of constants C'. Consider a set of distinct constants $\{c_m^i | i \geq 0, m \geq 0\}$ disjoint with C'. A denumerable family of denumerable sets of constants is given by $C^i = \{c_m^i | m \geq 0\}$. We will construct a Kripke model over the poset of all finite sequences of natural numbers, including the empty sequence $\langle \rangle$, with their natural ordering, "initial segment of".

Define $C(\langle \rangle) := C'$ and $C(\vec{n}) = C(\langle \rangle) \cup C^0 \cup \cdots \cup C^{k-1}$ for $\vec{n}$ of positive length k. $L(\vec{n})$ is the extension of L with the set of atoms $At(\vec{n})$, obtained from the constants from $C(\vec{n})$. Note that $L(\vec{n})$ only depends on the length $|\vec{n}|$ of $\vec{n}$; for convenience we denote it by $L_{|\vec{n}|}$. Now put $D(\vec{n}) := C(\vec{n})$.

We define[1] $\Sigma(\vec{n})$ by induction on the length of $\vec{n}$. We construct simultaneously a collection of (prime) theories $\Gamma(\vec{n})$.

1. $\Sigma(\langle \rangle) := \Gamma' \cap At(\langle \rangle)$, $\Gamma(\langle \rangle) = \Gamma'$.
2. Suppose $\Sigma(\vec{n})$ and $\Gamma(\vec{n})$ are given. Consider an enumeration $\langle \sigma_0, \tau_0 \rangle$, $\langle \sigma_1, \tau_1 \rangle$, $\langle \sigma_2, \tau_2 \rangle, \ldots$ of all pairs of sentences in $L_{|\vec{n}|+1}$ such that $\Gamma(\vec{n}), \sigma_i \nvdash \tau_i$ in the language $L_{|\vec{n}|+1}$. The sentences σ_i and τ_i are in the language $L_{|\vec{n}|+1}$, they involve only finitely many constants of $L_{|\vec{n}|+1}$. Hence we may apply Lemma 6.3.8 to $\Gamma(\vec{n}) \cup \{\sigma_i\}$ and τ_i. That is, taking away the fresh constants from σ_i and τ_i, there remain denumerably many fresh constants to carry out the construction described in the lemma. This yields a prime theory $\Gamma(\vec{n}, i)$ with the same language $L(\vec{n}, i)$ ($= L_{|\vec{n}|+1}$) such that $\sigma_i \in \Gamma(\vec{n}, i)$ and $\tau_i \notin \Gamma(\vec{n}, i)$.

Now put $\Sigma(\vec{n}, i) := \Gamma(\vec{n}, i) \cap At(\vec{n}, i)$. We observe that all conditions for a Kripke model are met. The model very much reflects (like the model of Lemma 4.1.11) the nature of the prime theories involved.

Claim $\vec{n} \Vdash \psi \Leftrightarrow \Gamma(\vec{n}) \vdash \psi$.

We prove the claim by induction on ψ.

- For atomic ψ the equivalence holds by definition.
- $\psi = \psi_1 \wedge \psi_2$—immediate.
- $\psi = \psi_1 \vee \psi_2$.

  (a) $\vec{n} \Vdash \psi_1 \vee \psi_2 \Leftrightarrow \vec{n} \Vdash \psi_1$ or $\vec{n} \Vdash \psi_2 \Rightarrow$ (ind. hyp.) $\Gamma(\vec{n}) \vdash \psi_1$ or $\Gamma(\vec{n}) \vdash \psi_2 \Rightarrow$ $\Gamma(\vec{n}) \vdash \psi_1 \vee \psi_2$.
  (b) $\Gamma(\vec{n}) \vdash \psi_1 \vee \psi_2 \Rightarrow \Gamma(\vec{n}) \vdash \psi_1$ or $\Gamma(\vec{n}) \vdash \psi_2$, since $\Gamma(\vec{n})$ is a prime theory (in the right language $L(\vec{n})$). So, by the induction hypothesis, $\vec{n} \Vdash \psi_1$ or $\vec{n} \Vdash \psi_2$, and hence $\vec{n} \Vdash \psi_1 \vee \psi_2$.

- $\psi = \psi_1 \rightarrow \psi_2$.

---

[1] I am indebted to Masahiko Rokuyama for noting a gap in the following construction and argument, and to Katsuhiko Sano for its correction.

(a) $\vec{n} \Vdash \psi_1 \to \psi_2$. Suppose $\Gamma(\vec{n}) \nvdash \psi_1 \to \psi_2$, then $\Gamma(\vec{n}), \psi_1 \nvdash \psi_2$. By the definition of the model there is an extension $\vec{m} = \langle n_0, \dots, n_{k-1}, i \rangle$ of $\vec{n}$ such that $\Gamma(\vec{n}) \cup \{\psi_1\} \subseteq \Gamma(\vec{m})$ and $\Gamma(\vec{m}) \nvdash \psi_2$. By the induction hypothesis $\vec{m} \Vdash \psi_1$ and by $\vec{m} \geq \vec{n}$ and $\vec{n} \Vdash \psi_1 \to \psi_2$, $\vec{m} \Vdash \psi_2$. Applying the induction hypothesis once more we get $\Gamma(\vec{m}) \vdash \psi_2$. Contradiction. Hence $\Gamma(\vec{n}) \vdash \psi_1 \to \psi_2$.

(b) The converse is simple; left to the reader.

- $\psi = \forall x \psi(x)$.

(a) Let $\vec{n} \Vdash \forall x \varphi(x)$, and assume $\Gamma(\vec{n}) \nvdash \forall x \varphi(x)$ in $L_{|\vec{n}|}$, hence, by conservativity, $\Gamma(\vec{n}) \nvdash \forall x \varphi(x)$ in $L_{|\vec{n}|+1}$ (see Exercise 6).

Now pick a constant c in $L_{|\vec{n}|+1}$ but not in $L_{|\vec{n}|}$.

If $\Gamma(\vec{n}) \vdash \varphi(c)$ in $L_{|\vec{n}|+1}$, then $\Gamma(\vec{n}) \vdash \forall x \varphi(x)$ in $L_{|\vec{n}|+1}$.

So we conclude $\Gamma(\vec{n}) \nvdash \varphi(c)$ in $L_{|\vec{n}|+1}$. Then $\Gamma(\vec{n}, i) \nvdash \varphi(c)$ for a suitable i (take $\top$ for σ_i and $\varphi(c)$ for τ_i in the above construction).

Now, by the induction hypothesis $\vec{n}, i \nVdash \varphi(c)$, which contradicts $\vec{n} \Vdash \forall x \varphi(x)$.

(b) $\Gamma(\vec{n}) \vdash \forall x \varphi(x)$. Suppose $\vec{n} \nVdash \forall x \varphi(x)$, then $\vec{m} \nVdash \varphi(c)$ for some $\vec{m} \geq \vec{n}$ and for some $c \in L(\vec{m})$, hence $\Gamma(\vec{m}) \nvdash \varphi(c)$ and therefore $\Gamma(\vec{m}) \nvdash \forall x \varphi(x)$. Contradiction.

- $\psi = \exists x \varphi(x)$.

The implication from left to right is obvious. For the converse we use the fact that $\Gamma(\vec{n})$ is a prime theory. The details are left to the reader.

We can now finish our proof. The bottom node forces Γ and φ is not forced.   □

We can get some extra information from the proof of the Model Existence Lemma: (i) the underlying partially ordered set is a *tree*, (ii) all sets $D(\vec{m})$ are denumerable.

From the Model Existence Lemma we easily derive the following.

Theorem 6.3.10 (Completeness Theorem—Kripke) $\Gamma \vdash_i \varphi \Leftrightarrow \Gamma \Vdash \varphi$ (Γ *and* φ *closed*).

Proof We have already shown $\Rightarrow$. For the converse we assume $\Gamma \nvdash_i \varphi$ and apply Lemma 6.3.9, which yields a contradiction.   □

Actually we have proved the following refinement: intuitionistic logic is complete for countable models over trees.

The above results are completely general (safe for the cardinality restriction on L), so we may as well assume that Γ contains the identity axioms $I_1, \dots, I_4$ (3.6). May we also assume that the identity predicate is interpreted by the real equality in each world? The answer is no; this assumption constitutes a real restriction, as the following theorem shows.

Theorem 6.3.11 *If for all* $k \in K$ $k \Vdash \overline{a} = \overline{b} \Rightarrow a = b$ *for* $a, b \in D(k)$ *then* $\mathcal{K} \Vdash \forall x y (x = y \vee x \neq y)$.

Proof Let $a, b \in D(k)$ and $k \Vdash \overline{a} = \overline{b}$, then $a \neq b$, not only in $D(k)$, but in all $D(l)$ for $l \geq k$, hence for all $l \geq k$, $l \Vdash \overline{a} = \overline{b}$, so $k \Vdash \overline{a} \neq \overline{b}$. $\qquad\square$

For a kind of converse, cf. Exercise 18.

The fact that the relation $a \sim_k b$ in $\mathfrak{A}(k)$, given by $k \Vdash \overline{a} = \overline{b}$, is not the identity relation is definitely embarrassing for a language with function symbols. So let us see what we can do about it. We assume that a function symbol F is interpreted in each k by a function F_k. We require $k \leq l \Rightarrow F_k \subseteq F_l$. F has to obey $I_4 : \forall \vec{x}\vec{y}(\vec{x} = \vec{y} \rightarrow F(\vec{x}) = F(\vec{y}))$. For more about functions see Exercise 34.

Lemma 6.3.12 *The relation $\sim_k$ is a congruence relation on $\mathfrak{A}(k)$, for each k.*

Proof Straightforward, by interpreting $I_1 - I_4$. $\qquad\square$

We may drop the index k; this means that we consider a relation $\sim$ on the whole model, which is interpreted node-wise by the local $\sim_k$'s.

We now define new structures by taking equivalence classes: $\mathfrak{A}^\star(k) := \mathfrak{A}(k)/\sim_k$, i.e. the elements of $|\mathfrak{A}^\star(k)|$ are equivalence classes $a/\sim_k$ of elements $a \in D(k)$, and the relations are canonically determined by

$R_k^\star(a/\sim, \ldots) \Leftrightarrow R_k(a, \ldots)$, similarly for the functions $F_k^\star(a/\sim, \ldots) = F_k(a, \ldots)/\sim$.

The inclusion $\mathfrak{A}(k) \subseteq \mathfrak{A}(l)$, for $k \leq l$, is now replaced by a map $f_{kl} : \mathfrak{A}^\star(k) \rightarrow \mathfrak{A}^\star(l)$, where f_{kl} is defined by $f_{kl}(a) = a^{\mathfrak{A}(l)}$ for $a \in |\mathfrak{A}^\star(k)|$. To be precise:

$a/\sim_k \longmapsto a/\sim_l$, so we have to show $a \sim_k a' \Rightarrow a \sim_l a'$ to ensure the well-definedness of f_{kl}. This, however, is obvious, since $k \Vdash \overline{a} = \overline{a'} \Rightarrow l \Vdash \overline{a} = \overline{a'}$.

Claim 6.3.13 *f_{kl} is a homomorphism.*

Proof Let us look at a binary relation. $R_k^\star(a/\sim, b/\sim) \Leftrightarrow R_k(a, b) \Leftrightarrow k \Vdash R(a, b) \Rightarrow l \Vdash R(a, b) \Leftrightarrow R_l(a, b) \Leftrightarrow R_l^\star(a/\sim, b/\sim)$.

The case of an operation is left to the reader. $\qquad\square$

The upshot is that we can define a modified notion of a Kripke model.

Definition 6.3.14 A modified Kripke model for a language L is a triple $\mathcal{K} = \langle K, \mathfrak{A}, f \rangle$ such that K is a partially ordered set, $\mathfrak{A}$ and f are mappings such that for $k \in K$, $\mathfrak{A}(k)$ is a structure for L and for $k, l \in K$ with $k \leq l$ $f(k, l)$ is a homomorphism from $\mathfrak{A}(k)$ to $\mathfrak{A}(l)$ and $f(l, m) \circ f(k, l) = f(k, m)$, $f(k, k) = id$.

Notation We write f_{kl} for $f(k, l)$, and $k \Vdash^\star \varphi$ for $\mathfrak{A}(k) \models \varphi$, for atomic φ.

Now one may mimic the development presented for the original notion of Kripke semantics.

In particular the connection between the two notions is given by the next lemma.

Lemma 6.3.15 *Let $\mathcal{K}^{\star}$ be the modified Kripke model obtained from $\mathcal{K}$ by dividing out $\sim$. Then $k \Vdash \varphi(\vec{a}) \Leftrightarrow k \Vdash^{\star} \varphi(\vec{a}/\sim)$ for all $k \in K$.*

Proof Left to the reader.   □

Corollary 6.3.16 *Intuitionistic logic (with identity) is complete with respect to modified Kripke semantics.*

Proof Apply Theorem 6.3.10 and Lemma 6.3.15.   □

We will usually work with ordinary Kripke models, but for convenience we will often replace inclusions of structures $\mathfrak{A}(k) \subseteq \mathfrak{A}(l)$ by inclusion mappings $\mathfrak{A}(k) \hookrightarrow \mathfrak{A}(l)$.

6.4 Some Model Theory

We will give some simple applications of Kripke's semantics. The first ones concern the *disjunction* and *existence properties*.

Definition 6.4.1 A set of sentences Γ has the

  (i) *Disjunction property (DP)* if $\Gamma \vdash \varphi \vee \psi \Rightarrow \Gamma \vdash \varphi$ or $\Gamma \vdash \psi$.
 (ii) *Existence property (EP)* if $\Gamma \vdash \exists x \varphi(x) \Rightarrow \Gamma \vdash \varphi(t)$ for some closed term t (where $\varphi \vee \psi$ and $\exists x \varphi(x)$ are closed).

In a sense *DP* and *EP* reflect the constructive character of the theory Γ (in the frame of intuitionistic logic), since it makes explicit the clause "if we have a proof of $\exists x \varphi(x)$, then we have a proof of a particular instance", and similarly for disjunction.

Classical logic does not have *DP* or *EP*, for consider in propositional logic $p_0 \vee \neg p_0$. Clearly $\vdash_c p_0 \vee \neg p_0$, but neither $\vdash_c p_0$ nor $\vdash_c \neg p_0$!

Theorem 6.4.2 *Intuitionistic propositional and predicate logic without function symbols have DP.*

Proof Let $\vdash \varphi \vee \psi$, and suppose $\nvdash \varphi$ and $\nvdash \psi$, then there are Kripke models $\mathcal{K}_1$ and $\mathcal{K}_2$ with bottom nodes k_1 and k_2 such that $k_1 \nVdash \varphi$ and $k_2 \nVdash \psi$.

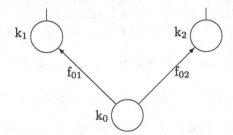

It is no restriction to suppose that the partially ordered sets K_1, K_2 of $\mathcal{K}_1$ and $\mathcal{K}_2$ are disjoint.

We define a new Kripke model with $K = K_1 \cup K_2 \cup \{k_0\}$ where $k_0 \notin K_1 \cup K_2$ (see the picture for the ordering).

$$\text{We define } \mathfrak{A}(k) = \begin{cases} \mathfrak{A}_1(k) & \text{for } k \in K_1, \\ \mathfrak{A}_2(k) & \text{for } k \in K_2, \\ |\mathfrak{A}| & \text{for } k = k_0, \end{cases}$$

where $|\mathfrak{A}|$ consists of all the constants of L, if there are any, otherwise $|\mathfrak{A}|$ contains only one element a. The inclusion mapping for $\mathfrak{A}(k_0) \hookrightarrow \mathfrak{A}(k_i)$ ($i = 1, 2$) is defined by $c \mapsto c^{\mathfrak{A}(k_i)}$ if there are constants; if not, we pick $a_i \in \mathfrak{A}(k_i)$ arbitrarily and define $f_{01}(a) = a_1$, $f_{02}(a) = a_2$. $\mathfrak{A}$ satisfies the definition of a Kripke model.

The models $\mathcal{K}_1$ and $\mathcal{K}_2$ are "submodels" of the new model in the sense that the forcing induced on $\mathcal{K}_i$ by that of $\mathcal{K}$ is exactly its old forcing, cf. Exercise 13. By the Completeness Theorem $k_0 \vdash \varphi \vee \psi$, so $k_0 \Vdash \varphi$ or $k_0 \Vdash \psi$. If $k_0 \Vdash \varphi$, then $k_1 \Vdash \varphi$. Contradiction. If $k_0 \vdash \psi$, then $k_2 \vdash \psi$. Contradiction. So $\nvdash \varphi$ and $\nvdash \psi$ are not true, hence $\vdash \varphi$ or $\vdash \psi$. $\square$

Observe that this proof can be considerably simplified for propositional logic. All we have to do is place an extra node under k_1 and k_2 in which no atom is forced (cf. Exercise 19).

Theorem 6.4.3 *Let the language of intuitionistic predicate logic contain at least one constant and no function symbols; then EP holds.*

Proof Let $\vdash \exists x \varphi(x)$ and $\nvdash \varphi(c)$ for all constants c. Then for each c there is a Kripke model $\mathcal{K}_c$ with bottom node k_c such that $k_c \nVdash \varphi(c)$. Now mimic the argument of Theorem 6.4.2 above, by taking the disjoint union of the $\mathcal{K}_c$'s and adding a bottom node k_0. Use the fact that $k_0 \Vdash \exists x \varphi(x)$. $\square$

The reader will have observed that we reason about our intuitionistic logic and model theory in a classical meta-theory. In particular we use the principle of the excluded third in our meta-language. This indeed detracts from the constructive nature of our considerations. For the present we will not bother to make our arguments constructive; it may suffice to remark that classical arguments can often be circumvented, cf. Chap. 7.

In constructive mathematics one often needs stronger notions than the classical ones. A paradigm is the notion of *inequality*. For example, in the case of the real numbers it does not suffice to know that a number is unequal (i.e. not equal) to 0 in order to invert it. The procedure that constructs the inverse for a given Cauchy sequence requires that there exists a number n such that the distance of the given number to zero is greater than 2^{-n}. Instead of a negative notion we need a positive one. This was introduced by Brouwer, and formalized by Heyting.

Definition 6.4.4 A binary relation # is called an *apartness relation* if

  (i) $\forall xy(x = y \leftrightarrow \neg x \# y)$,
  (ii) $\forall xy(x \# y \leftrightarrow y \# x)$,
 (iii) $\forall xyz(x \# y \to x \# z \vee y \# z)$.

Examples

1. For rational numbers the inequality is an apartness relation.
2. If the equality relation on a set is decidable (i.e. $\forall xy(x = y \vee x \neq y)$), then $\neq$ is an apartness relation (Exercise 22).
3. For real numbers the relation $|a - b| > 0$ is an apartness relation (cf. *Troelstra–van Dalen*, 2.7, 2.8).

We call the theory with axioms (i), (ii), (iii) of Definition 6.4.4 **AP**, the theory of apartness (obviously, the identity axiom $x_1 = x_2 \wedge y_1 = y_2 \wedge x_1 \# y_1 \to x_2 \# y_2$ is included).

Theorem 6.4.5 $\mathbf{AP} \vdash \forall xy(\neg\neg x = y \to x = y)$.

Proof Observe that $\neg\neg x = y \leftrightarrow \neg\neg\neg x \# y \leftrightarrow \neg x \# y \leftrightarrow x = y$. $\square$

We call an equality relation that satisfies the condition $\forall xy(\neg\neg x = y \to x = y)$ *stable*. Note that *stable* is essentially weaker than *decidable* (Exercise 23).

In the passage from intuitionistic theories to classical ones by adding the principle of the excluded third usually many notions are collapsed, e.g. $\neg\neg x = y$ and $x = y$. Or conversely, when passing from classical theories to intuitionistic ones (by deleting the principle of the excluded third) there is a choice of the right notions. Usually (but not always) the strongest notions fare best. An example is the notion of *linear order*.

The theory of linear order, **LO**, has the following axioms:

  (i) $\forall xyz(x < y \wedge y < z \to x < z)$,
  (ii) $\forall xyz(x < y \to z < y \vee x < z)$,
 (iii) $\forall xy(x = y \leftrightarrow \neg x < y \wedge \neg y < x)$.

One might wonder why we did not choose the axiom $\forall xyz(x < y \vee x = y \vee y < x)$ instead of (ii), it certainly would be stronger! There is a simple reason: the axiom is *too* strong, it does not hold, e.g. for the reals.

We will next investigate the relation between linear order and apartness.

Theorem 6.4.6 *The relation $x < y \vee y < x$ is an apartness relation.*

Proof An exercise in logic. $\square$

Conversely, Smoryński has shown how to introduce an order relation in a Kripke model of **AP**: let $\mathcal{K} \Vdash \mathbf{AP}$, then in each $D(k)$ the following is an equivalence relation: $k \nVdash a \# b$.

(a) $k \Vdash a = a \leftrightarrow \neg a \# a$, since $k \Vdash a = a$ we get $k \Vdash \neg a \# a$ and hence $k \nVdash a \# a$.

(b) $k \Vdash a \# b \leftrightarrow b \# a$, so obviously $k \nVdash a \# b \Leftrightarrow k \nVdash b \# a$.

(c) Let $k \nVdash a \# b$, $k \nVdash b \# c$ and suppose $k \Vdash a \# c$, then by axiom (iii) $k \Vdash a \# b$ or $k \Vdash c \# b$, which contradicts the assumptions. So $k \nVdash a \# c$.

Observe that this equivalence relation contains the one induced by the identity; $k \Vdash a = b \Rightarrow k \nVdash a \# b$. The domains $D(k)$ are thus split up in equivalence classes, which can be linearly ordered in the classical sense. Since we want to end up with a Kripke model, we have to be a bit careful. Observe that equivalence classes may be split by passing to a higher node, e.g. if $k < l$ and $k \nVdash a \# b$ then $l \Vdash a \# b$ is very well possible, but $l \nVdash a \# b \Rightarrow k \nVdash a \# b$. We take an arbitrary ordering of the equivalence classes of the bottom node (using the axiom of choice in our meta-theory if necessary). Next we indicate how to order the equivalence classes in an immediate successor l of k.

The "new" elements of $D(l)$ are indicated by the shaded part.

(i) Consider an equivalence class $[a_0]_k$ in $D(k)$, and look at the corresponding set $\hat{a}_0 := \bigcup \{[a]_l \mid a \in [a_0]_k\}$.
This set splits in a number of classes; we order those linearly. Denote the equivalence classes of $\hat{a}_0$ by $a_0 b$ (where b is a representative). Now the classes belonging to the b's are ordered, and we order all the classes on $\bigcup \{\hat{a}_0 \mid a_0 \in D(k)\}$ lexicographically according to the representation $a_0 b$.

(ii) Finally we consider the new equivalence classes, i.e. of those that are not equivalent to any b in $\bigcup \{\hat{a}_0 \mid a_0 \in D(k)\}$. We order those classes and put them in that order behind the classes of case (i).

Under this procedure we order all equivalence classes in all nodes.

We now define a relation R_k for each k: $R_k(a, b) := [a]_k < [b]_k$, where $<$ is the ordering defined above. By our definition $k < l$ and $R_k(a, b) \Rightarrow R_l(a, b)$.

We leave it to the reader to show that I_4 is valid, i.e. in particular $k \Vdash \forall xyz(x = z \wedge x < y \rightarrow z < y)$, where $<$ is interpreted by R_k.

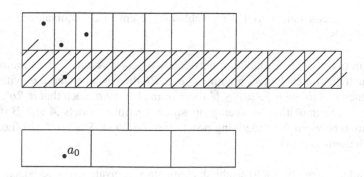

Observe that in this model the following holds:

$$(\#) \quad \forall xy(x \# y \leftrightarrow x < y \vee y < x),$$

for in all nodes $k, k \Vdash a \# b \leftrightarrow k \Vdash a < b$ or $k \Vdash b < a$.

Now we must check the axioms of linear order.

(i) *Transitivity.* $k_0 \Vdash \forall xyz(x < y \wedge y < z \rightarrow x < z) \Leftrightarrow$ for all $k \geq k_0$, for all $a, b, c \in D(k) k \Vdash a < b \wedge b < c \rightarrow a < c \Leftrightarrow$ for all $k \geq k_0$, for all $a, b, c \in D(k)$ and for all $l \geq k \ l \Vdash a < b$ and $l \Vdash b < c \Rightarrow l \Vdash a < c$.

So we have to show $R_l(a, b)$ and $R_l(b, c) \Rightarrow R_l(a, c)$, but that is indeed the case by the linear ordering of the equivalence classes.

(ii) *(Weak) linearity.* We must show $k_0 \Vdash \forall xyz(x < y \rightarrow z < y \vee x < z)$. Since in our model $\forall xy(x \# y \leftrightarrow x < y \vee y < x)$ holds the problem is reduced to pure logic. Show: $\mathbf{AP} + \forall xyz(x < y \wedge y < z \rightarrow x < z) + \forall xy(x \# y \leftrightarrow x < y \vee y < x) \vdash \forall xyz(x < y \rightarrow z < y \vee x < z)$.

We leave the proof to the reader.

(iii) *Anti-symmetry.* We must show $k_0 \Vdash \forall xy(x = y \leftrightarrow \neg x < y \wedge \neg y < x)$. As before the problem is reduced to logic. Show:
$$\mathbf{AP} + \forall xy(x \# y \leftrightarrow x < y \vee y < x) \vdash \forall xy(x = y \leftrightarrow \neg x < y \wedge \neg y < x).$$

Now we have finished the job—we have put a linear order on a model with an apartness relation. We can now draw some conclusions.

Theorem 6.4.7 $\mathbf{AP} + \mathbf{LO} + (\#)$ *is conservative over* $\mathbf{LO}$.

Proof Immediate, by Theorem 6.4.6.   □

Theorem 6.4.8 (van Dalen–Statman) $\mathbf{AP} + \mathbf{LO} + (\#)$ *is conservative over* $\mathbf{AP}$.

Proof Suppose $\mathbf{AP} \nvdash \varphi$, then by the Model Existence Lemma there is a tree model $\mathcal{K}$ of $\mathbf{AP}$ such that the bottom node k_0 does not force φ.

We now carry out the construction of a linear order on K, the resulting model $\mathcal{K}^*$ is a model of $\mathbf{AP} + \mathbf{LO} + (\#)$, and, since φ does not contain $<, k_0 \nVdash \varphi$. Hence $\mathbf{AP} + \mathbf{LO} + (\#) \nvdash \varphi$. This shows the conservative extension result:
$$\mathbf{AP} + \mathbf{LO} + (\#) \vdash \varphi \Rightarrow \mathbf{AP} \vdash \varphi, \quad \text{for } \varphi \text{ in the language of } \mathbf{AP}.$$
□

There is a convenient tool for establishing elementary equivalence between Kripke models.

Definition 6.4.9 (i) A bisimulation between two posets A and B is a relation $R \subseteq A \times B$ such that for each a, a', b with $a \leq a', aRb$ there is an element b' with $a'Rb'$ and for each a, b, b' with $aRb, b \leq b'$ there is an element a' such that $a'Rb'$.

(ii) R is a bisimulation between propositional Kripke models $\mathcal{A}$ and $\mathcal{B}$ if it is a bisimulation between the underlying posets and if $aRb \Rightarrow \Sigma(a) = \Sigma(b)$ (i.e. a and b force the same atoms).

Bisimulations are useful to establish elementary equivalence node-wise.

Lemma 6.4.10 *Let R be a bisimulation between $\mathcal{A}$ and $\mathcal{B}$, then for all $a, b, \varphi, aRb \Rightarrow (a \Vdash \varphi \leftrightarrow b \Vdash \varphi)$.*

Proof Induction on φ. For atoms and conjunctions and disjunctions the result is obvious.

Consider $\varphi = \varphi_1 \to \varphi_2$.

Let aRb and $a \Vdash \varphi_1 \to \varphi_2$. Suppose $b \not\Vdash \varphi_1 \to \varphi_2$, then for some $b' \geq b$ $b' \Vdash \varphi_1$ and $b' \not\Vdash \varphi_2$. By definition, there is an $a' \geq a$ such that $a'Rb'$. By the induction hypothesis $a' \Vdash \varphi_1$ and $a' \not\Vdash \varphi_2$. Contradiction.

The converse is completely similar.   □

Corollary 6.4.11 *If R is a total bisimulation between $\mathcal{A}$ and $\mathcal{B}$, i.e. $dom\, R = A$, $ran\, R = B$, then $\mathcal{A}$ and $\mathcal{B}$ are elementarily equivalent ($\mathcal{A} \Vdash \varphi \Leftrightarrow \mathcal{B} \Vdash \varphi$).*

We end this chapter by giving some examples of models with unexpected properties.

1.

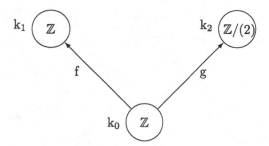

f is the identity and g is the canonical ring homomorphism $\mathbb{Z} \to \mathbb{Z}/(2)$.

$\mathcal{K}$ is a model of the ring axioms (p. 82).

Note that $k_0 \Vdash 3 \neq 0$, $k_0 \Vdash 2 = 0$, $k_0 \not\Vdash 2 \neq 0$ and $k_0 \not\Vdash \forall x(x \neq 0 \to \exists y(xy = 1))$, but also $k_0 \not\Vdash \exists x(x \neq 0 \land \forall y(xy \neq 1))$. We see that $\mathcal{K}$ is a commutative ring in which not all non-zero elements are invertible, but in which it is impossible to exhibit a non-invertible, non-zero element.

2.

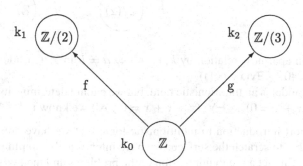

Again f and g are the canonical homomorphisms. $\mathcal{K}$ is an intuitionistic, commutative ring, as one easily verifies.

$\mathcal{K}$ has no zero divisors: $k_0 \Vdash \neg\exists xy(x \neq 0 \wedge y \neq 0 \wedge xy = 0) \Leftrightarrow$ for all i $k_i \nVdash \exists xy(x \neq 0 \wedge y \neq 0 \wedge xy = 0)$. (1)

For $i = 1, 2$ this is obvious, so let us consider $i = 0$. $k_0 \Vdash \exists xy(x \neq 0 \wedge y \neq 0 \wedge xy = 0) \Leftrightarrow k_0 \Vdash m \neq 0 \wedge n \neq 0 \wedge mn = 0$ for some m, n. So $m \neq 0, n \neq 0, mn = 0$. Contradiction. This proves (1).

The cardinality of the model is rather undetermined. We know $k_0 \Vdash \exists xy(x \neq y)$—take 0 and 1, and $k_0 \Vdash \neg\exists x_1 x_2 x_3 x_4 \bigwedge_{1 \le i < j \le 4} x_i \neq x_j$. But note that $k_0 \nVdash \exists x_1 x_2 x_3 \bigwedge_{1 \le i < j \le 3} x_i \neq x_j, k_0 \nVdash \forall x_1 x_2 x_3 x_4 \bigvee_{1 \le i < j \le 4} x_i = x_j$ and $k_0 \nVdash \neg\exists x_1 x_2 x_3 \bigwedge_{1 \le i < j \le 3} x_i \neq x_j$.

Observe that the equality relation in $\mathcal{K}$ is not stable: $k_0 \Vdash \neg\neg 0 = 6$, but $k_0 \nVdash 0 = 6$.

3.

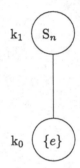

k_1 (S_n)

k_0 ($\{e\}$)

S_n is the (classical) symmetric group on n elements. Choose $n \ge 3$. k_0 forces the group axioms (p. 80). $k_0 \Vdash \neg\forall xy(xy = yx)$, but $k_0 \nVdash \exists xy(xy \neq yx)$, and $k_0 \nVdash \forall xy(xy = yx)$. So this group is not commutative, but one cannot indicate non-commuting elements.

4.

k_1 ($\mathbb{Z}/(2)$) k_2 ($\mathbb{Z}/(3)$)

Define an apartness relation by $k_1 \Vdash a\#b \Leftrightarrow a \neq b$ in $\mathbb{Z}/(2)$, idem for k_2. Then $\mathcal{K} \Vdash \forall x(x\#0 \to \exists y(xy = 1))$.

This model is an intuitionistic field, but we cannot determine its characteristic. $k_1 \Vdash \forall x(x + x = 0), k_2 \Vdash \forall x(x + x + x = 0)$. All we know is $\mathcal{K} \Vdash \forall x(6 \cdot x = 0)$.

In the short introduction to intuitionistic logic that we have presented we have only been able to scratch the surface. We have intentionally simplified the issues so that a reader can get a rough impression of the problems and methods without going into the finer foundational details. In particular we have treated intuitionistic logic in a classical meta-mathematics, e.g. we have freely applied proof by contradiction (cf. Theorem 6.3.10). Obviously this does not do justice to constructive mathematics as an alternative mathematics in its own right. For this and related issues the reader is referred to the literature. A more constructive approach is presented in the next chapter.

Exercises

1. (Informal mathematics). Let $\varphi(n)$ be a decidable property of natural numbers such that neither $\exists n\varphi(n)$ nor $\forall n\neg\varphi(n)$ has been established (e.g. "n is the largest number such that n and $n+2$ are prime"). Define a real number a by the Cauchy sequence:

$$a_n := \begin{cases} \sum_{i=1}^{n} 2^{-i} & \text{if } \forall k < n\neg\varphi(k) \\ \sum_{i=1}^{k} 2^{-i} & \text{if } k < n \text{ and } \varphi(k) \text{ and } \neg\varphi(i) \text{ for } i < k. \end{cases}$$

   Show that (a_n) is a Cauchy sequence and that "$\neg\neg a$ is rational", but there is no evidence for "a is rational".

2. Prove
$$\vdash \neg\neg(\varphi \to \psi) \to (\varphi \to \neg\neg\psi), \quad \vdash \neg\neg(\varphi \vee \neg\varphi),$$
$$\vdash \neg(\varphi \wedge \neg\varphi), \quad \vdash \neg\neg(\neg\neg\varphi \to \varphi),$$
$$\neg\neg\varphi, \neg\neg(\varphi \to \psi) \vdash \neg\neg\psi, \quad \vdash \neg\neg(\varphi \to \psi) \leftrightarrow \neg(\varphi \wedge \neg\psi),$$
$$\vdash \neg(\varphi \vee \psi) \leftrightarrow \neg(\neg\varphi \to \psi).$$

3. (a) $\varphi \vee \neg\varphi, \psi \vee \neg\psi \vdash (\varphi \square \psi) \vee \neg(\varphi \square \psi)$, where $\square \in \{\wedge, \vee, \to\}$.
   (b) Let the proposition φ have atoms $p_0, \ldots, p_n$, show $\bigwedge (p_i \vee \neg p_i) \vdash \varphi \vee \neg\varphi$.

4. Define the double negation translation $\varphi^{\neg\neg}$ of φ by placing $\neg\neg$ in front of each subformula. Show $\vdash_i \varphi^{\circ} \leftrightarrow \varphi^{\neg\neg}$ and $\vdash_c \varphi \Leftrightarrow \vdash_i \varphi^{\neg\neg}$.

5. Show that for propositional logic $\vdash_i \neg\varphi \Leftrightarrow \vdash_c \neg\varphi$.

6. Intuitionistic arithmetic **HA** (Heyting's arithmetic) is the first-order intuitionistic theory with the axioms of p. 82 as mathematical axioms. Show $\mathbf{HA} \vdash \forall xy(x = y \vee x \neq y)$ (use the principle of induction). Show that the Gödel translation works for arithmetic, i.e. $\mathbf{PA} \vdash \varphi \Leftrightarrow \mathbf{HA} \vdash \varphi^{\circ}$ (where **PA** is Peano (classical) arithmetic). Note that we need not doubly negate the atoms.

7. Show that **PA** is conservative over **HA** with respect to formulas not containing $\vee$ and $\exists$.

8. Show that $\mathbf{HA} \vdash \varphi \vee \psi \leftrightarrow \exists x((x = 0 \to \varphi) \wedge (x \neq 0 \to \psi))$.

9. (a) Show $\nvdash (\varphi \to \psi) \vee (\psi \to \varphi)$; $\nvdash (\neg\neg\varphi \to \varphi) \to (\varphi \vee \neg\varphi)$;
   $\nvdash \neg\varphi \vee \neg\neg\varphi$; $\nvdash (\neg\varphi \to \psi \vee \sigma) \to [(\neg\varphi \to \psi) \vee (\neg\varphi \to \sigma)]$;
   $\nvdash \bigvee_{1 \leq i < j \leq n}(\varphi_i \leftrightarrow \varphi_j)$, for all $n > 2$.
   (b) Use the Completeness Theorem to establish the following theorems:
   (i) $\varphi \to (\psi \to \varphi)$
   (ii) $(\varphi \vee \varphi) \to \varphi$
   (iii) $\forall xy\varphi(x, y) \to \forall yx\varphi(x, y)$
   (iv) $\exists x\forall y\varphi(x, y) \to \forall y\exists x\varphi(x, y)$.
   (c) Show $k \Vdash \forall xy\varphi(x, y) \Leftrightarrow \forall l \geq k\forall a, b \in D(l) \, l \Vdash \varphi(\bar{a}, \bar{b})$.
   $k \nVdash \varphi \to \psi \Leftrightarrow \exists l \geq k(l \Vdash \varphi \text{ and } l \nVdash \psi)$.

10. Give the simplified definition of a Kripke model for (the language of) propositional logic by considering the special case of Definition 6.3.1 with $\Sigma(k)$ consisting of propositional atoms only, and $D(k) = \{0\}$ for all k.

11. Give an alternative definition of a Kripke model based on the "structure map" $k \mapsto \mathfrak{A}(k)$ and show the equivalence with Definition 6.3.1 (without propositional atoms).

12. Prove the Soundness Theorem using Lemma 6.3.5.

13. A subset K' of a partially ordered set K is closed (under $\leq$) if $k \in K', k \leq l \Rightarrow l \in K'$. If K' is a closed subset of the underlying partially ordered set K of a Kripke model $\mathcal{K}$, then K' determines a Kripke model $\mathcal{K}'$ over K' with $D'(k) = D(k)$ and $k \Vdash' \varphi \Leftrightarrow k \Vdash \varphi$ for $k \in K'$ and φ atomic. Show $k \Vdash' \varphi \Leftrightarrow k \Vdash \varphi$ for all φ with parameters in $D(k)$, for $k \in K'$ (i.e. it is the future that matters, not the past).

14. Give a modified proof of the Model Existence Lemma by taking as nodes of the partially ordered set prime theories that extend Γ and that have a language with constants in some set $C^0 \cup C^1 \cup \cdots \cup C^{k-1}$ (cf. the proof of Lemma 6.3.9). Note that the resulting partially ordered set need not (and, as a matter of fact, is not) a tree, so we lose something. However compare Exercise 16.

15. Consider a propositional Kripke model $\mathcal{K}$, where the Σ function assigns only subsets of a finite set Γ of the propositions, which is closed under subformulas. We may consider the sets of propositions forced at a node instead of the node: define $[k] = \{\varphi \in \Gamma | k \Vdash \varphi\}$. The set $\{[k] | k \in K\}$ is partially ordered by inclusion. Define $\Sigma_\Gamma([k]) := \Sigma(k) \cap At$, show that the conditions of a Kripke model are satisfied; call this model $\mathcal{K}_\Gamma$, and denote the forcing by $\Vdash_\Gamma$. We say that $\mathcal{K}_\Gamma$ is obtained by *filtration* from $\mathcal{K}$.
    (a) Show $[k] \Vdash_\Gamma \varphi \Leftrightarrow k \Vdash \varphi$, for $\varphi \in \Gamma$.
    (b) Show that $\mathcal{K}_\Gamma$ has an underlying finite partially ordered set.
    (c) Show that $\vdash \varphi \Leftrightarrow \varphi$ holds in all finite Kripke models.
    (d) Show that intuitionistic propositional logic is *decidable* (i.e. there is a decision method for $\vdash \varphi$), apply Lemma 4.3.17.

16. Each Kripke model with bottom node k_0 can be turned into a model over a tree as follows: K_{tr} consists of all finite increasing sequences $\langle k_0, k_1, \ldots, k_n \rangle, k_i < k_{i+1} (0 \leq i < n)$, and $\mathfrak{A}_{tr}(\langle k_0, \ldots, k_n \rangle) := \mathfrak{A}(k_n)$. Show $\langle k_0, \ldots, k_n \rangle, \Vdash_{tr} \varphi \Leftrightarrow k_n \Vdash \varphi$, where $\Vdash_{tr}$ is the forcing relation in the tree model.

17. (a) Show that $(\varphi \to \psi) \vee (\psi \to \varphi)$ holds in all linearly ordered Kripke models for propositional logic.
    (b) Show that $\mathbf{LC} \nvdash \sigma \Rightarrow$ there is a linear Kripke model of $\mathbf{LC}$ in which σ fails, where $\mathbf{LC}$ is the propositional theory axiomatized by the schema $(\varphi \to \psi) \vee (\psi \to \varphi)$. (Hint: apply Exercise 15.) Hence $\mathbf{LC}$ is complete for linear Kripke models (Dummett).

18. Consider a Kripke model $\mathcal{K}$ for decidable equality (i.e. $\forall xy(x = y \vee x \neq y)$). For each k the relation $k \Vdash \overline{a} = \overline{b}$ is an equivalence relation. Define a new model $\mathcal{K}'$ with the same partially ordered set as $\mathcal{K}$, and $D'(k) = \{[a]_k | a \in D(k)\}$, where $[a]$ is the equivalence class of a. Replace the inclusion of $D(k)$ in $D(l)$, for $k < l$, by the corresponding canonical embedding $[a]_k \mapsto [a]_l$. Define for atomic φ $k \Vdash' \varphi := k \Vdash \varphi$ and show $k \Vdash' \varphi \Leftrightarrow k \Vdash \varphi$ for all φ.

19. Prove *DP* for propositional logic directly by simplifying the proof of Theorem 6.4.2.

20. Show that **HA** has *DP* and *EP*, the latter in the form: **HA** $\vdash \exists x \varphi(x) \Rightarrow$ **HA** $\vdash$ $\varphi(\bar{n})$ for some $n \in N$. (Hint: show that the model constructed in Theorems 6.4.2 and 6.4.3 is a model of **HA**.)

21. Consider predicate logic in a language without function symbols and constants. Show $\vdash \exists x \varphi(x) \Rightarrow \vdash \forall x \varphi(x)$, where $FV(\varphi) \subseteq \{x\}$. (Hint: add an auxiliary constant c, apply Theorem 6.4.3, and replace it by a suitable variable.)

22. Show $\forall xy(x = y \lor x \neq y) \vdash \bigwedge$ **AP**, where **AP** consists of the three axioms of the apartness relation, with $x \# y$ replaced by $\neq$.

23. Show $\forall xy(\neg\neg x = y \to x = y) \nvdash \forall xy(x = y \lor x \neq y)$.

24. Show that $k \Vdash \varphi \lor \neg\varphi$ for maximal nodes k of a Kripke model, so $\Sigma(k) = Th(\mathfrak{A}(k))$ (in the classical sense). That is, "the logic in maximal node is classical."

25. Give an alternative proof of Glivenko's theorem using Exercises 15 and 24.

26. Consider a Kripke model with two nodes $k_0, k_1; k_0 < k_1$ and $\mathfrak{A}(k_0) = \mathbb{R}$, $\mathfrak{A}(k_1) = \mathbb{C}$. Show $k_0 \nVdash \neg\forall x(x^2 + 1 \neq 0) \to \exists x(x^2 + 1 = 0)$.

27. Let $\mathbb{D} = \mathbb{R}[X]/X^2$ be the ring of dual numbers. $\mathbb{D}$ has a unique maximal ideal, generated by X. Consider a Kripke model with two nodes $k_0, k_1; k_0 < k_1$ and $\mathfrak{A}(k_0) = \mathbb{D}$, $\mathfrak{A}(k_1) = \mathbb{R}$, with $f : \mathbb{D} \to \mathbb{R}$ the canonical map $f(a + bX) = a$. Show that the model is an intuitionistic field, define the apartness relation.

28. Show that $\forall x(\varphi \lor \psi(x)) \to (\varphi \lor \forall x \psi(x))(x \notin FV(\varphi))$ holds in all Kripke models with constant domain function (i.e. $\forall kl(D(k) = D(l))$).

29. This exercise will establish the undefinability of propositional connectives in terms of other connectives. To be precise the connective $\square_1$ is not definable in (or "by") $\square_2, \ldots, \square_n$ if there is no formula φ, containing only the connectives $\square_2, \ldots, \square_n$ and the atoms p_0, p_1, such that $\vdash p_0 \square_1 p_1 \leftrightarrow \varphi$.

   (i) $\lor$ is not definable in $\to, \land, \bot$. Hint: suppose φ defines $\lor$, apply the Gödel translation.

   (ii) $\land$ is not definable in $\to, \lor, \bot$. Consider the Kripke model with three nodes k_1, k_2, k_3 and $k_1 < k_3, k_2 < k_3, k_1 \Vdash p, k_2 \Vdash q, k_3 \Vdash p, q$. Show that all $\land$-free formulas are either equivalent to $\bot$ or are forced in k_1 or k_2.

   (iii) $\to$ is not definable in $\land, \lor, \neg, \bot$. Consider the Kripke model with three nodes k_1, k_2, k_3 and $k_1 < k_3, k_2 < k_3, k_1 \Vdash p, k_3 \Vdash p, q$. Show for all $\to$-free formulas $k_2 \Vdash \varphi \Rightarrow k_1 \Vdash \varphi$.

30. In this exercise we now consider only propositions with a single atom p. Define a sequence of formulas by $\varphi_0 := \bot, \varphi_1 := p, \varphi_2 := \neg p, \varphi_{2n+3} := \varphi_{2n+1} \lor \varphi_{2n+2}, \varphi_{2n+4} := \varphi_{2n+2} \to \varphi_{2n+1}$ and an extra formula $\varphi_\infty := \top$. There is a specific set of implications among the φ_i, indicated in the diagram on the left.

   (i) Show that the following implications hold:
   $\vdash \varphi_{2n+1} \to \varphi_{2n+3}, \vdash \varphi_{2n+1} \to \varphi_{2n+4}, \vdash \varphi_{2n+2} \to \varphi_{2n+3},$
   $\vdash \varphi_0 \to \varphi_n, \vdash \varphi_n \to \varphi_\infty.$

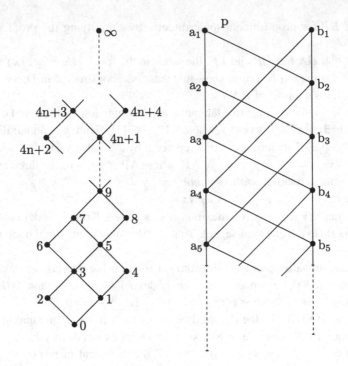

(ii) Show that the following "identities" hold:

$\vdash (\varphi_{2n+1} \to \varphi_{2n+2}) \leftrightarrow \varphi_{2n+2}, \ \vdash (\varphi_{2n+2} \to \varphi_{2n+4}) \leftrightarrow \varphi_{2n+4},$
$\vdash (\varphi_{2n+3} \to \varphi_{2n+1}) \leftrightarrow \varphi_{2n+4}, \ \vdash (\varphi_{2n+4} \to \varphi_{2n+1}) \leftrightarrow \varphi_{2n+6},$
$\vdash (\varphi_{2n+5} \to \varphi_{2n+1}) \leftrightarrow \varphi_{2n+1}, \ \vdash (\varphi_{2n+6} \to \varphi_{2n+1}) \leftrightarrow \varphi_{2n+4},$
$\vdash (\varphi_k \to \varphi_{2n+1}) \leftrightarrow \varphi_{2n+1}$ for $k \geq 2n + 7,$
$\vdash (\varphi_k \to \varphi_{2n+2}) \leftrightarrow \varphi_{2n+2}$ for $k \geq 2n + 3.$

Determine identities for the implications not covered above.

(iii) Determine all possible identities for conjunctions and disjunctions of φ_i's (look at the diagram).

(iv) Show that each formula in p is equivalent to some φ_i.

(v) In order to show that there are no other implications than those indicated in the diagram (and the compositions of course) it suffices to show that no φ_n is derivable. Why?

(vi) Consider the Kripke model indicated in the diagram on the right. $a_1 \Vdash p$ and no other node forces p. Show: $\forall a_n \exists \varphi_i \forall k (k \Vdash \varphi_i \Leftrightarrow k \geq a_n)$, $\forall b_n \exists \varphi_j \forall k (k \Vdash \varphi_j \Leftrightarrow k \geq b_n)$.

Clearly the $\varphi_i(\varphi_j)$ is uniquely determined, call it $\varphi(a_n)$, resp. $\varphi(b_n)$. Show $\varphi(a_1) = \varphi_1$, $\varphi(b_1) = \varphi_2$, $\varphi(a_2) = \varphi_4$, $\varphi(b_2) = \varphi_6$, $\varphi(a_{n+2}) = [(\varphi(a_{n+1}) \vee \varphi(b_n)) \to (\varphi(a_n) \vee \varphi(b_n))] \to (\varphi(a_{n+1}) \vee \varphi(b_n))$, $\varphi(b_{n+2}) = [(\varphi(a_{n+1}) \vee \varphi(b_{n+1})) \to (\varphi(a_{n+1}) \vee \varphi(b_n))] \to (\varphi(a_{n+1}) \vee \varphi(b_{n+1})).$

(vii) Show that the diagram on the left contains all provable implications.

Remark The diagram of the implications is called the *Rieger-Nishimura lattice* (it actually is the free Heyting algebra with one generator).

31. Consider intuitionistic predicate logic without function symbols. Prove the following extension of the existence property: $\vdash \exists y \varphi(x_1, \ldots, x_n, y) \Leftrightarrow \vdash \varphi(x_1, \ldots, x_n, t)$, where t is a constant or one of the variables $x_1, \ldots, x_n$. (Hint: replace $x_1, \ldots, x_n$ by new constants $a_1, \ldots, a_n$.)

32. Let $Q_1 x_1 \ldots Q_n x_n \varphi(\vec{x}, \vec{y})$ be a prenex formula (without function symbols), then we can find a suitable substitution instance φ' of φ obtained by replacing the existentially quantified variables by certain universally quantified variables or by constants, such that $\vdash Q_1 x_1 \ldots Q_n x_n \varphi(\vec{x}, \vec{y}) \Leftrightarrow \vdash \varphi'$ (use Exercise 31).

33. Show that $\vdash \varphi$ is decidable for prenex φ (use Lemma 4.3.17 and Exercise 32).
    Remark. Combined with the fact that intuitionistic predicate logic is undecidable, this shows that not every formula is equivalent to one in prenex normal form.

34. Consider a language with identity and function symbols, and interpret an n-ary symbol F by a function $F_k : D(k)^n \rightarrow D(k)$ for each k in a given Kripke model $\mathcal{K}$. We require *monotonicity*: $k \leq l \Rightarrow F_k \subseteq F_l$, and *preservation of equality*: $\vec{a} \sim_k \vec{b} \Rightarrow F_k(\vec{a}) \sim_k F_k(\vec{b})$, where $a \sim_k b \Leftrightarrow k \Vdash \vec{a} = \vec{b}$.
    (i) Show $\mathcal{K} \Vdash \forall \vec{x} \exists! y (F(\vec{x}) = y)$.
    (ii) Show $\mathcal{K} \Vdash I_4$.
    (iii) Let $\mathcal{K} \Vdash \forall \vec{x} \exists! y \varphi(\vec{x}, y)$, show that we can define for each k an F_k satisfying the above requirements such that $\mathcal{K} \Vdash \forall \vec{x} \varphi(\vec{x}, F(\vec{x}))$.
    (iv) Show that one can conservatively add definable Skolem functions.
    Note that we have shown how to introduce functions in Kripke models, when they are given by "functional" relations. So, strictly speaking, Kripke models with just relations are good enough.

35. Γ is a prime theory in L. $L(c)$ is obtained by adding a constant c. $\Gamma(c) = \{\varphi | \Gamma \vdash \varphi \in L(c)\}$. Show that $\Gamma(c)$ is prime in $L(c)$.

Chapter 7
Normalization

7.1 Cuts

Anyone with a reasonable experience in making natural deduction derivations observes that one somehow gets fairly efficient derivations. The worst that can happen is that a number of steps end up with what was already derived or given, but then one can obviously shorten the derivation. Here is an example:

$$\dfrac{\dfrac{[\sigma \wedge \varphi]^2}{\varphi}\, \wedge E \qquad [\varphi \to \psi]^1}{\psi} \to E \qquad \dfrac{\dfrac{\dfrac{[\sigma \wedge \varphi]^2}{\sigma}\, \wedge E}{\psi \to \sigma}\to I}{} $$

$$\dfrac{\dfrac{\dfrac{\dfrac{[\sigma\wedge\varphi]^2}{\varphi}\wedge E \quad [\varphi\to\psi]^1}{\psi}\to E \quad \dfrac{\dfrac{[\sigma\wedge\varphi]^2}{\sigma}\wedge E}{\psi\to\sigma}\to I}{\sigma}\to E}{\dfrac{\dfrac{(\varphi\to\psi)\to\sigma}{}\to I_1}{(\sigma\wedge\varphi)\to((\varphi\to\psi)\to\sigma)}\to I_2}$$

σ occurs twice; the first time it is a premise for a $\to I$, and the second time the result of a $\to E$. We can shorten the derivation as follows:

$$\dfrac{\dfrac{\dfrac{[\sigma \wedge \varphi]^1}{\sigma}\, \wedge E}{(\varphi \to \psi) \to \sigma}\to I}{(\sigma \wedge \varphi) \to ((\varphi \to \psi) \to \sigma)}\to I_1$$

It is apparently not a good idea to introduce something and to eliminate it right away. This indeed is the key idea for simplifying derivations: avoid eliminations after introductions. If a derivation contains an introduction followed by an elimination, then one can, as a rule, easily shorten the derivation, the question is, can one get rid of *all* those unfortunate steps? The answer is yes, but the proof is not trivial.

The topic of this chapter belongs to proof theory; the system of natural deduction was introduced by Gentzen, who also showed that "detours" in derivations can be

D. van Dalen, *Logic and Structure*, Universitext, DOI 10.1007/978-1-4471-4558-5_7,
© Springer-Verlag London 2013

eliminated. The subject was revived again by Prawitz, who considerable extended Gentzen's techniques and results.

We will introduce a number of notions in order to facilitate the treatment.

Definition 7.1.1 The formulas directly above the line in a derivation rule are called the *premises*, the formula directly below the line, the *conclusion*. In elimination rules a premise not containing the connective is called a *minor premise*. All other premises are called the *major premises*.

Convention The major premises will from now on appear on the left-hand side.

Definition 7.1.2 A formula occurrence γ is a *cut* in a derivation when it is the conclusion of an introduction rule and the major premise of an elimination rule. γ is called the *cut formula* of the cut.

In the above example $\psi \to \sigma$ is a cut formula.

We will adopt a slightly modified $\forall I$-rule, this will help to streamline the system.

$$\forall I \qquad \frac{\varphi}{\forall x\ \varphi[x/y]}\ \forall I$$

where y does not occur free in a hypothesis of the derivation of φ, and x is free for y in φ.

The old version of $\forall I$ is clearly a special case of the new rule. We will use the familiar notation, e.g.

$$\forall I \qquad \frac{\varphi(y)}{\forall x\ \varphi(x)}\ \forall I$$

Note that with the new rule we get a shorter derivation for

$$\begin{array}{c}\mathcal{D}\\[4pt]\dfrac{\varphi(x)}{\forall x\varphi(x)}\ \forall I\\[6pt]\dfrac{}{\varphi(y)}\ \forall E\\[6pt]\hline \forall y\varphi(y)\end{array}\ \forall I \qquad \text{namely} \qquad \begin{array}{c}\mathcal{D}\\[4pt]\dfrac{\varphi(x)}{\forall y\varphi(y)}\ \forall I\end{array}$$

The adoption of the new rule is not necessary, but rather convenient.

We will first look at predicate calculus with $\wedge, \to, \bot, \forall$.

Derivations will systematically be converted into simpler ones by "elimination of cuts"; here is an example:

$$\begin{array}{c}\mathcal{D}\\[4pt]\dfrac{\sigma}{\psi \to \sigma}\ \to I \qquad \begin{array}{c}\mathcal{D}'\\ \psi\end{array}\\[6pt]\hline \sigma\end{array}\ \to E \qquad \text{converts to} \qquad \begin{array}{c}\mathcal{D}\\ \sigma\end{array}$$

In general, when the tree under consideration is a subtree of a larger derivation the whole subtree ending with σ is replaced by the second one. The rest of the derivation remains unaltered. This is one of the features of natural deduction derivations: for a formula σ in the derivation only the part above σ is relevant to σ. Therefore we will only indicate conversions as far as required, but the reader will do well to keep in mind that we make the replacement inside a given bigger derivation.

We list the possible conversions:

$$
\begin{array}{c}
\mathcal{D}_1 \quad\ \mathcal{D}_2 \\
\dfrac{\varphi_1 \qquad \varphi_2}{\varphi_1 \wedge \varphi_2}\ \wedge I \\
\dfrac{}{\varphi_i}\ \wedge E
\end{array}
\qquad \text{is converted to} \qquad
\begin{array}{c}
\mathcal{D}_i \\
\varphi_i
\end{array}
$$

$$
\begin{array}{c}
\qquad\ [\psi] \\
\qquad\ \mathcal{D}_2 \\
\mathcal{D}_1 \qquad \varphi \\
\dfrac{\qquad\quad}{}\ \\
\psi \quad \dfrac{}{\psi \to \varphi}\ \to I \\
\dfrac{\qquad\qquad\qquad}{\varphi}\ \to E
\end{array}
\qquad \text{is converted to} \qquad
\begin{array}{c}
\mathcal{D}_1 \\
\psi \\
\mathcal{D}_2 \\
\varphi
\end{array}
$$

$$
\begin{array}{c}
\mathcal{D} \\
\dfrac{\varphi}{\forall x \varphi[x/y]}\ \forall I \\
\dfrac{}{\varphi[t/y]}\ \forall E
\end{array}
\qquad \text{is converted to} \qquad
\begin{array}{c}
\mathcal{D}[t/y] \\
\varphi[t/y]
\end{array}
$$

It is not immediately clear that this conversion is a legitimate operation on derivations. For example, consider the following derivation, where we assume that $z \in FV(\varphi(u, z))$ and $v \notin FV(\varphi(u, z))$. The elimination of the lower cut yields the conversion.

$$
\begin{array}{c}
\qquad\qquad \dfrac{\forall u \varphi(u, z)}{\varphi(v, z)} \\
\mathcal{D} \\
\dfrac{\varphi(z, z)}{\forall x \varphi(x, x)} = \dfrac{\forall v \varphi(v, z)}{\varphi(z, z)} \\
\dfrac{}{\varphi(v, v)} \qquad \dfrac{\forall x \varphi(x, x)}{\varphi(v, v)}\ \forall I \\
\qquad\qquad\qquad\qquad \forall E
\end{array}
\quad \text{to} \quad
\begin{array}{c}
\forall u \varphi(u, v) \\
\dfrac{\varphi(v, v)}{\forall v \varphi(v, v)}\ \forall I = \mathcal{D}[v/z] \\
\dfrac{}{\varphi(v, v)}
\end{array}
$$

The inadvertent substitution of v for z in $\mathcal{D}$ is questionable because v is not free for z in the third line of the right-hand derivation and we see that in the resulting derivation $\forall I$ violates the condition on the variable involved in the $\forall I$ application.

Proper variable   In order to avoid confusion of the above kind, we have to look a bit closer at the way we handle our variables in derivations. There is, of course, the obvious distinction between free and bound variables, but even the free variables do not all play the same role. Some of them are "the variable" involved in a $\forall I$. We call these occurrences *proper variables* and we extend the name to all occurrences that are "related" to them. The notion "related" is the transitive closure of the relation that two occurrences of the same variable have if one occurs in a conclusion and the other in a premise of a rule in "related" formula occurrences. It is simplest to define "related" as the reflexive, symmetric, transitive closure of the "direct relative" relation which is given by checking all derivation rules, e.g. in $\dfrac{\varphi(x) \wedge \psi(x, y)}{\psi(x,y)} \wedge E$ the top occurrence and bottom occurrence of $\psi(x, y)$ are directly related, and so are the corresponding occurrences of x and y. This applies similarly to the φ at the top and the one at the bottom in

$$[\varphi]$$
$$\mathcal{D}$$
$$\frac{\psi}{\varphi \to \psi} \to I$$

The details are left to the reader.

Dangerous clashes of variables can always be avoided, it just requires a routine renaming of variables. Since these syntactic matters present notorious pitfalls, we will exercise some care. Recall that we have shown earlier that bound variables may be renamed while retaining logical equivalence. We will also use this expedient trick in derivations.

Lemma 7.1.3 *In a derivation the bound variables can be renamed so that no variable occurs both free and bound.*

Proof By induction on $\mathcal{D}$. Actually it is better to do some "induction loading", in particular to prove that the bound variables can be chosen outside a given set of variables (including the free variables under consideration). The proof is simple, and hence left to the reader.   □

Note that the formulation of the lemma is rather cryptic. We mean of course that the resulting configuration is again a derivation.

It also expedient to rename some of the free variables in a derivation; in particular we want to keep the proper and the non-proper free variables separated.

Lemma 7.1.4 *In a derivation the free variables may be renamed, so that unrelated proper variables are distinct and each one is used exactly once in its inference rule. Moreover, no variable occurs as a proper and a non-proper variable.*

Proof Induction on $\mathcal{D}$. Always choose a fresh variable for a proper variable. Note that the renaming of the proper variables does not influence the hypotheses and the conclusion.   □

In practice it may be necessary to keep renaming variables in order to satisfy the results of the preceding lemmas.

From now on we assume that our derivations satisfy the above condition, i.e.

(i) bound and free variables are distinct,
(ii) proper and non-proper variables are distinct and each proper variable is used in precisely one $\forall I$.

Lemma 7.1.5 *The conversions for $\rightarrow, \wedge, \forall$ yield derivations.*

Proof The only difficult case is the $\forall$-conversion. But according to our variables condition $\mathcal{D}[t/u]$ is a derivation when $\mathcal{D}$ is one, for the variables in t do not act as proper variables in $\mathcal{D}$.   □

Exercise 7.1.6 *Carry out suitable renamings in the derivation on the previous page so that the conversion again yields a derivation.*

Remark There is an alternative practice for formulating the rules of logic, which is handy indeed for proof theoretical purposes: make a typographical distinction between bound and free variables (a distinction in the alphabet). Free variables are called *parameters* in that notation. We have seen that the same effect can be obtained by the syntactical transformations described above. It is then necessary, of course, to formulate the $\forall$-introduction in the liberal form!

7.2 Normalization for Classical Logic

Definition 7.2.1 A string of conversions is called a *reduction sequence*. A derivation $\mathcal{D}$ is called an *irreducible derivation* if there is no $\mathcal{D}'$ such that $\mathcal{D} >_1 \mathcal{D}'$.

Notation $\mathcal{D} >_1 \mathcal{D}'$ stands for "$\mathcal{D}$ is converted to $\mathcal{D}'$". $\mathcal{D} > \mathcal{D}'$ stands for "there is a finite sequence of conversions $\mathcal{D} = \mathcal{D}_0 >_1 \mathcal{D}_1 >_1 \cdots >_1 \mathcal{D}_{n-1} = \mathcal{D}$ and $\mathcal{D} \geq \mathcal{D}'$ stands for $\mathcal{D} > \mathcal{D}'$ or $\mathcal{D} = \mathcal{D}'$. ($\mathcal{D}$ *reduces to* $\mathcal{D}'$.)

The basic question is of course "does every sequence of conversions terminate in finitely many steps?", or equivalently, "is $>$ well-founded?" The answer turns out to be yes, but we will first look at a simpler question: "does every derivation reduce to an irreducible derivation?"

Definition 7.2.2 If there is no $\mathcal{D}_1'$ such that $\mathcal{D}_1 >_1 \mathcal{D}_1'$ (i.e. if $\mathcal{D}_1$ does not contain cuts), then we call $\mathcal{D}_1$ a *normal derivation*, or we say that $\mathcal{D}_1$ is *in normal form*, and if $\mathcal{D} \geq \mathcal{D}'$ where $\mathcal{D}'$ is normal, then we say that $\mathcal{D}$ *normalizes to* $\mathcal{D}'$.

We say that $>$ has the *strong normalization property* if $>$ is well-founded, i.e. there are no infinite reduction sequences, and we say that it has the *weak normalization property* if every derivation normalizes.

Popularly speaking, strong normalization tells you that no matter how you choose your conversions, you will ultimately find a normal form; weak normalization tells you that if you choose your conversions in a particular way, you will find a normal form.

Before discussing the normalization proofs, we remark that the $\bot$-rule can be restricted to instances where the conclusion is atomic. This is achieved by lowering the rank of the conclusion step by step.

Example

$$
\begin{array}{c}
\mathcal{D} \\
\bot \\
\hline
\varphi \wedge \psi
\end{array}
\quad \text{is replaced by} \quad
\begin{array}{c}
\dfrac{\dfrac{\mathcal{D}}{\bot}}{\varphi} \quad \dfrac{\dfrac{\mathcal{D}}{\bot}}{\psi} \\
\hline
\varphi \wedge \psi
\end{array} \wedge I
$$

$$
\begin{array}{c}
\mathcal{D} \\
\bot \\
\hline
\varphi \rightarrow \psi
\end{array}
\quad \text{is replaced by} \quad
\begin{array}{c}
\dfrac{\dfrac{\mathcal{D}}{\bot}}{\psi} \\
\hline
\varphi \rightarrow \psi
\end{array} \rightarrow I
\qquad \text{etc.}
$$

(Note that in the right-hand derivation some hypothesis may be canceled, this is, however, not necessary; if we want to get a derivation from the same hypotheses, then it is wiser not to cancel the φ at that particular $\forall I$.) A similar fact holds for *RAA*: it suffices to apply *RAA* to atomic instances. The proof is again a matter of reducing the complexity of the relevant formula.

$$
\begin{array}{c}
[\neg(\varphi \wedge \psi)] \\
\mathcal{D} \\
\bot \\
\hline
\varphi \wedge \psi
\end{array}
\quad \text{is replaced by} \quad
\begin{array}{cc}
\dfrac{\dfrac{[\neg\varphi] \quad \dfrac{[\varphi \wedge \psi]}{\varphi}}{\bot}}{\dfrac{\neg(\varphi \wedge \psi)}{\dfrac{\mathcal{D}}{\dfrac{\bot}{\varphi}\, RAA}}}
&
\dfrac{\dfrac{[\neg\psi] \quad \dfrac{[\varphi \wedge \psi]}{\psi}}{\bot}}{\dfrac{\neg(\varphi \wedge \psi)}{\dfrac{\mathcal{D}}{\dfrac{\bot}{\psi}\, RAA}}}
\end{array}
$$
$$
\begin{array}{c}
\hline
\varphi \wedge \psi
\end{array} \wedge I
$$

$$
\begin{array}{c}
\dfrac{[\varphi] \quad [\varphi \to \psi]}{\psi} \qquad [\neg\psi] \\[2mm]
\hline
\bot \\[1mm]
\hline
\neg(\varphi \to \psi) \\[1mm]
\mathcal{D} \\[1mm]
\dfrac{\bot}{\psi} \; RAA \\[1mm]
\hline
\varphi \to \psi
\end{array}
$$

$$
\begin{array}{c}
[\neg(\varphi \to \psi)] \\
\mathcal{D} \\
\dfrac{\bot}{\varphi \to \psi}
\end{array}
\qquad \text{is replaced by} \qquad
$$

$$
\begin{array}{c}
\dfrac{[\neg\varphi(x)] \quad [\forall x\, \varphi(x)] \;\; \varphi(x)}{\bot} \\[2mm]
\hline
\neg\forall x\, \varphi(x) \\[1mm]
\mathcal{D} \\[1mm]
\dfrac{\bot}{\varphi(x)} \; RAA \\[1mm]
\hline
\forall x\, \varphi(x)
\end{array}
$$

$$
\begin{array}{c}
[\neg\forall x\, \varphi(x)] \\
\mathcal{D} \\
\dfrac{\bot}{\forall x\, \varphi(x)}
\end{array}
\qquad \text{is replaced by} \qquad
$$

Some definitions are in order now.

Definition 7.2.3

(i) A maximal cut formula is a cut formula with maximal rank.

(ii) $d = \max\{r(\varphi)|\varphi \text{ cut formula in } \mathcal{D}\}$ (observe that max $\emptyset = 0$).
    n = number of maximal cut formulas and $cr(\mathcal{D}) = (d, n)$, the *cut rank* of $\mathcal{D}$.

If $\mathcal{D}$ has no cuts, put $cr(\mathcal{D}) = (0, 0)$. We will systematically lower the cut rank of a derivation until all cuts have been eliminated. The ordering on cut ranks is lexicographic:

$$(d, n) < (d', n') := d < d' \lor (d = d' \land n < n').$$

Fact 7.2.4 $<$ *is a well-ordering (actually $\omega \cdot \omega$) and hence has no infinite descending sequences.*

Lemma 7.2.5 *Let $\mathcal{D}$ be a derivation with a cut at the bottom, let this cut have rank n while all other cuts have rank $< n$, then the conversion of $\mathcal{D}$ at this lowest cut yields a derivation with only cuts of rank $< n$.*

Proof Consider all the possible cuts at the bottom and check the ranks of the cuts after the conversion.

(i) $\rightarrow$-cut

$$
\begin{array}{ccc}
[\varphi] & & \mathcal{D}_2 \\
\mathcal{D}_1 & & \varphi \\
\dfrac{\psi}{\varphi \rightarrow \psi} \quad \varphi & \mathcal{D}_2 = \mathcal{D}. \quad \text{Then } \mathcal{D} >_1 \mathcal{D}' = & \begin{array}{c}\mathcal{D}_1\\ \\ \psi\end{array} \\
\psi
\end{array}
$$

Observe that nothing in $\mathcal{D}_1$ and $\mathcal{D}_2$ was changed in the process of conversion, so all the cuts in $\mathcal{D}'$ have rank $< n$.

(ii) $\forall$-cut

$$
\begin{array}{c}
\mathcal{D} \\
\dfrac{\varphi(x)}{\forall y \, \varphi(y)} = \mathcal{D}. \quad \text{Then } \mathcal{D} >_1 \mathcal{D}' = \left(\begin{array}{c}\mathcal{D}\\ \varphi\end{array}\right)[t/x] \\
\varphi(t)
\end{array}
$$

The substitution of a term does not affect the cut rank of a derivation, so in $\mathcal{D}'$ all cuts have rank $< n$.

(iii) $\wedge$-cut. Similar.   □

Observe that in the $\wedge, \rightarrow, \bot, \forall$-language the reductions are fairly simple, i.e. parts of derivations are replaced by parts of the old derivation (forgetting for a moment about the terms)—things get smaller!

Lemma 7.2.6 *If* $cr(\mathcal{D}) > (0,0)$, *then there is a* $\mathcal{D}'$ *with* $\mathcal{D} >_1 \mathcal{D}'$ *and* $cr(\mathcal{D}') < cr(\mathcal{D})$.

Proof Select a maximal cut formula in $\mathcal{D}$ such that all cuts above it have lower rank. Apply the appropriate reduction to this maximal cut, then the part of the derivation $\mathcal{D}$ ending in the conclusion σ of the cut is replaced, by Lemma 7.2.5, by a (sub)derivation in which all cut formulas have lower rank. If the maximal cut formula was the only one, then d is lowered by 1, otherwise n is lowered by 1 and d remains unchanged. In both cases $cr(\mathcal{D})$ gets smaller. Note that in the first case n may become much larger, but that does not matter in the lexicographic order.

Observe that the elimination of a cut (here!) is a local affair, i.e. it only affects the part of the derivation tree above the conclusion of the cut.

Theorem 7.2.7 (Weak Normalization) *All derivations normalize.*

Proof By Lemma 7.2.6 the cut rank can be lowered to $(0,0)$ in a finite number of steps, hence the last derivation in the reduction sequence has no more cuts.   □

Normal derivations have a number of convenient properties, which can be read off from their structure. In order to formulate these properties and the structure, we introduce some more terminology.

Definition 7.2.8 (i) A *path* in a derivation is a sequence of formulas $\varphi_0, \dots, \varphi_n$, such that φ_0 is a hypothesis, φ_n is the conclusion and φ_i is a premise immediately above $\varphi_{i+1} (0 \leq i \leq n - 1)$. (ii) A *track* is an initial part of a path which stops at the first minor premise or at the conclusion. In other words, a track can only pass through introduction rules and through major premises of elimination rules.

Example

$$\frac{\dfrac{[\varphi \rightarrow (\psi \rightarrow \sigma)] \quad \dfrac{[\varphi \wedge \psi]}{\varphi}}{\psi \rightarrow \sigma} \quad \dfrac{[\varphi \wedge \psi]}{\psi}}{\dfrac{\dfrac{\sigma}{\varphi \wedge \psi \rightarrow \sigma}}{\varphi \rightarrow (\psi \rightarrow \sigma) \rightarrow (\varphi \wedge \psi \rightarrow \sigma)}}$$

The underlying tree is labeled with numbers:

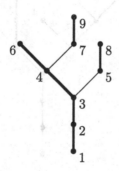

and the tracks are $(6, 4, 3, 2, 1)$, $(9, 7)$ and $(8, 5)$.

Fact 7.2.9 *In a normal derivation no introduction rule (application) can precede an elimination rule (application) in a track.*

Proof Suppose an introduction rule precedes an elimination rule in a track, then there is a last introduction rule that precedes the first elimination rule. Because the derivation is normal, one cannot immediately precede the other. So there has to be a rule in between, which must be the $\bot$-rule or the *RAA*, but that clearly is impossible, since $\bot$ cannot be the conclusion of an introduction rule. □

Fact 7.2.10 *A track in a normal derivation is divided into (at most) three parts: an elimination part, followed by a ⊥-part, followed by an introduction part. Each of the parts may be empty.*

Proof By Fact 7.2.9 we know that if the first rule is an elimination, then all eliminations come first. Look at the last elimination, it results (1) in the conclusion of $\mathcal{D}$, or (2) in ⊥, in which case the ⊥-rule or *RAA* may be applied, or (3) it is followed by an introduction. In the last case only introductions can follow. If we applied the ⊥- or *RAA*-rule, then an atom appears, which can only be the premise of an introduction rule (or the conclusion of $\mathcal{D}$).   □

Fact 7.2.11 *Let $\mathcal{D}$ be a normal derivation. Then $\mathcal{D}$ has at least one maximal track, ending in the conclusion.*

The underlying tree of a normal derivation looks like the following diagram:

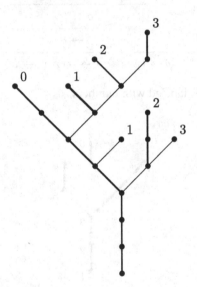

The picture suggests that the tracks are classified as to "how far" they are from the maximal track. We formalize this in the notion of *order*.

Definition 7.2.12 Let $\mathcal{D}$ be a normal derivation.

$o(t_m) = 0$ for a maximal track t_m.

$o(t) = o(t') + 1$ if the end formula of track t is a minor premise

belonging to a major premise in t'.

The orders of the various tracks are indicated in the picture.

Theorem 7.2.13 (Subformula Property)   *Let $\mathcal{D}$ be a normal derivation of $\Gamma \vdash \varphi$, then each formula (occurrence) ψ of $\mathcal{D}$ is a subformula of φ or of a formula in Γ, unless ψ is canceled by an application of RAA or when it is the $\bot$ immediately following such a canceled hypothesis.*

Proof Consider a formula ψ in $\mathcal{D}$; if it occurs in the elimination part of its track t, then it evidently is a subformula of the hypothesis at the top of t. If not, then it is a subformula of the end formula ψ_1 of t. Hence ψ_1 is a subformula of a formula ψ_2 of a track t_1 with $o(t_1) < o(t)$. Repeating the argument we find that ψ is a subformula of a hypothesis or of the conclusion.

So far we have considered all hypotheses, but we can do better. If φ is a subformula of a canceled hypothesis, it must be a subformula of the resulting implicational formula in case of an $\rightarrow I$ application, or of the resulting formula in case of an RAA-application, or (and these are the only exceptions) it is itself canceled by an RAA-application or it is a $\bot$ immediately following such a hypothesis.   □

One can draw some immediate corollaries from our results so far.

Corollary 7.2.14   *Predicate logic is consistent.*

Proof Suppose $\vdash \bot$, then there is a normal derivation ending in $\bot$ with all hypotheses canceled. There is a track through the conclusion; in this track there are no introduction rules, so the top (hypothesis) is not canceled. Contradiction.   □

Note that Corollary 7.2.14 does not come as a surprise. We already knew that predicate logic is consistent on the basis of the Soundness Theorem. The nice point of the above proof is that it uses only syntactical arguments.

Corollary 7.2.15   *Predicate logic is conservative over propositional logic.*

Proof Let $\mathcal{D}$ be a normal derivation of $\Gamma \vdash \varphi$, where Γ and φ contain no quantifiers; then by the subformula property $\mathcal{D}$ contains only quantifier-free formulas, hence $\mathcal{D}$ is a derivation in propositional logic.   □

7.3 Normalization for Intuitionistic Logic

When we consider the full language, including $\vee$ and $\exists$, some of the notions introduced above have to be reconsidered. We briefly mention them:

- In the $\exists E$
$$\frac{\exists x\, \varphi(x) \qquad \begin{array}{c} \varphi(u) \\ \mathcal{D} \\ \sigma \end{array}}{\sigma} \qquad u \text{ is called the proper variable.}$$

- The lemmas on bound variables, proper variables and free variables remain correct.
- Cuts and cut formulas are more complicated, they will be dealt with below.

As before we assume that our derivations satisfy the conditions on free and bound variables and on proper variables.

Intuitionistic logic adds certain complications to the technique developed above. We can still define all conversions:

$$
\vee\text{-conversion} \qquad
\begin{array}{c}
\mathcal{D} \\[2pt]
\varphi_i \\ \hline
\varphi_1 \vee \varphi_2
\end{array} \vee I
\quad
\begin{array}{cc}
[\varphi_1] & [\varphi_2] \\
\mathcal{D}_1 & \mathcal{D}_2 \\
\sigma & \sigma
\end{array}
\;\; \vee E
\qquad \text{converts to} \qquad
\begin{array}{c}
\mathcal{D}_i \\
\varphi_i \\
\mathcal{D}_1 \\
\sigma
\end{array}
$$

$$
\text{(conclusion } \sigma \text{)}
$$

$$
\exists\text{-conversion} \qquad
\begin{array}{c}
\mathcal{D} \\
\varphi(t) \\ \hline
\exists x\, \varphi(x)
\end{array}
\quad
\begin{array}{c}
\varphi(y) \\
\mathcal{D}' \\
\sigma
\end{array}
\qquad \text{converts to} \qquad
\begin{array}{c}
\mathcal{D} \\
\varphi(t) \\
\mathcal{D}'[t/y] \\
\sigma
\end{array}
$$

Lemma 7.3.1 *For any derivation* $\begin{smallmatrix}\varphi(y)\\ \mathcal{D}'\\ \sigma\end{smallmatrix}$ *with* y *not free in* σ *and* t *free for* y *in* $\varphi(y)$, $\begin{smallmatrix}\varphi(t)\\ \mathcal{D}'[t/y]\\ \sigma\end{smallmatrix}$ *is also a derivation.*

Proof Induction on $\mathcal{D}'$.   □

It becomes somewhat harder to define tracks; recall that tracks were introduced in order to formalize something like "essential successor". In $\dfrac{\varphi \to \psi \quad \varphi}{\psi}$ we did not consider φ to be an "essential successor" of φ (the minor premise) since ψ has nothing to do with φ.

In $\vee E$ and $\exists E$ the canceled hypotheses have something to do with the major premise, so we deviate from the geometric idea of going down in the tree and we make a track that ends in $\varphi \vee \psi$ both continuing through (the canceled) φ and ψ; similarly a track that gets to $\exists x \varphi(x)$ continues through (the canceled) $\varphi(y)$.

The old clauses are still observed, except that tracks are not allowed to start at hypotheses, canceled by $\vee E$ or $\exists E$. Moreover, a track (naturally) ends in a major premise of $\vee E$ or $\exists E$ if no hypotheses are canceled in these rule applications.

Example

$$\frac{[\exists x(\varphi(x) \vee \psi(x))] \quad \dfrac{[\varphi(y) \vee \psi(y)] \quad \dfrac{[\varphi(y)]}{\dfrac{\exists x\varphi(x)}{\exists x\varphi(x) \vee \exists x\psi(x)}} \quad \dfrac{[\psi(y)]}{\dfrac{\exists x\psi(x)}{\exists x\varphi(x) \vee \exists x\psi(x)}}}{\exists x\varphi(x) \vee \exists x\psi(x)} \exists E}{\dfrac{\exists x\varphi(x) \vee \exists x\psi(x)}{\exists x(\varphi(x) \vee \psi(x)) \to \exists x\varphi(x) \vee \exists x\psi(x)}}$$

In tree form:

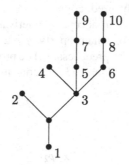

The derivation contains the following tracks:

$$(2, 4, 9, 7, 5, 3, 1), (2, 4, 10, 8, 6, 3, 1).$$

There are still more problems to be faced in the intuitionistic case.

(i) There may be superfluous applications of $\vee E$ and $\exists E$ in the sense that "nothing is canceled".

That is, in $\dfrac{\begin{array}{cc} \mathcal{D} & \mathcal{D}' \\ \exists x\varphi(x) & \sigma \end{array}}{\sigma}$ no hypotheses $\varphi(y)$ are canceled in $\mathcal{D}'$.

We add extra conversions to get rid of those elimination rule applications:

$$\frac{\begin{array}{ccc} \mathcal{D} & \mathcal{D}_1 & \mathcal{D}_2 \\ \varphi \vee \psi & \sigma & \sigma \end{array}}{\sigma} \quad \text{converts to} \quad \begin{array}{c} \mathcal{D}_i \\ \sigma \end{array}$$

if φ and ψ are not canceled in resp. $\mathcal{D}_1, \mathcal{D}_2$.

$$\frac{\begin{array}{cc} \mathcal{D} & \mathcal{D}' \\ \exists x\varphi(x) & \sigma \end{array}}{\sigma} \quad \text{converts to} \quad \begin{array}{c} \mathcal{D}' \\ \sigma \end{array}$$

if $\varphi(y)$ is not canceled in $\mathcal{D}'$.

(ii) An introduction may be followed by an elimination in a track without giving
rise to a conversion.

Example

$$\frac{\dfrac{[\varphi]\quad[\varphi]}{\varphi\wedge\varphi}\wedge I \qquad \dfrac{[\varphi]\quad[\varphi]}{\varphi\wedge\varphi}\wedge I}{\dfrac{\varphi\wedge\varphi}{\varphi}\wedge E}\vee E$$

with $\varphi\vee\varphi$ on the left below the first $\wedge I$.

In each track there is an $\wedge$-introduction and two steps later an $\wedge$-elimination, but
we are not in a position to apply a reduction.

We would still not be willing to accept this derivation as "normal", if only because nothing is left of the subformula property: $\varphi\wedge\varphi$ is neither a subformula of its
predecessor in the track, nor of its predecessor. The problem is caused by the repeti-
tions that may occur because of $\vee E$ and $\exists E$, e.g. one may get a string of occurrences
of the same formula:

$$\begin{array}{c}
\mathcal{D}_1\\
\dfrac{\exists x_1\varphi_1(x_1)\qquad\sigma}{\ }\\
\dfrac{\exists x_2\varphi_2(x_2)\qquad\qquad\sigma}{\ }\\
\dfrac{\exists x_3\varphi_3(x_3)\qquad\qquad\qquad\sigma}{\sigma}
\end{array}$$

$$\vdots$$

$$\frac{\exists x_n\varphi_n(x_n)\qquad\sigma}{\sigma}$$

Clearly the formulas that would have to interact in a reduction may be too far
apart. The solution is to change the order of the rule applications; we call this a
permutation conversion.

Our example is converted by "pulling" the $\wedge E$ upwards:

$$\frac{\varphi\vee\varphi \qquad \dfrac{\dfrac{[\varphi]\quad[\varphi]}{\varphi\wedge\varphi}\wedge I}{\varphi}\wedge E \qquad \dfrac{\dfrac{[\varphi]\quad[\varphi]}{\varphi\wedge\varphi}\wedge I}{\varphi}\wedge E}{\varphi}\vee E$$

Now we can apply the $\wedge$-conversion:

$$\frac{\varphi\vee\varphi \quad [\varphi]\quad[\varphi]}{\varphi}\vee E$$

In view of the extra complications we have to extend our notion of *cut*.

Definition 7.3.2 A string of occurrences of a formula σ in a track which starts with the result of an introduction and ends with an elimination is called a *cut segment*. A maximal cut segment is one with a cut formula of maximal rank.

We have seen that the elimination at the bottom of the cut segment can be permuted upwards:

Example

$$
\begin{array}{ccc}
\cfrac{\cfrac{\begin{array}{c}[\psi]\\ \mathcal{D}\\ \sigma\end{array}}{\cfrac{\exists x\varphi_1(x) \quad \psi \to \sigma}{\cfrac{\exists y\varphi_2(y) \qquad \psi \to \sigma}{\psi \to \sigma} \qquad \psi}}}{\sigma}
& \text{converts to} &
\cfrac{\cfrac{\exists x\varphi_1(x) \quad \cfrac{\begin{array}{c}[\psi]\\ \mathcal{D}\\ \sigma\end{array}}{\psi \to \sigma}}{\cfrac{\exists y\varphi_2(y) \qquad \cfrac{\psi \to \sigma \qquad \psi}{\sigma}}{\sigma}}}{}
\end{array}
$$

and then to

$$
\cfrac{\exists y\varphi_2(y) \qquad \cfrac{\exists x\varphi_1(x) \qquad \cfrac{\cfrac{\begin{array}{c}[\psi]\\ \mathcal{D}\\ \sigma\end{array}}{\psi \to \sigma} \quad \psi}{\sigma}}{\sigma}}{\sigma}
$$

Now we can eliminate the cut formula $\psi \to \sigma$:

$$
\cfrac{\exists y\varphi_2(y) \qquad \cfrac{\exists x\varphi_1(x) \qquad \cfrac{\begin{array}{c}\psi\\ \mathcal{D}\\ \sigma\end{array}}{\sigma}}{\sigma}}{\sigma}
$$

So a cut segment may be eliminated by applying a series of permutation conversions followed by a "connective conversion".

As in the smaller language, we can restrict our attention to applications of the $\bot$-rule for atomic instances.

We just have to consider the extra connectives:

$$\frac{\dfrac{\mathcal{D}}{\bot}}{\varphi \vee \psi} \quad \text{can be replaced by} \quad \frac{\dfrac{\dfrac{\mathcal{D}}{\bot}}{\varphi}}{\varphi \vee \psi}$$

$$\frac{\dfrac{\mathcal{D}}{\bot}}{\exists x \varphi(x)} \quad \text{can be replaced by} \quad \frac{\dfrac{\dfrac{\mathcal{D}}{\bot}}{\varphi(y)}}{\exists x \varphi(x)}$$

We will show that in intuitionistic logic derivations can be normalized. We define the cut rank as before, but now for cut segments.

Definition 7.3.3 (i) The rank of a cut segment is the rank of its formula.

(ii) $d = \max\{r(\varphi) | \varphi \text{ cut formula in } \mathcal{D}\}$, $n = $ number of maximal cut segments, $cr(\mathcal{D}) = (d, n)$ with the same lexicographical ordering.

Lemma 7.3.4 *If $\mathcal{D}$ is a derivation ending with a cut segment of maximal rank such that all cut segments distinct from this segment have a smaller rank, then a number of permutation conversions and a conversion reduce $\mathcal{D}$ to a derivation with smaller cut rank.*

Proof (i) Carry out the permutation conversions on a maximal segment, so that an elimination immediately follows an introduction. For example,

$$\cfrac{\cfrac{\cdots \quad \cfrac{\varphi \quad \psi}{\varphi \wedge \psi}}{\cfrac{\varphi \wedge \psi}{\cfrac{\cdots \quad \varphi \wedge \psi}{\varphi}}} }{\varphi} \quad > \quad \cfrac{\cfrac{\cdots \quad \cfrac{\varphi \quad \psi}{\cfrac{\varphi \wedge \psi}{\varphi}}}{\cfrac{\cdots \quad \varphi}{\varphi}}}{\varphi}$$

Observe that the cut rank is not raised. We now apply the "connective" conversion to the remaining cut. The result is a derivation with a lower d.   □

Lemma 7.3.5 *If $cr(\mathcal{D}) > (0, 0)$, then there is a $\mathcal{D}'$ such that $\mathcal{D} > \mathcal{D}'$ and $cr(\mathcal{D}') < cr(\mathcal{D})$.*

Proof Let s be a maximal segment such that in the subderivation $\hat{\mathcal{D}}$ ending with s no other maximal segments occur. Apply the reduction steps indicated in Lemma 7.3.4; then $\mathcal{D}$ is replaced by $\mathcal{D}'$ and either the d is not lowered, but n is lowered, or d is lowered. In both cases $cr(\mathcal{D}') < cr(\mathcal{D})$.   □

Theorem 7.3.6 (Weak Normalization) *Each intuitionistic derivation normalizes.*

Proof Apply Lemma 7.3.5. □

Observe that the derivation may grow in size during the reductions, e.g.

$$
\cfrac{
\cfrac{
\varphi \vee \varphi \quad \cfrac{[\varphi]^1 \quad [\varphi]^1}{\varphi \vee \psi \quad \varphi \vee \psi}}{\varphi \vee \psi} 1 \quad \quad \cfrac{\varphi \to \sigma \quad [\varphi]^2}{\sigma} \quad \cfrac{\psi \to \sigma \quad [\psi]^2}{\sigma}
}{\sigma} 2
$$

is reduced by a permutation conversion to

$$
\cfrac{
\varphi \vee \varphi \quad \cfrac{\cfrac{[\varphi]^1}{\varphi \vee \psi} \quad \cfrac{[\varphi]^2 \quad \varphi \to \sigma}{\sigma} \quad \cfrac{[\psi]^2 \quad \psi \to \sigma}{\sigma}}{\sigma} 2 \quad \mathcal{D}
}{\sigma} 1
$$

where

$$
\mathcal{D} = \cfrac{\cfrac{[\varphi]^1}{\varphi \vee \psi} \quad \cfrac{[\varphi]^3 \quad \varphi \to \sigma}{\sigma} \quad \cfrac{[\psi]^3 \quad \psi \to \sigma}{\sigma}}{\sigma} 3
$$

In general, parts of derivations may be duplicated.

The structure theorem for normal derivations holds for intuitionistic logic as well; note that we have to use the extended notion of *track* and that *segments* may occur.

Fact 7.3.7 (i) *In a normal derivation, no application of an introduction rule can precede an application of an elimination rule.*

(ii) *A track in a normal derivation is divided into (at most) three parts: an elimination part, followed by a* ⊥ *part, followed by an introduction part. These parts contain segments, the last formula of which is resp. the major premise of an elimination rule, the falsum rule or (an introduction rule or the conclusion).*

(iii) *In a normal derivation the conclusion is in at least one maximal track.*

Theorem 7.3.8 (Subformula Property) *In a normal derivation of $\Gamma \vdash \varphi$, each formula is a subformula of a hypothesis in Γ, or of φ.*

Proof Left to the reader. □

Definition 7.3.9 The relation "φ is a *strictly positive subformula occurrence* of ψ" is inductively defined by:

(1) φ is a strictly positive subformula occurrence of φ,
(2) ψ is a strictly positive subformula occurrence of $\varphi \wedge \psi, \psi \wedge \varphi, \varphi \vee \psi, \psi \vee \varphi,$
$\varphi \rightarrow \psi$,
(3) ψ is a strictly positive subformula occurrence of $\forall x\psi, \exists x\psi$.

Note that here we consider occurrences; as a rule this will be tacitly understood. We will also say, for short, φ is strictly positive in ψ, or φ occurs strictly positive in ψ. The extension to connectives and terms is obvious, e.g. "$\forall$ is strictly positive in ψ".

Lemma 7.3.10 (i) *The immediate successor of the major premise of an elimination rule is strictly positive in this premise (for $\rightarrow E, \wedge E, \forall E$ this actually is the conclusion).*

(ii) *A strictly positive part of a strictly positive part of φ is a strictly positive part of φ.*

Proof Immediate.   □

We now give some applications of the Normal Form Theorem.

Theorem 7.3.11 *Let $\Gamma \vdash \varphi \vee \psi$, where Γ does not contain $\vee$ in strictly positive subformulas, then $\Gamma \vdash \varphi$ or $\Gamma \vdash \psi$.*

Proof Consider a normal derivation $\mathcal{D}$ of $\varphi \vee \psi$ and a maximal track t. If the first occurrence $\varphi \vee \psi$ of its segment belongs to the elimination part of t, then $\varphi \vee \psi$ is a strictly positive part of the hypothesis in t, which has not been canceled. Contradiction.

Hence $\varphi \vee \psi$ belongs to the introduction part of t, and thus $\mathcal{D}$ contains a subderivation of φ or of ψ.

$\mathcal{D}$ looks like

The last k steps are $\exists E$ or $\vee E$. If any of them were an $\vee$-elimination then the disjunction would be in the elimination part of a track and hence a $\vee$ would occur strictly positive in some hypothesis of Γ. Contradiction.

Hence all the eliminations are $\exists E$. Replace the derivation now by:

$$
\begin{array}{c}
\mathcal{D}' \\
\dfrac{\mathcal{D}_1 \quad \varphi}{} \\
\dfrac{\mathcal{D}_2 \quad \varphi}{} \\
\varphi \\
\vdots \\
\dfrac{\mathcal{D}_k \quad \varphi}{} \\
\varphi
\end{array}
$$

In this derivation exactly the same hypotheses have been canceled, so $\Gamma \vdash \varphi$. $\square$

Consider a language without function symbols (i.e. all terms are variables or constants).

Theorem 7.3.12 *If $\Gamma \vdash \exists x \varphi(x)$, where Γ does not contain an existential formula as a strictly positive part, then $\Gamma \vdash \varphi(t_1) \vee \cdots \vee \varphi(t_n)$, where the terms $t_1, \ldots, t_n$ occur in the hypotheses or in the conclusion.*

Proof Consider an end segment of a normal derivation $\mathcal{D}$ of $\exists x \varphi(x)$ from Γ. End segments run through minor premises of $\vee E$ and $\exists E$. In this case an end segment cannot result from $\exists E$, since then some $\exists u \varphi(u)$ would occur strictly positive in Γ. Hence the segment runs through minor premises of $\vee E$'s. That is, we get:

$$
\begin{array}{c}
\quad\quad\quad [\alpha_1] \quad\quad [\beta_1] \\
\quad\quad\quad \mathcal{D}_1 \quad\quad \mathcal{D}_2 \\
\dfrac{\alpha_1 \vee \beta_1 \quad \exists x \varphi(x) \quad \exists x \varphi(x)}{\quad\quad\quad \exists x \varphi(x)} \quad\quad \vdots \\
\dfrac{\alpha_2 \vee \beta_2 \quad\quad\quad\quad\quad\quad\quad\quad\quad\quad \exists x \varphi(x)}{\exists x \varphi(x)} \\
\vdots \\
\dfrac{}{\exists x \varphi(x)}
\end{array}
$$

$\exists x \varphi(x)$ at the beginning of an end segment results from an introduction (else it would occur strictly positive in Γ), say from $\varphi(t_i)$. It could also result from a $\bot$ rule, but then we could conclude a suitable instance of $\varphi(x)$.

We now replace the parts of $\mathcal{D}$ yielding the tops of the end segments by parts yielding disjunctions:

$$
\begin{array}{ccc}
& [\alpha_1] & [\beta_1] \\
& \mathcal{D}_1 & \mathcal{D}_2 \\
& \varphi(t_1) & \varphi(t_2)
\end{array}
$$

$$
\alpha_1 \vee \beta_1 \quad \underline{\varphi(t_1) \vee \varphi(t_2) \quad \varphi(t_1) \vee \varphi(t_2)} \qquad \vdots
$$

$$
\alpha_2 \vee \beta_2 \quad \underline{\qquad \varphi(t_1) \vee \varphi(t_2) \qquad} \qquad \varphi(t_3)
$$

$$
\iddots
$$

$$
\underline{\qquad\qquad\qquad\qquad}
$$

$$
\varphi(t_1) \vee \varphi(t_2) \vee \ldots \vee \varphi(t_n)
$$

So $\Gamma \vdash \bigvee \varphi(t_i)$. Since the derivation was normal the various t_i's are subterms of Γ or $\exists x \varphi(x)$. $\qquad\square$

Corollary 7.3.13 *If in addition $\vee$ does not occur strictly positive in Γ, then $\Gamma \vdash \varphi(t)$ for a suitable t.*

Corollary 7.3.14 *If the language does not contain constants, then we get $\Gamma \vdash \forall x \varphi(x)$.*

We have obtained here constructive proofs of the disjunction and existence properties, which had already been proved by classical means in Chap. 6.

Exercises

1. Show that there is no formula φ with atoms p and q without $\vee$ so that $\vdash \varphi \leftrightarrow p \vee q$ (hence $\vee$ is not definable from the remaining connectives).
2. If φ does not contain $\to$ then $\nvdash_i \varphi$. Use this to show that $\to$ is not definable by the remaining connectives.
3. If $\wedge$ does not occur in φ and p and q are distinct atoms, then $\varphi \vdash p$ and $\varphi \vdash q$ $\Rightarrow \varphi \vdash \bot$.
4. Eliminate the cut segment $(\sigma \vee \tau)$ from

$$
\begin{array}{ccccc}
& & \mathcal{D}_3 & & \\
& \mathcal{D}_2 & \sigma & & \\
\mathcal{D}_1 & \exists x \varphi_2(x) & \sigma \vee \tau & [\sigma] & [\tau] \\
\exists y \varphi_1(y) & \underline{\qquad \sigma \vee \tau \qquad} & & \mathcal{D}_4 & \mathcal{D}_5 \\
& \sigma \vee \tau & & \rho & \rho \\
\underline{\qquad\qquad\qquad\qquad\qquad\qquad\qquad\qquad\qquad\qquad} \\
\rho
\end{array}
$$

5. Show that a prenex formula $(Q_1 x_1) \cdots (Q_n x_n)\varphi$ is derivable if and only if a suitable quantifier-free formula, obtained from φ, is derivable. This, in combination with Exercise 15 of Chap. 6, yields another proof of Exercise 33 of Chap. 6.

7.4 Additional Remarks: Strong Normalization and the Church–Rosser Property

As we have already mentioned, there is a stronger result for natural deduction: every reduction sequence terminates (i.e. $<_1$ is well-founded). For proofs see Girard (1987) and Girard et al. (1989). Indeed, one can also show for $>$ the *Church–Rosser property* (or *confluence property*): if $\mathcal{D} \geq \mathcal{D}_1$, $\mathcal{D} \geq \mathcal{D}_2$ then there is a $\mathcal{D}_3$ such that $\mathcal{D}_1 \geq \mathcal{D}_3$ and $\mathcal{D}_2 \geq \mathcal{D}_3$. As a consequence each $\mathcal{D}$ has a *unique normal form*. One easily shows, however, that a given φ may have more than one normal derivation.

Normalization is an integral part of proof theory; a full treatment is definitely beyond this exposition. In particular we have suppressed the technicalities inherent to the reduction process, the reader may consult the treatment in Prawitz (1965), Troelstra and Schwichtenberg (1996), Troelstra and van Dalen (1988), and Schwichtenberg and Wainer (2012).

Chapter 8
Gödel's Theorem

8.1 Primitive Recursive Functions

We will introduce a class of numerical functions which evidently are effectively computable. The procedure may seem rather ad hoc, but it gives us a surprisingly rich class of algorithms. We use the inductive method, that is, we fix a number of initial functions which are as effective as one can wish; after that we specify certain ways to manufacture new algorithms out of old ones.

The initial algorithms are extremely simple indeed: the successor function, the constant functions and the projection functions. It is obvious that composition (or substitution) of algorithms yields algorithms. The use of recursion was as a device to obtain new functions already known to Dedekind; that recursion produces algorithms from given algorithms is also easily seen. In logic the study of primitive recursive functions was initiated by Skolem, Herbrand, Gödel and others.

We will now proceed with a precise definition, which will be given in the form of an inductive definition. First we present a list of initial functions of an unmistakably algorithmic nature, and then we specify how to get new algorithms from old ones. All functions have their own arity, that is to say, they map $\mathbb{N}^k$ to $\mathbb{N}$ for a suitable k. We will in general not specify the arities of the functions involved, and assume that they are chosen correctly.

The so-called *initial functions* are

- the *constant functions* C_m^k with $C_m^k(n_0, \ldots, n_{k-1}) = m$,
- the *successor function* S with $S(n) = n + 1$,
- the *projection functions* P_i^k with $P_i^k(n_0, \ldots, n_{k-1}) = n_i$ $(i < k)$.

New algorithms are obtained from old ones by *substitution* or *composition* and *primitive recursion*.

- A class $\mathcal{F}$ of functions is *closed under substitution* if $g, h_0, \ldots, h_{p-1} \in \mathcal{F} \Rightarrow f \in \mathcal{F}$, where $f(\vec{n}) = g(h_0(\vec{n}), \ldots, h_{p-1}(\vec{n}))$.
- $\mathcal{F}$ is *closed under primitive recursion* if $g, h \in \mathcal{F} \Rightarrow f \in \mathcal{F}$, where

$$\begin{cases} f(0, \vec{n}) = g(\vec{n}) \\ f(m+1, \vec{n}) = h(f(m, \vec{n}), \vec{n}, m). \end{cases}$$

D. van Dalen, *Logic and Structure*, Universitext, DOI 10.1007/978-1-4471-4558-5_8,
© Springer-Verlag London 2013

Definition 8.1.1 The class of primitive recursive functions is the smallest class of functions containing the constant functions, the successor function, and the projection functions, which is closed under substitution and primitive recursion.

Remark Substitution has been defined in a particular way: the functions that are substituted all have the same string of inputs. In order to make arbitrary substitutions one has to do a little bit of extra work. Consider for example the function $f(x, y)$ in which we want to substitute $g(z)$ for x and $f(z, x)$ for y: $f(g(z), f(z, x))$. This is accomplished as follows: put $h_0(x, z) = g(z) = g(P_1^2(x, z))$ and $h(x, z) = f(z, x) = f(P_1^2(x, z), P_0^2(x, z))$. Then the required $f(g(z), f(z, x))$ is obtained as $f(h_0(x, z), h_1(x, z))$. The reader is expected to handle cases of substitution that will come up.

Let us start by building up a stock of primitive recursive functions. The technique is not difficult at all; most readers will have used it on numerous occasions. The surprising fact is that so many functions can be obtained by these simple procedures. Here is a first example:
$x + y$, defined by

$$\begin{cases} x + 0 = x \\ x + (y + 1) = (x + y) + 1. \end{cases}$$

We will reformulate this definition so that the reader can see that it indeed fits the prescribed format:

$$\begin{cases} +(0, x) = P_0^1(x) \\ +(y + 1, x) = S(P_0^3(+(y, x), P_0^2(x, y), P_1^2(x, y))). \end{cases}$$

As a rule we will we will stick to traditional notation, so we will simply write $x + y$ for $+(y, x)$. We will also tacitly use the traditional abbreviations from mathematics, e.g. we will mostly drop the multiplication dot.

There are two convenient tricks to add or delete variables. The first one is the introduction of dummy variables.

Lemma 8.1.2 (Dummy Variables) *If f is primitive recursive, then so is g with* $g(x_0, \ldots, x_{n-1}, z_0, \ldots, z_{m-1}) = f(x_0, \ldots, x_{n-1})$.

Proof Put $g(x_0, \ldots, x_{n-1}, z_0, \ldots, z_{m-1}) = f(P_0^{n+m}(\vec{x}, \vec{z}), \ldots, P_n^{n+m}(\vec{x}, \vec{z}))$. □

Lemma 8.1.3 (Identification of Variables) *If f is primitive recursive, then so is* $f(x_0, \ldots, x_{n-1})[x_i/x_j]$, *where $i, j \leq n$.*

Proof We need only consider the case $i \neq j$. $f(x_0, \ldots, x_{n-1})[x_i/x_j] = f(P_0^n(x_0, \ldots, x_{n-1}), \ldots, P_i^n(x_0, \ldots, x_{n-1}), \ldots, P_i^n(x_0, \ldots, x_{n-1}), \ldots, P_{n-1}^n(x_0, \ldots, x_{n-1}))$, where the second P_i^n is at the jth entry. □

A more pedestrian notation is $f(x_0, \ldots, x_i, \ldots, x_i, \ldots, x_{n-1})$.

Lemma 8.1.4 (Permutation of Variables) *If f is primitive recursive, then so is g
with $g(x_0, \ldots, x_{n-1}) = f(x_0, \ldots, x_{n-1})[x_i, x_j/x_j, x_i]$, where $i, j \leq n$.*

Proof Use substitution and projection functions. □

From now on we will use the traditional informal notation, e.g. $g(x) = f(x, x, x)$, or $g(x, y) = f(y, x)$. For convenience we have used and will use, when no confusion can arise, the vector notation for strings of inputs.

The reader can easily verify that the following examples can be cast in the required format of the primitive recursive functions.

1. $x + y$

$$\begin{cases} x + 0 = x \\ x + (y + 1) = (x + y) + 1 \end{cases}$$

2. $x \cdot y$

$$\begin{cases} x \cdot 0 = 0 \\ x \cdot (y + 1) = x \cdot y + x \quad \text{(we use (1))} \end{cases}$$

3. x^y

$$\begin{cases} x^0 = 1 \\ x^{y+1} = x^y \cdot x \end{cases}$$

4. *Predecessor function*

$$p(x) = \begin{cases} x - 1 & \text{if } x > 0 \\ 0 & \text{if } x = 0 \end{cases}$$

Apply recursion:

$$\begin{cases} p(0) = 0 \\ p(x + 1) = x \end{cases}$$

5. *Cut-off subtraction* (*monus*)

$$x \dot{-} y = \begin{cases} x - y & \text{if } x \geq y \\ 0 & \text{else} \end{cases}$$

Apply recursion:

$$\begin{cases} x \dot{-} 0 = x \\ x \dot{-} (y + 1) = p(x \dot{-} y) \end{cases}$$

6. *Factorial function*

$$n! = 1 \cdot 2 \cdot 3 \cdots (n - 1) \cdot n$$

7. *Signum function*

$$sg(x) = \begin{cases} 0 & \text{if } x = 0 \\ 1 & \text{otherwise} \end{cases}$$

Apply recursion:

$$\begin{cases} sg(0) = 0 \\ sg(x+1) = 1 \end{cases}$$

8. $\overline{sg}(x) = 1 \dotminus sg(x)$.

$$\overline{sg}(x) = \begin{cases} 1 & \text{if } x = 0 \\ 0 & \text{otherwise} \end{cases}$$

9. $|x - y|$.

Observe that $\quad |x - y| = (x \dotminus y) + (y \dotminus x)$

10. $f(\vec{x}, y) = \sum_{i=0}^{y} g(\vec{x}, i)$, where g is primitive recursive.

$$\begin{cases} \sum_{i=0}^{0} g(\vec{x}, i) = g(\vec{x}, 0) \\ \sum_{i=0}^{y+1} g(\vec{x}, i) = \sum_{i=0}^{y} g(\vec{x}, i) + g(\vec{x}, y+1) \end{cases}$$

11. $\prod_{i=0}^{y} g(\vec{x}, i)$, idem.
12. If f is primitive recursive and π is a permutation of the set $\{0, \dots, n-1\}$, then g with $g(\vec{x}) = f(x_{\pi 0}, \dots, x_{\pi(n-1)})$ is also primitive recursive.
13. If $f(\vec{x}, y)$ is primitive recursive, so is $f(\vec{x}, k)$.

The definition of primitive recursive functions by direct means is a worthwhile challenge, and the reader will find interesting cases among the exercises. For an efficient and quick access to a large stock of primitive recursive functions there are, however, techniques that cut a number of corners. We will present them here.

In the first place we can relate sets and functions by means of *characteristic functions*. In the setting of number theoretic functions, we define characteristic functions as follows: for $A \subseteq \mathbb{N}^k$ the characteristic function $K_A : \mathbb{N}^k \to \{0, 1\}$ of A is given by $\vec{n} \in A \Leftrightarrow K_A(\vec{n}) = 1$ (and hence $\vec{n} \notin A \Leftrightarrow K_A(\vec{n}) = 0$). Warning: in logic the characteristic function is sometimes defined with 0 and 1 interchanged. For the theory that does not make any difference. Note that a subset of $\mathbb{N}^k$ is also called a *k-ary relation*. When dealing with relations we tacitly assume that we have the correct number of arguments, e.g. when we write $A \cap B$ we suppose that A, B are subsets of the same $\mathbb{N}^k$.

Definition 8.1.5 A relation R is primitive recursive if its characteristic function is so.

Note that this corresponds to the idea of using K_R as a test for membership.
The following sets (relations) are primitive recursive:

1. $\emptyset$: $K_\emptyset(n) = 0$ for all n.
2. The set of even numbers, E:

$$\begin{cases} K_E(0) = 1 \\ K_E(x+1) = \overline{sg}(K_E(x)) \end{cases}$$

3. The equality relation: $K_=(x, y) = \overline{sg}(|x - y|)$.
4. The order relation: $K_<(x, y) = \overline{sg}((x+1) \dotminus y)$.

Lemma 8.1.6 *The primitive recursive relations are closed under* $\cup, \cap, {}^c$ *and bounded quantification.*

Proof Let $C = A \cap B$, then $x \in C \Leftrightarrow x \in A \wedge x \in B$, so $K_C(x) = 1 \Leftrightarrow K_A(x) = 1 \wedge K_B(x) = 1$. Therefore we put $K_C(x) = K_A(x) \cdot K_B(x)$. Hence the intersection of primitive recursive sets is primitive recursive. For union take $K_{A \cup B}(x) = sg(K_A(x) + K_B(x))$, and for the complement $K_{A^c}(x) = \overline{sg}(K_A(x))$.

We say that R is *obtained by bounded quantification from* S if $R(\vec{n}, m) := Qx \leq m \, S(\vec{n}, x)$, where Q is one of the quantifiers $\forall, \exists$.

Consider the bounded existential quantification: $R(\vec{x}, n) := \exists y \leq n \, S(\vec{x}, y)$, then $K_R(\vec{x}, n) = sg(\sum_{y \leq n} K_S(\vec{x}, y))$, so if S is primitive recursive, then R is so.

The $\forall$ case is similar; it is left to the reader. □

Lemma 8.1.7 *The primitive recursive relations are closed under primitive recursive substitutions, i.e. if* $f_0, \ldots, f_{n-1}$ *and* R *are primitive recursive, then so is* $S(\vec{x}) := R(f_0(\vec{x}), \ldots, f_{n-1}(\vec{x}))$.

Proof $K_S(\vec{x}) = K_R(f_1(\vec{x}), \ldots, f_{n-1}(\vec{x}))$. □

Lemma 8.1.8 (Definition by Cases) *Let* $R_1, \ldots, R_p$ *be mutually exclusive primitive recursive predicates, such that* $\forall \vec{x}(R_1(\vec{x}) \vee R_2(\vec{x}) \vee \cdots \vee R_p(\vec{x}))$ *and let* $g_1, \ldots, g_p$ *be primitive recursive functions, then* f *with*

$$f(\vec{x}) = \begin{cases} g_1(\vec{x}) & \text{if } R_1(\vec{x}) \\ g_2(\vec{x}) & \text{if } R_2(\vec{x}) \\ \quad \vdots \\ g_p(\vec{x}) & \text{if } R_p(\vec{x}) \end{cases}$$

is primitive recursive.

Proof If $K_{R_i}(\vec{x}) = 1$, then all the other characteristic functions yield 0, so we put $f(\vec{x}) = g_1(\vec{x}) \cdot K_{R_1}(\vec{x}) + \cdots + g_p(\vec{x}) \cdot K_{R_p}(\vec{x})$. □

The natural numbers are well-ordered, that is, each non-empty subset has a least element. If we can test the subset for membership, then we can always find this least element effectively. This is made precise for primitive recursive sets.

Some notation—the minimization operator μ is defined as follows: $(\mu y)R(\vec{x}, y)$ is the least number y such that $R(\vec{x}, y)$ if there is one. Bounded minimization is given by: $(\mu y < m)R(\vec{x}, y)$ is the least number $y < m$ such that $R(\vec{x}, y)$ if such a number exists; if not, we simply take it to be m.

Lemma 8.1.9 (Bounded Minimization) R *is primitive recursive* $\Rightarrow (\mu y < m) R(\vec{x}, y)$ *is primitive recursive.*

Proof Consider the following table:

R	$R(\vec{x},0)$	$R(\vec{x},1)$	...	$R(\vec{x}i),$	$R(\vec{x},i+1)$	...	$R(\vec{x},m)$
K_R	0	0	...	1	0	...	1
g	0	0	...	1	1	...	1
h	1	1	...	0	0	...	0
f	1	2	...	i	i	...	i

In the first line we write the values of $K_R(\vec{x},i)$ for $0 \le i \le m$, in the second line we make the sequence monotone, e.g. take $g(\vec{x},i) = sg \sum_{j=0}^{i} K_R(\vec{x},j)$. Next we switch 0 and 1: $h(\vec{x},i)=\overline{sg}g(\vec{x},i)$ and finally we sum the $h : f(\vec{x},i) = \sum_{j=0}^{i} h(\vec{x},j)$. If $R(\vec{x},j)$ holds for the first time in i, then $f(\vec{x},m-1)=i$, and if $R(\vec{x},j)$ does not hold for any $j < m$, then $f(\vec{x},m-1)=m$.

So $(\mu y < m)R(\vec{x},y) = f(\vec{x},m-1)$, and thus *bounded minimization* yields a primitive recursive function.   □

We put $(\mu y \le m)R(\vec{x},y) := (\mu y < m+1)R(\vec{x},y)$.

Now it is time to apply our arsenal of techniques to obtain a large variety of primitive recursive relations and functions.

Theorem 8.1.10 *The following are primitive recursive.*

1. *The set of primes*:

$$Prime(x) \Leftrightarrow x \text{ is a prime} \Leftrightarrow x \ne 1 \land \forall yz \le x(x = yz \to y = 1 \lor z = 1).$$

2. *The divisibility relation*:

$$x \mid y \Leftrightarrow \exists z \le y(x \cdot z = y)$$

3. *The exponent of the prime p in the factorization of x*:

$$(\mu y \le x)[p^y \mid x \land \neg \ p^{y+1} \mid x]$$

4. *The "nth prime" function*:

$$\begin{cases} p(0) = 2 \\ p(n+1) = (\mu x \le p(n)^{n+2})(x \text{ is prime} \land x > p(n)). \end{cases}$$

Note that we start to count the prime numbers from zero, and we use the notation $p_n = p(n)$. So $p_0 = 2$, $p_1 = 3$, $p_2 = 5,\dots$. The first prime is p_0, and the ith prime is p_{i-1}.

Proof One easily checks that the defining predicates are primitive recursive by applying the above lemmas.   □

Coding of Finite Sequences One of the interesting features of the natural number system is that it allows a fairly simple coding of pairs of numbers, triples, ..., and n-tuples in general. There are quite a number of these codings, each having its own strong points. The two best known ones are those of Cantor and of Gödel. Cantor's coding is given in Exercise 6, Gödel's coding will be used here. It is based on the well-known fact that numbers have a unique (up to order) prime factorization.

The idea is to associate to a sequence $(n_0, \ldots, n_{k-1})$ the number $2^{n_0+1} \cdot 3^{n_1+1} \cdot \ldots \cdot p_i^{n_i+1} \cdot \ldots \cdot p_{k-1}^{n_{k-1}+1}$. The extra $+1$ in the exponents is used to take into account that the coding has to show the zeros that occur in a sequence. From the prime factorization of a coded sequence we can effectively extract the original sequence. The way we have introduced these codes makes the coding unfortunately not a bijection; for example, 10 is not a coded sequence, whereas 6 is. This is not a terrible drawback; there are remedies, which we will not consider here.

Recall that, in the framework of set theory a sequence of length n is a mapping from $\{0, \ldots, n-1\}$ to $\mathbb{N}$, so we define the *empty sequence* as the unique sequence of length 0, i.e. the unique map from $\emptyset$ to $\mathbb{N}$, which is the empty function (i.e. set). We put the code of the empty sequence 1.

Definition 8.1.11

1. $Seq(n) := \forall p, q \leq n (Prime(p) \land Prime(q) \land q < p \land p \mid n \to q \mid n) \land n \neq 0$
   (*sequence number*)
   In words: n is a sequence number if it is a product of consecutive positive prime powers.
2. $lth(n) := (\mu x \leq n + 1)[\neg\, p_x \mid n]$ (*length*)
3. $(n)_i = (\mu x < n)[p_i^x \mid n \land \neg\, p_i^{x+1} \mid n] \dotminus 1$ (*decoding* or *projection*)
   In words: the exponent of the ith prime in the factorization of n, minus 1. $(n)_i$ extracts the ith element of the sequence.
4. $n * m = n \cdot \prod_{i=0}^{lth(m)-1} p_{lth(n)+i}^{(m)_i+1}$ (*concatenation*)
   In words: if m, n are codes of two sequences $\vec{m}, \vec{n}$, then the code of the concatenation of $\vec{m}$ and $\vec{n}$ is obtained by the product of n and the prime powers that one gets by "moving up" all primes in the factorization of m by the length of n.

Remark 1 is trivially a sequence number. The length function only yields the correct output for sequence numbers, e.g. $lth(10) = 1$. Furthermore the length of 1 is indeed 0, and the length of a 1-tuple is 1.

Notation We will use abbreviations for the iterated decoding functions: $(n)_{i,j} = ((n)_i)_j$, etc.

Sequence numbers are from now on written as $\langle n_0, \ldots, n_{k-1} \rangle$. So, for example, $\langle 5, 0 \rangle = 2^6 \cdot 3^1$. We write $\langle\,\rangle$ for the code of the empty sequence. The binary coding, $\langle x, y \rangle$, is usually called a *pairing function*.

So far we have used a straightforward form of recursion, each next output depends on the parameters and on the previous output. But the *Fibonacci sequence* shows us that there are more forms of recursion that occur in practice:

$$\begin{cases} F(0) = 1 \\ F(1) = 1 \\ F(n+2) = F(n) + F(n+1). \end{cases}$$

The obvious generalization is a function, where each output depends on the parameters and all the preceding outputs. This is called *course-of-value recursion.*

Definition 8.1.12 For a given function $f(y, \vec{x})$ its "course-of-value" function $\bar{f}(y, \vec{x})$ is given by

$$\begin{cases} \bar{f}(0, \vec{x}) = 1 \\ \bar{f}(y+1, \vec{x}) = \bar{f}(y \cdot \vec{x}) \cdot p_y^{f(y, \vec{x})+1}. \end{cases}$$

Example If $f(0) = 1$, $f(1) = 0$, $f(2) = 7$, then $\bar{f}(0) = 1$, $\bar{f}(1) = 2^{1+1}$, $\bar{f}(2) = 2^{1+1} \cdot 3^1$, $\bar{f}(3) = 2^2 \cdot 3 \cdot 5^8 = \langle 1, 0, 7 \rangle$.

Lemma 8.1.13 *If f is primitive recursive, then so is $\bar{f}$.*

Proof Obvious. $\square$

Since $\bar{f}(n+1)$ "codes" so to speak all information on f up to the nth value, we can use $\bar{f}$ to formulate course-of-value recursion.

Theorem 8.1.14 *If g is primitive recursive and $f(y, \vec{x}) = g(\bar{f}(y, \vec{x}), y, \vec{x})$, then f is primitive recursive.*

Proof We first define $\bar{f}$.

$$\begin{cases} \bar{f}(0, \vec{x}) = 1 \\ \bar{f}(y+1, \vec{x}) = \bar{f}(y, \vec{x}) * \langle g(\bar{f}(y, \vec{x}), y, \vec{x}) \rangle. \end{cases}$$

$\bar{f}$ is obviously primitive recursive. Since $f(y, \vec{x}) = (\bar{f}(y+1, \vec{x}))_y$ we see that f is primitive recursive. $\square$

By now we have collected enough facts for future use about the primitive recursive functions. We might ask if there are more algorithms than just the primitive recursive functions. The answer turns out to be yes. Consider the following construction: each primitive recursive function f is determined by its definition, which consists of a string of functions $f_0, f_1, \ldots, f_{n-1} = f$ such that each function is either an initial function, or obtained from earlier ones by substitution or primitive recursion.

It is a matter of routine to code the whole definition into a natural number such that all information can be effectively extracted from the code (see Hinman

1978, p. 34). The construction shows that we may define a function F such that $F(x, y) = f_x(y)$, where f_x is the primitive recursive function with code x. Now consider $D(x) = F(x, x) + 1$. Suppose that D is primitive recursive, i.e. $D = f_n$ for a certain n, but then $D(n) = F(n, n) + 1 = f_n(n) + 1 \neq f_n(n)$. Contradiction. It is clear, however, from the definition of D that it is effective, i.e. we have an (informal) algorithm to compute $D(n)$ for each n; so we have indicated how to get an effective function which is not primitive recursive. The above result can also be given the following formulation: there is no binary primitive recursive function $F(x, y)$ such that each unary primitive function is $F(n, y)$ for some n. In other words, the primitive functions cannot be primitive recursively enumerated.

The argument is in fact completely general; suppose we have a class of effective functions that can enumerate itself in the manner considered above, then we can always "diagonalize out of the class" by the D function. We call this *diagonalization*. The moral of this observation is that we have little hope of obtaining *all* effective functions in an effective way. The diagonalization technique goes back to Cantor, who introduced it to show that the reals are not denumerable. In general he used diagonalization to show that the cardinality of a set is less than the cardinality of its power set.

Exercises

1. If h_1 and h_2 are primitive recursive, then so are f and g, where

$$\begin{cases} f(0) = a_1 \\ g(0) = a_2 \\ f(x+1) = h_1(f(x), g(x), x) \\ g(x+1) = h_2(f(x), g(x), x). \end{cases}$$

2. Show that the Fibonacci series is primitive recursive, where

$$\begin{cases} f(0) = f(1) = 1 \\ f(x+2) = f(x) + f(x+1). \end{cases}$$

3. Let $[a]$ denote the integer part of the real number a (i.e. the greatest integer $\leq a$). Show that $[\frac{x}{y+1}]$, for natural numbers x and y, is primitive recursive.
4. Show that $\max(x, y)$ and $\min(x, y)$ are primitive recursive.
5. Show that the gcd (greatest common divisor) and lcm (least common multiple) are primitive recursive.
6. Cantor's pairing function is given by $P(x, y) = \frac{1}{2}((x + y)^2 + 3x + y)$. Show that P is primitive recursive, and that P is a bijection of $\mathbb{N}^2$ onto $\mathbb{N}$. (Hint: consider in the plane a walk along all lattice points as follows: $(0, 0) \to (0, 1) \to (1, 0) \to (0, 2) \to (1, 1) \to (2, 0) \to (0, 3) \to (1, 2) \to \cdots$.) Define the "inverses" L and R such that $P(L(z), R(z)) = z$ and show that they are primitive recursive.
7. Show $p_n \leq 2^{2^n}$.

8.2 Partial Recursive Functions

Given the fact that the primitive recursive functions do not exhaust the numerical
algorithms, we extend in a natural way the class of effective functions. As we have
seen that an effective generation of all algorithms invariably brings us in conflict
with diagonalization, we will widen our scope by allowing partial functions. In this
way the conflicting situation $D(n) = D(n) + 1$ for a certain n only tells us that D is
not defined for n.

In the present context functions have natural domains, i.e. sets of the form $\mathbb{N}^n$
($= \{(m_0, \ldots, m_{n-1}) \mid m_i \in \mathbb{N}\}$, called Cartesian products), a partial function has a
domain that is a subset of $\mathbb{N}^n$. If the domain is all of $\mathbb{N}^n$, then we call the function
total.

Example $f(x) = x^2$ is total, $g(x) = \mu y[y^2 = x]$ is partial and not total ($g(x)$ is the
square root of x if it exists).

The algorithms that we are going to introduce are called *partial recursive functions*; perhaps recursive, partial functions would have been a better name. However,
the name has come to be generally accepted. The particular technique for defining
partial recursive functions that we employ here goes back to Kleene. As before,
we use an inductive definition; apart from clause R7 below, we could have used
a formulation almost identical to that of the definition of the primitive recursive
functions. Since we want a built-in universal function, that is a function that effectively enumerates the functions, we have to employ a more refined technique that
allows explicit reference to the various algorithms. The trick is not esoteric at all,
we simply give each algorithm a code number, called its *index*. We fix these indices
in advance so that we can speak of the "algorithm with index e yields output y on
input $(x_0, \ldots, x_{n-1})$", symbolically represented as $\{e\}(x_0, \ldots, x_{n-1}) \simeq y$.

The heuristics of this "index applied to input" is that an index is viewed as
a description of an abstract machine that operates on inputs of a fixed arity. So
$\{e\}(\vec{n}) \simeq m$ must be read as "the machine with index e operates on $\vec{n}$ and yields
output m". It may very well be the case that the machine does not yield an output;
in that case we say that $\{e\}(\vec{n})$ diverges. If there is an output, we say that $\{e\}(\vec{n})$ converges. That the abstract machine is an algorithm will appear from the specification
in the definition below.

Note that we do not know in advance that the result is a function, i.e. that for
each input there is at most one output. However plausible that is, it has to be shown.
Kleene has introduced the symbol $\simeq$ for "equality" in contexts where terms may
be undefined. This happens to be useful in the study of algorithms that need not
necessarily produce an output. The abstract machines above may, for example, get
into a computation that runs on forever. For example, it might have an instruction
of the form "the output at n is the successor of the output at $n + 1$". It is easy to
see that for no n an output can be obtained. In this context the use of the existence
predicate would be useful, and $\simeq$ would be the $\equiv$ of the theory of partial objects
(cf. *Troelstra–van Dalen*, 2.2). The convention ruling $\simeq$ is: if $t \simeq s$ then t and s are
simultaneously defined and identical, or they are simultaneously undefined.

Definition 8.2.1 The relation $\{e\}(\vec{x}) \simeq y$ is inductively defined by

R1 $\{\langle 0, n, q \rangle\}(m_0, \dots, m_{n-1}) \simeq q$

R2 $\{\langle 1, n, i \rangle\}(m_0, \dots, m_{n-1}) \simeq m_i$, for $0 \leq i < n$

R3 $\{\langle 2, n, i \rangle\}(m_0, \dots, m_{n-1}) \simeq m_i + 1$, for $0 \leq i < n$

R4 $\{\langle 3, n+4 \rangle\}(p, q, r, s, m_0, \dots, m_{n-1}) \simeq p$ if $r = s$

 $\{\langle 3, n+4 \rangle\}(p, q, r, s, m_0, \dots, m_{n-1}) \simeq q$ if $r \neq s$

R5 $\{\langle 4, n, b, c_0, \dots, c_{k-1} \rangle\}(m_0, \dots, m_{n-1}) \simeq p$ if there are $q_0, \dots, q_{k-1}$ such that

 $\{c_i\}(m_0, \dots, m_{n-1}) \simeq q_i (0 \leq i < k)$ and $\{b\}(q_0, \dots, q_{k-1}) \simeq p$

R6 $\{\langle 5, n+2 \rangle\}(p, q, m_0, \dots, m_{n-1}) \simeq S_n^1(p, q)$

R7 $\{\langle 6, n+1 \rangle\}(b, m_0, \dots, m_{n-1}) \simeq p$ if $\{b\}(m_0, \dots, m_{n-1}) \simeq p$

The function S_n^1 in R6 will be specified in the S_n^m theorem below.

Keeping the above reading of $\{e\}(\vec{x})$ in mind, we can paraphrase the schemas as follows:

R1 the machine with index $\langle 0, n, q \rangle$ yields for input $(m_0, \dots, m_{n-1})$ output q (the *constant function*)

R2 the machine with index $\langle 1, n, i \rangle$ yields for input $\vec{m}$ output m_i (the *projection function P_i^n*)

R3 the machine with index $\langle 2, n, i \rangle$ yields for input $\vec{m}$ output $m_i + 1$ (the *successor function* on the ith argument)

R4 the machine with index $\langle 3, n+4 \rangle$ tests the equality of the third and fourth arguments of the input and yields the first argument in the case of equality, and the second argument otherwise (the *discriminator function*)

R5 the machine with index $\langle 4, n, b, c_0, \dots, c_{k-1} \rangle$ first simulates the machines with index $c_0, \dots, c_{k-1}$ with input $\vec{m}$, then uses the output sequence $(q_0, \dots, q_{k-1})$ as input and simulates the machine with index b (*substitution*)

R7 the machine with index $\langle 6, n+1 \rangle$ simulates for a given input $b, m_0, \dots, m_{n-1}$, the machine with index b and input $m_0, \dots, m_{n-1}$ (*reflection*)

Another way to view R7 is that it provides a *universal machine* for all machines with n-argument inputs, that is, it accepts as an input the indices of machines, and then simulates them. This is the kind of machine required for the diagonalization process. If one thinks of idealized abstract machines, then R7 is quite reasonable. One would expect that if indices can be "deciphered", a universal machine can be constructed. This was indeed accomplished by Alan Turing, who constructed (abstractly) a universal Turing machine.

The scrupulous might call R7 a case of cheating, since it does away with all the hard work one has to do in order to obtain a universal machine, for example in the case of Turing machines.

As $\{e\}(\vec{x}) \simeq y$ is inductively defined, everything we proved about inductively defined sets applies here. For example, if $\{e\}(\vec{x}) \simeq y$ is the case, then we know that there is a formation sequence (see p. 9) for it. This sequence specifies how $\{e\}$ is built up from simpler partial recursive functions.

Note that we could also have viewed the above definition as an inductive definition of the set of indices (of partial recursive functions).

Lemma 8.2.2 *The relation $\{e\}(\vec{x}) \simeq y$ is functional.*

Proof We have to show that $\{e\}$ behaves as a function, that is $\{e\}(\vec{x}) \simeq y$, $\{e\}(\vec{x}) \simeq z \Rightarrow y = z$. This is done by induction on the definition of $\{e\}$. We leave the proof to the reader.   □

The definition of $\{e\}(\vec{n}) \simeq m$ has a computational content, it tells us what to do. When presented with $\{e\}(\vec{n})$, we first look at e; if the first "entry" of e is 0, 1 or 2, then we compute the output via the corresponding initial function. If the first "entry" is 3, then we determine the output "by cases". If the first entry is 4, we first do the subcomputations indicated by $\{c_i\}(\vec{m})$, then we use the outputs to carry out the subcomputation for $\{b\}(\vec{n})$. And so on.

If R7 is used in such a computation, we are no longer guaranteed that it will stop; indeed, we may run into a loop, as the following simple example shows.

From R7 it follows, as we will see below, that there is an index e such that $\{e\}(x) = \{x\}(x)$. To compute $\{e\}$ for the argument e we pass, according to R7, to the right-hand side, i.e. we must compute $\{e\}(e)$, since e was introduced by R7, we must repeat the transitions to the right-hand side, etc. Evidently our procedure does not get us anywhere!

Conventions. The relation $\{e\}(\vec{x}) \simeq y$ defines a function on a domain, which is a subset of the "natural domain", i.e. a set of the form $\mathbb{N}^n$. Such functions are called *partial recursive functions*; they are traditionally denoted by symbols from the Greek alphabet, φ, ψ, σ, etc. If such a function is total on its natural domain, it is called *recursive*, and denoted by a roman symbol, f, g, h, etc. The use of the equality symbol "=" is proper in the context of total functions. However, in practice when no confusion arises, we will often use it instead of "$\simeq$". The reader should take care not to confuse formulas and partial recursive functions; it will always be clear from the context what a symbol stands for. Sets and relations will be denoted by roman capitals. When no confusion can arise, we will sometimes drop brackets, as in $\{e\}x$ for $\{e\}(x)$. Some authors use a "bullet" notation for partial recursive functions: $e \bullet \vec{x}$. We will stick to "Kleene brackets": $\{e\}(\vec{x})$.

The following terminology is traditionally used in recursion theory.

Definition 8.2.3

1. If for a partial function φ $\exists y (\varphi(\vec{x}) = y)$, then we say that φ *converges at* $\vec{x}$, otherwise φ *diverges at* $\vec{x}$.
2. If a partial function converges for all (proper) inputs, it is called *total*.
3. A total partial recursive function (sic!) will be called a *recursive function*.
4. A set (relation) is called *recursive* if its characteristic function (which, by definition, is total) is recursive.

Observe that it is an important feature of computations, as defined in Definition 8.2.1, that $\{e\}(\psi_0(\vec{n}), \psi_1(\vec{n}), \ldots, \psi_{k-1}(\vec{n}))$ diverges if one of its arguments $\psi_i(\vec{n})$ diverges. So, for example, the partial recursive function $\{e\}(x) - \{e\}(x)$ need not converge for all e and x, we first must know that $\{e\}(x)$ converges!

This feature is sometimes inconvenient and slightly paradoxical, e.g. in direct applications of the discriminator scheme R4, $\{\langle 3, 4 \rangle\}(\varphi(x), \psi(x), 0, 0)$ is undefined when the (seemingly irrelevant) function $\psi(x)$ is undefined.

With a bit of extra work, we can get an index for a partial recursive function that does *definition by cases* on partial recursive functions:

$$\{e\}(\vec{x}) = \begin{cases} \{e_1\}(\vec{x}) & \text{if } g_1(\vec{x}) = g_2(\vec{x}) \\ \{e_2\}(\vec{x}) & \text{if } g_1(\vec{x}) \neq g_2(\vec{x}) \end{cases}$$

for recursive g_1, g_2.

Define

$$\varphi(\vec{x}) = \begin{cases} e_1 & \text{if } g_1(\vec{x}) = g_2(\vec{x}) \\ e_2 & \text{if } g_1(\vec{x}) \neq g_2(\vec{x}) \end{cases}$$

by $\varphi(\vec{x}) = \{\langle 3, 4 \rangle\}(e_1, e_2, g_1(\vec{x}), g_2(\vec{x}))$. So $\{e\}(\vec{x}) = \{\psi(\vec{x})\}(\vec{x}) =$ [by R7] $\{\langle 6, n+1 \rangle\}(\psi(\vec{x}), \vec{x})$. Now use R5 (substitution) to get the required index.

Since the primitive recursive functions form such a natural class of algorithms, it will be our first goal to show that they are included in the class of recursive functions.

The following important theorem has a neat machine motivation. Consider a machine with index e operating on two arguments x and y. Keeping x fixed, we have a machine operating on y. So we get a sequence of machines, one for each x. Does the index of each such machine depend in a decent way on x? The plausible answer seems "yes". The following theorem confirms this.

Theorem 8.2.4 (The S_n^m Theorem) *For every m, n with $0 < m < n$ there exists a primitive recursive function S_n^m such that*

$$\{S_n^m(e, x_0, \ldots, x_{m-1})\}(x_m, \ldots, x_{n-1}) = \{e\}(\vec{x}).$$

Proof The first function, S_n^1, occurs in R6. We have postponed the precise definition, here it is:

$$S_n^1(e, y) = \langle 4, (e)_1 \dot{-} 1, e, \langle 0, (e)_1 \dot{-} 1, y \rangle,$$
$$\langle 1, (e)_1 \dot{-} 1, 0 \rangle, \ldots, \langle 1, (e)_1 \dot{-} 1, n \dot{-} 2 \rangle \rangle.$$

Note that the arities are correct, $\{e\}$ has one argument more than the constant function and the projection functions involved.

Now $\{S_n^1(e, y)\}(\vec{x}) = z \Leftrightarrow$ there are $q_0 \cdots q_{n-1}$ such that

$$\{\langle 0, (e)_1 \dot{-} 1, y \rangle\}(\vec{x}) = q_0$$
$$\{\langle 1, (e)_1 \dot{-} 1, 0 \rangle\}(\vec{x}) = q_1$$
$$\vdots$$
$$\{\langle 1, (e)_1 \dot{-} 1, n \dot{-} 2 \rangle\}(\vec{x}) = q_{n-1}$$
$$\{e\}(q_0, \ldots, q_{n-1}) = z.$$

By the clauses R1 and R2 we get $q_0 = y$ and $q_{i+1} = x_i$ $(0 \leq i \leq n - 1)$, so $\{S_n^1(e, y)\}(\vec{x}) = \{e\}(y, \vec{x})$. Clearly, S_n^1 is primitive recursive.

The primitive recursive function S_n^m is obtained by applying S_n^1 m times. From our definition it follows that S_n^m is also recursive.   □

The S_n^m function allows us to consider some inputs as parameters, and the rest as proper inputs. This is a routine consideration in everyday mathematics: "consider $f(x, y)$ as a function of y". The logical notation for this specification of inputs makes use of the *lambda operator*. Say $t(x, y, z)$ is a term (in some language), then $\lambda x \cdot t(x, y, z)$ is for each choice of y, z the function $x \mapsto t(x, y, z)$. We say that y and z are parameters in this function. The evaluation of these lambda terms is simple: $\lambda x \cdot t(x, y, z)(n) = t(n, y, z)$. This topic belongs to the *lambda calculus*; for us the notation is just a convenient tool to express ourselves succinctly.

The S_n^m theorem expresses a *uniformity* property of the partial recursive functions. It is obvious indeed that, say for a partial recursive function $\varphi(x, y)$, each individual $\varphi(n, y)$ is partial recursive (substitute the constant n function for x), but this does not yet show that the index of $\lambda y \cdot \varphi(x, y)$ is in a systematic, uniform way computable from the index of φ and x. By the S_n^m theorem, we know that the index of $\{e\}(x, y, z)$, considered as a function of, say, z depends primitive recursively on x and y: $\{h(x, y)\}(z) = \{e\}(x, y, z)$. We will see a number of applications of the S_n^m theorem.

Next we will prove a powerful theorem about partial recursive functions that allows us to introduce partial recursive functions by inductive definitions, or by implicit definitions. Partial recursive functions can, by this theorem, be given as solutions of certain equations.

Example

$$\varphi(n) = \begin{cases} 0 & \text{if } n \text{ is a prime, or } 0, \text{ or } 1 \\ \varphi(2n + 1) + 1 & \text{otherwise.} \end{cases}$$

Then $\varphi(0) = \varphi(1) = \varphi(2) = \varphi(3) = 0$, $\varphi(4) = \varphi(9) + 1 = \varphi(19) + 2 = 2$, $\varphi(5) = 0$, and, e.g. $\varphi(85) = 6$. Prima facie, we cannot say much about such a sequence. The following theorem of Kleene shows that we can always find a partial recursive solution to such an equation for φ.

Theorem 8.2.5 (The Recursion Theorem) *There exists a primitive recursive function rc such that for each e and $\vec{x}$ $\{rc(e)\}(\vec{x}) = \{e\}(rc(e), \vec{x})$.*

Let us note first that the theorem indeed gives the solution r of the following equation: $\{r\}(\vec{x}) = \{e\}(r, \vec{x})$. Indeed the solution depends primitive recursively on the given index e: $\{f(e)\}(x) = \{e\}(f(e), x)$. If we are not interested in the (primitive recursive) dependence of the index of the solution on the old index, we may even be content with the solution of $\{f\}(x) = \{e\}(f, x)$.

Proof Let $\varphi(m, e, \vec{x}) = \{e\}(S_{n+2}^2(m, m, e), \vec{x})$ and let p be an index of φ. Put $rc(e) = S_{n+2}^2(p, p, e)$, then

$$\{rc(e)\}(\vec{x}) = \{S_{n+2}^2(p, p, e)\}(\vec{x}) = \{p\}(p, e, \vec{x}) = \varphi(p, e, \vec{x})$$
$$= \{e\}(S_{n+2}^2(p, p, e), \vec{x}) = \{e\}(rc(e), \vec{x}).$$
□

As a special case we get the following.

Corollary 8.2.6 *For each e there exists an n such that* $\{n\}(\vec{x}) = \{e\}(n, \vec{x})$.

Corollary 8.2.7 *If* $\{e\}$ *is primitive recursive, then the solution of the equation* $\{f(e)\}(\vec{x}) = \{e\}(f(e), \vec{x})$ *given by the recursion theorem is also primitive recursive.*

Proof Immediate from the explicit definition of the function *rc*.   □

We will give a number of examples as soon as we have shown that we can obtain all primitive recursive functions, for then we have an ample stock of functions to experiment on. First we have to prove some more theorems.

The partial recursive functions are closed under a general form of minimization, sometimes called *unbounded search*, which for a given recursive function $f(y, \vec{x})$ and arguments $\vec{x}$ runs through the values of y and looks for the first one that makes $f(y, \vec{x})$ equal to zero.

Theorem 8.2.8 *Let f be a recursive function, then* $\varphi(\vec{x}) = \mu y[f(y, \vec{x}) = 0]$ *is partial recursive.*

Proof The idea is to compute consecutively $f(0, \vec{x}), f(1, \vec{x}), f(2, \vec{x}), \ldots$ until we find a value 0. This need not happen at all, but if it does, we will get to it. While we are computing these values, we keep count of the number of steps. This is taken care of by a recursive function. So we want a function ψ with index e, operating on y and $\vec{x}$, that does the job for us, i.e. a ψ that after computing a positive value for $f(y, \vec{x})$ moves on to the next input y and adds a 1 to the counter. Since we have hardly any arithmetical tools at the moment, the construction is rather roundabout and artificial.

In the table below we compute $f(y, \vec{x})$ step by step (the outputs are in the third row), and in the last row we compute $\psi(y, \vec{x})$ backwards, as it were.

y	0	1	2	3	...	$k-1$	k
$f(y, \vec{x})$	$f(0, \vec{x})$	$f(1, \vec{x})$	$f(2, \vec{x})$	$f(3, \vec{x})$	...	$f(k-1, \vec{x})$	$f(k, \vec{x})$
	2	7	6	12	...	3	0
$\psi(y, \vec{x})$	k	$k-1$	$k-2$	$k-3$	...	1	0

$\psi(0, \vec{x})$ is the required k. The instruction for ψ is simple:

$$\psi(y, \vec{x}) = \begin{cases} 0 & \text{if } f(y, \vec{x}) = 0 \\ \psi(y+1, \vec{x}) + 1 & \text{else.} \end{cases}$$

In order to find an index for ψ, put $\psi(y, \vec{x}) = \{e\}(y, \vec{x})$ and look for a value for e. We introduce two auxiliary functions ψ_1 and ψ_2 with indices b and c such that $\psi_1(e, y, \vec{x}) = 0$ and $\psi_2(e, y, \vec{x}) = \psi(y + 1, \vec{x}) + 1 = \{e\}(y + 1, \vec{x}) + 1$. The index c follows easily by applying R_3, R_7 and the S_n^m Theorem. If $f(y, \vec{x}) = 0$ then we consider ψ_1, if not, ψ_2. Now we introduce, by clause R4, a new function χ_0 which computes an index:

$$\chi_0(e, y, \vec{x}) = \begin{cases} b & \text{if } f(y, \vec{x}) = 0 \\ c & \text{else} \end{cases}$$

and we put $\chi(e, y, \vec{x}) = \{\chi_0(e, y, \vec{x})\}(e, y, \vec{x})$. The recursion theorem provides us with an index e_0 such that $\chi(e_0, y, \vec{x}) = \{e_0\}(y, \vec{x})$.

We claim that $\{e_0\}(0, \vec{x})$ yields the desired value k, if it exists at all, i.e. e_0 is the index of the ψ we were looking for, and $\varphi(\vec{x}) = \{e\}(o, \vec{x})$.

If $f(y, \vec{x}) \neq 0$ then $\chi(e_0, y, \vec{x}) = \{c\}(e_0, y, \vec{x}) = \psi_2(e_0, y, \vec{x}) = \psi(y + 1, \vec{x}) + 1$, and if $f(y, \vec{x}) = 0$ then $\chi(e_0, y, \vec{x}) = \{b\}(e_0, y, \vec{x}) = 0$.

If, on the other hand, k is the first value y such that $f(y, \vec{x}) = 0$, then $\psi(0, \vec{x}) = \psi(1, \vec{x}) + 1 = \psi(2, \vec{x}) + 2 = \cdots = \psi(y_0, \vec{x}) + y_0 = k$. $\square$

Note that the given function need not be recursive, and that the above argument also works for partial recursive f. We then have to reformulate $\mu y[f(x, \vec{y}) = 0]$ as the y such that $f(y, \vec{x}) = 0$ and for all $z < y$ $f(z, \vec{x})$ is defined and positive.

Lemma 8.2.9 *The predecessor is recursive.*

Proof Define

$$x \dot{-} 1 = \begin{cases} 0 & \text{if } x = 0 \\ \mu y[y + 1 = x] & \text{else} \end{cases}$$

where $\mu y[y + 1 = x] = \mu y[f(y, x) = 0]$ with

$$f(y, x) = \begin{cases} 0 & \text{if } y + 1 = x \\ 1 & \text{else.} \end{cases}$$ $\square$

Theorem 8.2.10 *The recursive functions are closed under primitive recursion.*

Proof Let g and h be recursive, and let f be given by

$$\begin{cases} f(0, \vec{x}) = g(\vec{x}) \\ f(y + 1, \vec{x}) = h(f(y, \vec{x}), \vec{x}, y). \end{cases}$$

We rewrite the schema as

$$f(y, \vec{x}) = \begin{cases} g(\vec{x}) & \text{if } y = 0 \\ h(f(y \dot{-} 1, \vec{x}), \vec{x}, y \dot{-} 1) & \text{otherwise.} \end{cases}$$

On the right-hand side we have a definition by cases. So, it defines a partial recursive function with index, say, a of $y, \vec{x}$ and the index e of the function f we are looking for. This yields an equation $\{e\}(y, \vec{x}) = \{a\}(y, \vec{x}, e)$. By the recursion theorem the equation has a solution e_0. And an easy induction on y shows that $\{e_0\}$ is total, so f is a recursive function. $\qquad\square$

We now get the obligatory result.

Corollary 8.2.11 *All primitive recursive functions are recursive.*

Now that we have recovered the primitive recursive functions, we can get lots of partial recursive functions.

Examples

1. Define $\varphi(x) = \{e\}(x) + \{f\}(x)$, then by Corollary 8.2.11 and R5 φ is partial recursive and we would like to express the index of φ as a function of e and f. Consider $\psi(e, f, x) = \{e\}(x) + \{f\}(x)$. ψ is partial recursive, so it has an index n, i.e. $\{n\}(e, f, x) = \{e\}(x) + \{f\}(x)$. By the S_n^m Theorem there is a primitive recursive function h such that $\{n\}(e, f, x) = \{h(n, e, f)\}(x)$. Therefore, $g(e, f) = h(n, e, f)$ is the required function.
2. There is a partial recursive function φ such that $\varphi(n) = (\varphi(n + 1) + 1)^2$: consider $\{z\}(n) = \{e\}(z, n) = (\{z\}(n + 1) + 1)^2$. By the recursion theorem there is a solution $rc(e)$ for z, hence φ exists. A simple argument shows that φ cannot be defined for any n, so the solution is the empty function (the machine that never gives an output).
3. *The Ackermann function*, see Smoryński (1991, p. 70). Consider the following sequence of functions:

$$\varphi_0(m, n) = n + m$$
$$\varphi_1(m, n) = n \cdot m$$
$$\varphi_2(m, n) = n^m$$
$$\vdots$$
$$\begin{cases} \varphi_{k+1}(0, n) = n \\ \varphi_{k+1}(m + 1, n) = \varphi_k(\varphi_{k+1}(m, n), n) \quad (k \geq 2). \end{cases}$$

This sequence consists of faster and faster growing functions. We can lump all those functions together in one function

$$\varphi(k, m, n) = \varphi_k(m, n).$$

The above equations can be summarized as

$$\begin{cases} \varphi(0, m, n) = n + m \\ \varphi(k + 1, 0, n) = \begin{cases} 0 & \text{if } k = 0 \\ 1 & \text{if } k = 1 \\ n & \text{else} \end{cases} \\ \varphi(k + 1, m + 1, n) = \varphi(k, \varphi(k + 1, m, n), n). \end{cases}$$

Note that the second equation has to distinguish cases according to the φ_{k+1} being the multiplication, exponentiation or the general case ($k \geq 2$).

Using the fact that all primitive recursive functions are recursive, we rewrite
the three cases into one equation of the form $\{e\}(k, m, n) = f(e, k, m, n)$ for a
suitable recursive f (Exercise 3). Hence, by the Recursion Theorem there exists
a recursive function with index e that satisfies the equations above. Ackermann
has shown that the function $\varphi(n, n, n)$ eventually grows faster than any primitive
recursive function.

4. The Recursion Theorem can also be used for inductive definitions of sets or relations. This is seen by changing over to characteristic functions, e.g. suppose we
want a relation $R(x, y)$ such that

$$R(x, y) \quad \Leftrightarrow \quad (x = 0 \wedge y \neq 0) \vee (x \neq 0 \wedge y \neq 0) \wedge R(x \dot{-} 1, y \dot{-} 1).$$

Then we write

$$K_R(x, y) = sg\big(\overline{sg}(x) \cdot sg(y) + sg(x) \cdot sg(y) \cdot K_R(x \dot{-} 1, y \dot{-} 1)\big),$$

so there is an e such that

$$K_R(x, y) = \{e\}\big(K_R(x \dot{-} 1, y \dot{-} 1), x, y\big).$$

Now suppose K_R has index z, then we have

$$\{z\}(x, y) = \{e'\}(z, x, y).$$

The solution $\{n\}$ as provided by the Recursion Theorem is the required characteristic function. One immediately sees that R is the relation "less than". Therefore $\{n\}$ is total, and hence recursive; this shows that R is also recursive. Note
that by the remark following the Recursion Theorem we even get the primitive
recursiveness of R.

The following fundamental theorem is extremely useful for many applications.
Its theoretical importance is that it shows that all partial recursive functions can be
obtained from a primitive recursive relation by *one* minimization.

So minimization is the missing link between primitive recursive and (partial)
recursive.

Theorem 8.2.12 (Normal Form Theorem) *There is a primitive recursive predicate
T such that $\{e\}(\vec{x}) = ((\mu z)T(e, \langle \vec{x} \rangle, z))_1$.*

Proof Our heuristics for partial recursive functions was based on the machine
metaphor: think of an abstract machine with actions prescribed by the clauses R1
through R7. By retracing the index e of such a machine, we more or less give a
computation procedure. It now is a clerical task to specify all the steps involved in
such a "computation". Once we have accomplished this, we have made our notion
of "computation" precise, and from the form of the specification, we can immediately conclude that "c is the code of a computation" is indeed primitive recursive.

We look for a predicate $T(e, u, z)$ that formalizes the heuristic statement "z is a (coded) computation that is performed by a partial recursive function with index e on input u" (i.e. $\langle \vec{x} \rangle$). The "computation" has been arranged in such a way that the first projection of z is its output.

The proof is a matter of clerical perseverance—not difficult, but not exciting either. For the reader it is better to work out a few cases by himself and to leave the rest, than to spell out the following details.

First two examples.

(1) The successor function applied to $(1, 2, 3)$:

$$S_1^3(1, 2, 3) = 2 + 1 = 3.$$

Warning, here S_1^3 is used for the successor function operating on the second item of the input string of length 3. The notation is only used here.

The index is $e = \langle 2, 3, 1 \rangle$, the input is $u = \langle 1, 2, 3 \rangle$ and the step is the direct computation $z = \langle 3, \langle 1, 2, 3 \rangle, \langle 2, 3, 1 \rangle \rangle = \langle 3, u, e \rangle$.

(2) The composition of projection and constant functions.

$$P_2^3\big(C_0^2(7, 0), 5, 1\big) = 1.$$

By R5 the input of this function has to be a string of numbers, so we have to introduce a suitable input. The simplest solution is to use $(7, 0)$ as input and manufacture the remaining 5 and 1 out of them. So let us put

$$P_2^3\big(C_0^2(7, 0), 5, 1\big) = P_2^3\big(C_0^2(7, 0), C_5^2(7, 0), C_1^2(7, 0)\big).$$

In order to keep the notation readable, we will use variables instead of the numerical inputs.

$$\varphi(y_0, y_1) = P_2^3\big(C_0^2(y_0, y_1), C_5^2(y_0, y_1), C_1^2(y_0, y_1)\big) = P_2^3\big(C_0^2(y_0, y_1), x_1, x_2\big).$$

Let us first write down the data for the component functions:

	Index	Input	Step
C_0^2	$\langle 0, 2, 0 \rangle = e_0$	$\langle y_0, y_1 \rangle = u$	$\langle 0, u, e_0 \rangle = z_0$
$C_{x_1}^2$	$\langle 0, 2, x_1 \rangle = e_1$	$\langle y_0, y_1 \rangle = u$	$\langle x_1, u, e_1 \rangle = z_1$
$C_{x_2}^2$	$\langle 0, 2, x_2 \rangle = e_2$	$\langle y_0, y_1 \rangle = u$	$\langle x_2, u, e_2 \rangle = z_2$
P_2^3	$\langle 1, 3, 2 \rangle = e_3$	$\langle 0, x_1, x_2 \rangle = u'$	$\langle x_2, u', e_3 \rangle = z_3$

Now for the composition:

Index	Input	Step
$f(y_0, y_1)$ $\langle 4, 2, e_3, e_0, e_1, e_2 \rangle = e$	$\langle y_0, y_1 \rangle = u$	$\langle x_2, \langle y_0, y_1 \rangle, e, z_3, \langle z_0, z_1, z_2 \rangle \rangle = z$

As we see in this example, "step" means the last step in the chain of steps that leads to the output. Now for an actual computation on numerical inputs, all one has to do is to replace y_0, y_1, x_1, x_2 by numbers and write out the data for $\varphi(y_0, y_1)$.

We have tried to arrange the proof in a readable manner by providing a running commentary.

The ingredients for, and conditions on, computations are displayed below. The index contains the information given in the clauses Ri. The computation codes the following items:

(1) the output
(2) the input
(3) the index
(4) subcomputations.

Note that z in the table below is the "master number", i.e. we can read off the remaining data from z, e.g. $e = (z)_2$, $lth(u) = (e)_1 = (z)_{2,1}$ and the output (if any) of the computation, $(z)_0$. In particular we can extract the "master numbers" of the subcomputations. So, by decoding the code for a computation, we can effectively find the codes for the subcomputations, etc. This suggests a primitive recursive algorithm for the extraction of the total "history" of a computation from its code. As a matter of fact, that is essentially the content of the Normal Form Theorem.

	Index e	Input u	Step z	Conditions on subcomputations
R1	$\langle 0, n, q \rangle$	$\langle \vec{x} \rangle$	$\langle q, u, e \rangle$	
R2	$\langle 1, n, i \rangle$	$\langle \vec{x} \rangle$	$\langle x_i, u, e \rangle$	
R3	$\langle 2, n, i \rangle$	$\langle \vec{x} \rangle$	$\langle x_i + 1, u, e \rangle$	
R4	$\langle 3, n+4 \rangle$	$\langle p, q, r, s, \vec{x} \rangle$	$\langle p, u, e \rangle$ if $r = s$ $\langle q, u, e \rangle$ if $r \neq s$	
R5	$\langle 4, n, b, c_0, \ldots, c_{k-1} \rangle$	$\langle \vec{x} \rangle$	$\langle (z')_0, u, e, z', \langle z_0'', \ldots, z_{k-1}'' \rangle \rangle$	$z', z_0'', \ldots, z_{k-1}''$ are computations with indices $b, c_0, \ldots, c_{k-1}$. z' has input $\langle (z_0'')_0, \ldots, (z_{k-1}'')_0 \rangle$ (cf. Theorem 8.2.4)
R6	$\langle 5, n+2 \rangle$	$\langle p, q, \vec{x} \rangle$	$\langle s, u, e \rangle$	
R7	$\langle 6, n+1 \rangle$	$\langle b, \vec{x} \rangle$	$\langle (z')_0, u, e, z' \rangle$	z' is a computation with input $\langle \vec{x} \rangle$ and index b

We will now proceed in a (slightly) more formal manner, by defining a predicate $C(z)$ (for z is a computation), using the information of the preceding table. For convenience, we assume that in the clauses below, sequences u (in $Seq(u)$) have positive length.

$C(z)$ is defined by cases as follows:

$$
C(z) :=
\begin{cases}
\exists q, u, e < z[z = \langle q, u, e \rangle \wedge Seq(u) \wedge e = \langle 0, lth(u), q \rangle] & (1) \\
\text{or} \\
\exists u, e, i < z[z = \langle (u)_i, u, e \rangle \wedge Seq(u) \wedge e = \langle 1, lth(u), i \rangle] & (2) \\
\text{or} \\
\exists u, e, i < z[z = \langle (u)_i + 1, u, e \rangle \wedge Seq(u) \wedge e = \langle 2, lth(u), i \rangle] & (3) \\
\text{or} \\
\exists u, e < z[Seq(u) \wedge e = \langle 3, lth(u) \rangle \wedge lth(u) > 4 \wedge ([z = \langle (u)_0, u, e \rangle \\
\qquad \wedge (u)_2 = (u)_3] \vee [z = \langle (u)_1, u, e \rangle \wedge (u)_2 neq (u)_3])] & (4) \\
\text{or} \\
Seq(z) \wedge lth(z) = 5 \wedge Seq((z)_2) \wedge Seq((z)_4) \wedge lth((z)_2) \\
\quad = 3 + lth((z)_4) \wedge (z)_{2,0} = 4 \wedge C((z)_3) \wedge (z)_{3,0} = (z)_0 \wedge (z)_{3,1} \\
\quad = \langle (z)_{4,0,0}, \ldots, (z)_{4,lth((z)_4),0} \rangle \wedge (z)_{3,2} = (z)_{2,2} \\
\quad \wedge \bigwedge_{i=0}^{lth((z)_4)-1} [C((z)_{4,i}) \wedge (z)_{4,i,2} = (z)_{0,2+i} \wedge (z)_{4,i,1} = (z)_1] & (5) \\
\text{or} \\
\exists s, u, e < z[z = \langle s, u, e \rangle \wedge Seq(u) \wedge e = \langle 5, lth(u) \rangle \\
\quad \wedge s = \langle 4, (u)_{0,1} \dot{-} 1, (u)_0, \langle 0, (u)_{0,1} \dot{-} 1, (u)_1 \rangle, \langle 1, (u)_{0,1} \dot{-} 1, 0 \rangle, \\
\quad \ldots, \langle 1, (u)_{0,1} \dot{-} 1, (e)_1 \dot{-} 2 \rangle \rangle], & (6) \\
\text{or} \\
\exists u, e, w < z[Seq(u) \wedge e = \langle 6, lth(y) \rangle \wedge z = \langle (w)_0, u, e, w \rangle \wedge C(w) \\
\quad \wedge (w)_2 = (u)_0 \wedge (w)_1 = \langle (u)_1, \ldots, (u)_{lth(u)-1} \rangle]. & (7)
\end{cases}
$$

We observe that the predicate C occurs on the right-hand side only for smaller arguments, furthermore all operations involved in this definition of $C(z)$ are primitive recursive. We now apply the recursion theorem, as in the example on p. 226, and conclude that $C(z)$ is primitive recursive.

Now we put $T(e, \vec{x}, z) := C(z) \wedge e = (z)_2 \wedge \langle \vec{x} \rangle = (z)_1$. So the predicate $T(e, \vec{x}, z)$ formalizes the statement "z is the computation of the partial recursive function (machine) with index e operating on input $\langle \vec{x} \rangle$". The output of the computation, if it exists, is $U(z) = (z)_0$; hence we have $\{e\}(\vec{x}) = (\mu z T(e, \vec{x}, z))_0$.

For applications the precise structure of T is not important; it is good enough to know that it is primitive recursive. □

Exercises

1. Show that the empty function (that is the function that diverges for all inputs) is partial recursive. Indicate an index for the empty function.
2. Show that each partial recursive function has infinitely many indices.
3. Carry out the conversion of the three equations of the Ackermann function into one function, see p. 225.

8.3 Recursively Enumerable Sets

If a set A has a recursive characteristic function, then this function acts as an effective test for membership. We can decide which elements are in A and which not.

Decidable sets, convenient as they are, demand too much; it is usually not necessary
to decide what is in a set, as long as we can generate it effectively. Equivalently, as
we shall see, it is good enough to have an abstract machine that only accepts elements, and does not reject them. If you feed it an element, it may eventually show a
green light of acceptance, but there is no red light for rejection.

Definition 8.3.1

1. A set (relation) is (recursively) *decidable* if it is recursive.
2. A set is *recursively enumerable* (RE) if it is the domain of a partial recursive
   function.
3. $W_e^k = \{\vec{x} \in \mathbb{N}^k | \exists y(\{e\}(\vec{x}) = y)\}$, i.e. the domain of the partial recursive func-
   tion $\{e\}$. We call e the RE index of W_e^k. If no confusion arises we will delete
   the superscript.

Notation We write $\varphi(\vec{x}) \downarrow$ (resp. $\varphi(\vec{x}) \uparrow$) for $\varphi(\vec{x})$ converges (resp. $\varphi(\vec{x})$ diverges).

It is good heuristics to think of RE sets as being accepted by machines, e.g. if A_i
is accepted by machine M_i $(i = 0, 1)$, then we make a new machine that simulates
M_0 and M_1 simultaneously, e.g. you feed M_0 and M_1 an input, and carry out the
computation alternatingly—one step for M_0 and then one step for M_1, and so n is
accepted by M if it is accepted by M_0 or M_1. Hence the union of two RE sets is
also RE.

Example 8.3.2

1. $\mathbb{N} =$ *the domain of the constant 1 function.*
2. $\emptyset =$ *the domain of the empty function.* This function is partial recursive, as we
   have already seen.
3. *Every recursive set is RE.* Let A be recursive, put

$$\psi(\vec{x}) = \mu y\big[K_A(\vec{x}) = y \wedge y \neq 0\big].$$

Then $Dom(\psi) = A$.

The recursively enumerable sets derive their importance from the fact that they
are effectively given, in the precise sense of the following theorem. Furthermore it is
the case that the majority of important relations (sets) in logic are RE. For example
the set of (codes of) provable sentences of arithmetic or predicate logic is RE. The
RE sets represent the first step beyond the decidable sets, as we will show below.

Theorem 8.3.3 *The following statements are equivalent* $(A \subseteq \mathbb{N})$:

1. $A = Dom(\varphi)$ *for some partial recursive* φ,
2. $A = Ran(\varphi)$ *for some partial recursive* φ,
3. $A = \{x | \exists y R(x, y)\}$ *for some recursive* R.

Proof (1) $\Rightarrow$ (2). Define $\psi(x) = x \cdot sg(\varphi(x) + 1)$. If $x \in Dom(\varphi)$, then $\psi(x) = x$, so $x \in Ran(\psi)$, and if $x \in Ran(\psi)$, then $\varphi(x) \downarrow$, so $x \in Dom(\varphi)$.

(2) $\Rightarrow$ (3) Let $A = Ran(\varphi)$, with $\{g\} = \varphi$, then

$$x \in A \quad \Leftrightarrow \quad \exists w \big[T\big(g, (w)_0, (w)_1\big) \wedge x = (w)_{1,0} \big].$$

The relation in the scope of the quantifier is recursive.

Note that w "simulates" a pair: first coordinate—input, second coordinate—computation, all in the sense of the Normal Form Theorem.

(3) $\Rightarrow$ (1) Define $\varphi(x) = \mu y R(x, y)$. φ is partial recursive and $Dom(\varphi) = A$.
Observe that (1) $\Rightarrow$ (3) also holds for $A \subseteq \mathbb{N}^k$. □

Since we have defined recursive sets by means of characteristic functions, and since we have established closure under primitive recursion, we can copy all the closure properties of primitive recursive sets (and relations) for the recursive sets (and relations).

Next we list a number of closure properties of RE sets. We will also write sets and relations as predicates, when that turns out to be convenient.

Theorem 8.3.4

1. *If A and B are RE, then so are $A \cup B$ and $A \cap B$.*
2. *If $R(x, \vec{y})$ is RE, then so is $\exists x R(x, \vec{y})$.*
3. *If $R(x, \vec{y})$ is RE and φ partial recursive, then $R(\varphi(\vec{y}, \vec{z}), \vec{y})$ is RE.*
4. *If $R(x, \vec{y})$ is RE, then so are $\forall x < z R(x, \vec{y})$ and $\exists x < z R(x, \vec{y})$.*

Proof (1) There are recursive R and S such that

$$A\vec{y} \quad \Leftrightarrow \quad \exists x R(x, \vec{y}),$$
$$B\vec{y} \quad \Leftrightarrow \quad \exists x S(x, \vec{y}).$$

Then

$$A\vec{y} \wedge B\vec{y} \quad \Leftrightarrow \quad \exists x_1 x_2 \big(R(x_1, \vec{y}) \wedge S(x_2, \vec{y})\big)$$
$$\Leftrightarrow \quad \exists z \big(R((z)_0, \vec{y}) \wedge S((z)_1, \vec{y})\big).$$

The relation in the scope of the quantifier is recursive, so $A \cap B$ is RE. A similar argument establishes the recursive enumerability of $A \cup B$. The trick of replacing x_1 and x_2 by $(z)_0$ and $(z)_1$ and $\exists x_1 x_2$ by $\exists z$ is called *contraction of quantifiers*.

(2) Let $R(x, \vec{y}) \Leftrightarrow \exists z S(z, x, \vec{y})$ for a recursive S, then $\exists x R(x, \vec{y}) \Leftrightarrow \exists x \exists z$ $S(z, x, \vec{y}) \Leftrightarrow \exists u S((u)_0, (u)_1, \vec{y})$. So the *projection* $\exists x R(x, \vec{y})$ of R is RE.

Geometrically speaking, $\exists x R(x, \vec{y})$ is indeed a projection. Consider the two-dimensional case.

The vertical projection S of R is given by $Sx \Leftrightarrow \exists y R(x, y)$.

(3) Let R be the domain of a partial recursive ψ, then $R(\varphi(\vec{y}, \vec{z}), \vec{y})$ is the domain of $\psi(\varphi(\vec{y}, \vec{z}), \vec{y})$.

(4) Left to the reader. □

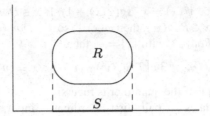

Theorem 8.3.5 *The graph of a partial function is RE iff the function is partial recursive.*

Proof $G = \{(\vec{x}, y) \mid y = \{e\}(\vec{x})\}$ is the graph of $\{e\}$. Now $(\vec{x}, y) \in G \Leftrightarrow \exists z(T(e, \langle \vec{x} \rangle, z) \wedge y = (z)_0)$, so G is RE.

Conversely, if G is RE, then $G(\vec{x}, y) \Leftrightarrow \exists z R(\vec{x}, y, z)$ for some recursive R. Hence $\varphi(\vec{x}) = (\mu w R(\vec{x}, (w)_0, (w)_1))_0$, so φ is partial recursive. □

We can also characterize recursive sets in terms of RE sets. Suppose both A and its complement A^c are RE, then (heuristically) we have two machines enumerating A and A^c. Now the test for membership of A is simple: turn both machines on and wait for n to turn up as output of the first or second machine. This must necessarily occur in finitely many steps since $n \in A$ or $n \in A^c$ (principle of the excluded third!). Hence, we have an effective test. We formalize the above.

Theorem 8.3.6 *A is recursive $\Leftrightarrow A$ and A^c are RE.*

Proof $\Rightarrow$ is trivial: $A(\vec{x}) \Leftrightarrow \exists y A(\vec{x})$, where y is a dummy variable. Similarly for A^c.

$\Leftarrow$ Let $A(\vec{x}) \Leftrightarrow \exists y R(\vec{x}, y)$, $\neg A(\vec{x}) \Leftrightarrow \exists z S(v, z)$. Since $\forall \vec{x}(A(\vec{x}) \vee \neg A(\vec{x}))$, we have $\forall \vec{x} \exists y (R(\vec{x}, y) \vee S(\vec{x}, y))$, so $f(\vec{x}) = \mu y[R(\vec{x}, y) \vee S(\vec{x}, y)]$ is recursive and if we put in the y that we found in $R(\vec{x}, y)$, then we know that if $R(\vec{x}, f(\vec{x}))$ is true, the $\vec{x}$ belongs to A. So $A(\vec{x}) \Leftrightarrow R(\vec{x}, f(\vec{x}))$, i.e. A is recursive. □

For partial recursive functions we have a strong form of definition by cases.

Theorem 8.3.7 *Let $\psi_1, \ldots, \psi_k$ be partial recursive, $R_0, \ldots, R_{k-1}$ mutually disjoint RE relations, then the following function is partial recursive:*

$$\varphi(\vec{x}) = \begin{cases} \psi_0(\vec{x}) & \text{if } R_0(\vec{x}) \\ \psi_1(\vec{x}) & \text{if } R_1(\vec{x}) \\ \vdots \\ \psi_{k-1}(\vec{x}) & \text{if } R_{k-1}(\vec{x}) \\ \uparrow & \text{else} \end{cases}$$

Proof We consider the graph of the function φ.

$$G(\vec{x}, y) \quad \Leftrightarrow \quad (R_0(\vec{x}) \wedge y = \psi_1(\vec{x})) \vee \cdots \vee (R_{k-1}(\vec{x}) \wedge y = \psi_{k-1}(\vec{x})).$$

By the properties of RE sets, $G(\vec{x}, y)$ is RE and, hence, $\varphi(\vec{x})$ is partial recursive. (Note that the last case in the definition of φ is just a bit of decoration.) ☐

Now we can show the existence of undecidable RE sets.

Examples

(1) (*The Halting Problem (Turing)*)

Consider $K = \{x \mid \exists z T(x, x, z)\}$. K is the projection of a recursive relation, so it is RE. Suppose that K^c is also RE, then $x \in K^c \Leftrightarrow \exists z T(e, x, z)$ for some index e. Now $e \in K \Leftrightarrow \exists z T(e, e, z) \Leftrightarrow e \in K^c$. Contradiction. Hence K is not recursive by Theorem 8.3.6. This tells us that there are recursively enumerable sets which are not recursive. In other words, the fact that one can effectively enumerate a set, does not guarantee that it is decidable.

The decision problem for K is called the *halting problem*, because it can be paraphrased as "decide if the machine with index x performs a computation that halts after a finite number of steps when presented with x as input". Note that it is *ipso facto* undecidable if "the machine with index x eventually halts on input y".

It is a characteristic feature of decision problems in recursion theory, that they concern tests for inputs out of some domain. It does not make sense to ask for a decision procedure for, say, the Riemann hypothesis, since trivially there is a recursive function f that tests the problem in the sense that $f(0) = 0$ if the Riemann hypothesis holds and $f(0) = 1$ if the Riemann hypothesis is false. Namely, consider the functions f_0 and f_1, which are the constant 0 and 1 functions respectively. Now logic tells us that one of the two is the required function (this is the law of the excluded middle); unfortunately we do not know which function it is. So for single problems (i.e. problems without a parameter), it does not make sense in the framework of recursion theory to discuss decidability. As we have seen, intuitionistic logic sees this "pathological example" in a different light.

(2) It is not decidable if $\{x\}$ is a total function.

Suppose it were decidable, then we would have a recursive function f such that $f(x) = 0 \Leftrightarrow \{x\}$ is total. Now consider

$$\varphi(x, y) := \begin{cases} 0 & \text{if } x \in K \\ \uparrow & \text{else} \end{cases}$$

By the S_n^m Theorem there is a recursive h such that $\{h(x)\}(y) = \varphi(x, y)$. Now $\{h(x)\}$ is total $\Leftrightarrow x \in K$, so $f(h(x)) = 0 \Leftrightarrow x \in K$, i.e. we have a recursive characteristic function $\overline{sg}(f(h(x)))$ for K. Contradiction. Hence such an f does not exist, that is $\{x \mid \{x\}$ is total$\}$ is not recursive.

(3) The problem "W_e is finite" is not recursively solvable.

In words, "it is not decidable whether a recursively enumerable set is finite".

Suppose that there was a recursive function f such that $f(e) = 0 \Leftrightarrow W_e$ is finite. Consider the $h(x)$ defined in example (2). Clearly $W_{h(x)} = Dom\{h(x)\} = \emptyset \Leftrightarrow$

$x \notin K$, and $W_{h(x)}$ is infinite for $x \in K$. $f(h(x)) = 0 \Leftrightarrow x \notin K$, and hence $sg(f(h(x)))$ is a recursive characteristic function for K. Contradiction.

Note that $x \in K \Leftrightarrow \{x\}x \downarrow$, so we can reformulate the above solutions as follows: in (2) take $\varphi(x, y) = 0 \cdot \{x\}(x)$ and in (3) $\varphi(x, y) = \{x\}(x)$.

(4) The equality of RE sets is undecidable.

That is, $\{(x, y) \mid W_x = W_y\}$ is not recursive. We reduce the problem to the solution of (3) by choosing $W_y = \emptyset$.

(5) It is not decidable if W_e is recursive.

Put $\varphi(x, y) = \{x\}(x) \cdot \{y\}(y)$, then $\varphi(x, y) = \{h(x)\}(y)$ for a certain recursive h, and

$$Dom\{h(x)\} = \begin{cases} K & \text{if } x \in K \\ \emptyset & \text{otherwise.} \end{cases}$$

Suppose there were a recursive function f such that $f(x) = 0 \Leftrightarrow W_x$ is recursive, then $f(h(x)) = 0 \Leftrightarrow x \notin K$ and, hence, K would be recursive. Contradiction.

There are several more techniques for establishing undecidability. We will treat here the method of inseparability.

Definition 8.3.8 Two disjoint RE sets W_m and W_n are *recursively separable* if there is a recursive set A such that $W_n \subseteq A$ and $W_m \subseteq A^c$. Disjoint sets A and B are *effectively inseparable* if there is a partial recursive φ such that for every m, n with $A \subseteq W_m, B \subseteq W_n, W_m \cap W_n = \emptyset$ we have $\varphi(m, n) \downarrow$ and $\varphi(m, n) \notin W_m \cup W_n$.

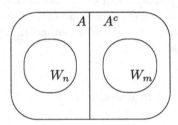   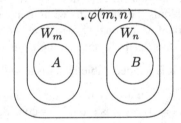

We immediately see that effectively inseparable RE sets are recursively inseparable, i.e. not recursively separable.

Theorem 8.3.9 *There exist effectively inseparable RE sets.*

Proof Define $A = \{x \mid \{x\}(x) = 0\}$, $B = \{x \mid \{x\}(x) = 1\}$. Clearly $A \cap B = \emptyset$ and both sets are RE.

Let $W_m \cap W_n = \emptyset$ and $A \subseteq W_m, B \subset W_n$. To define φ we start testing $x \in W_m$ or $x \in W_n$; if we first find $x \in W_m$, we put an auxiliary function $\sigma(x)$ equal to 1, if x turns up first in W_n then we put $\sigma(x) = 0$.

Formally

$$\sigma(m,n,x) = \begin{cases} 1 & \text{if } \exists z(T(m,x,z) \text{ and } \forall y < z\neg T(n,x,y)) \\ 0 & \text{if } \exists z(T(n,x,z) \text{ and } \forall y \leq z\neg T(m,x,y)) \\ \uparrow & \text{else.} \end{cases}$$

By the S_n^m Theorem $\{h(m,n)\}(x) = \sigma(m,n,x)$ for some recursive h.

$$\begin{aligned} h(m,n) \in W_m \quad &\Rightarrow \quad h(m,n) \notin W_n. \text{ So } \exists z\big(T\big(m, h(m,n), z\big) \\ &\qquad\qquad \wedge \forall y < z\neg T\big(n, h(m,n), y\big)\big) \\ &\Rightarrow \quad \sigma\big(m,n,h(m,n)\big) = 1 \Rightarrow \{h(m,n)\}\big(h(m,n)\big) = 1 \\ &\Rightarrow \quad h(m,n) \in B \Rightarrow h(m,n) \in W_n. \end{aligned}$$

Contradiction. Hence $h(m,n) \notin W_m$. Similarly $h(m,n) \notin W_n$. Thus h is the required φ.   □

Definition 8.3.10 A subset A of $\mathbb{N}$ is productive if there is a partial recursive function φ, such that for each $W_n \subseteq A$, $\varphi(n) \downarrow$ and $\varphi(n) \in A - W_n$.

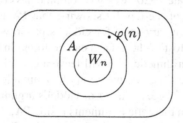

The theorem above gives us the following corollary.

Corollary 8.3.11 *There are productive sets.*

Proof The set A^c defined in the above proof is *productive*. Let $W_k \subseteq A^c$. Put $W_\ell = B \cup W_k = W_n \cup W_k = W_{h(n,k)}$ for a suitable recursive function h. Now apply the separating function from the proof of the preceding theorem to $A = W_m$ and $W_{h(n,k)}$: $\varphi(m, h(n,k)) \in A^c - W_m$.   □

Productive sets are in a strong sense not RE: no matter how one tries to fit an RE set into them, one can uniformly and effectively indicate a point that is missed by this RE set.

Exercises

1. The projection of an RE set is RE, i.e. if $R(\vec{x}, y)$ is RE then so is $\exists y R(\vec{x}, y)$.
2. (i) If A is enumerated by a strictly monotone function, then A is recursive.
   (ii) If A is infinite and recursive, then A is enumerated by a strictly increasing recursive function.
   (iii) An infinite RE set contains an infinite recursive subset.

3. Every non-empty RE set is enumerated by a total recursive function.
4. If A is an RE set and f a partially recursive function, then $f^{-1}(A)(= \{x \mid f(x) \in A\})$ and $f(A)$ are RE.
5. Show that the following are not recursive:
   (i) $\{(x, y) \mid W_x = W_y\}$
   (ii) $\{x \mid W_x \text{ is recursive}\}$
   (iii) $\{x \mid 0 \in W_x\}$.

8.4 Some Arithmetic

In the section on recursive functions we have been working in the standard model of arithmetic; as we are now dealing with provability *in* arithmetic we have to avoid semantical arguments and to rely solely on derivations inside the formal system of arithmetic. The generally accepted theory for arithmetic goes back to Peano, and thus we speak of *Peano arithmetic*, **PA** (cf. Sect. 3.7)

A major issue in the late 1920s was the completeness of **PA**. Gödel put an end to prevailing high hopes of the day by showing that **PA** is incomplete (1931). In order to carry out the necessary steps for Gödel's proof, we have to prove a number of theorems in **PA**. Most of these facts can be found in texts on number theory, or on the foundation of arithmetic. We will leave a considerable number of proofs to the reader. Most of the time one has to apply a suitable form of induction. Important as the actual proofs are, the heart of Gödel's argument lies in his ingenious incorporation of recursion theoretic arguments inside **PA**.

One of the obvious stumbling blocks for a straightforward imitation of "self-reference" is the apparent poverty of the language of **PA**. It does not allow us to speak of, e.g., a finite string of numbers. Once we have exponentiation we can simply code finite sequences of numbers. Gödel showed that one can indeed define the exponential (and much more) at the cost of some extra arithmetic, yielding his famous β-function. In 1971 Matijasevich showed by other means that the exponential is definable in **PA**, thus enabling us to handle coding of sequences in **PA** directly. Peano arithmetic plus exponentiation is prima facie stronger than **PA**, but the above mentioned results show that exponentiation can be eliminated. Let us call the extended system **PA**; no confusion will arise.

We repeat the axioms:

- $\forall x (S(x) \neq 0)$,
- $\forall xy (S(x) = S(y) \rightarrow x = y)$,
- $\forall x (x + 0 = x)$,
- $\forall xy (x + S(y) = S(x + y))$,
- $\forall x (x \cdot 0 = 0)$,
- $\forall xy (x \cdot S(y) = x \cdot y + x)$,
- $\forall x (x^0 = 1)$,
- $\forall xy (x^{Sy} = x^y \cdot x)$,
- $\varphi(0) \wedge \forall x (\varphi(x) \rightarrow \varphi(S(x))) \rightarrow \forall x \varphi(x)$.

Since $\vdash \bar{1} = S(0)$, we will use both $S(x)$ and $x + \bar{1}$, whichever is convenient. We will also use the usual abbreviations. In order to simplify the notation, we will tacitly drop the "**PA**" in front of "$\vdash$" whenever possible. As another harmless simplification of notation we will often simply write n for $\bar{n}$ when no confusion arises.

In the following we will give a number of theorems of **PA**; in order to improve the readability, we will drop the universal quantifiers preceding the formulas. The reader should always think of "the universal closure of . . .".

Furthermore we will use the standard abbreviations of algebra, i.e. leave out the multiplication dot, superfluous brackets, etc., when no confusion arises. We will also write "n" instead of "$\bar{n}$".

The basic operations satisfy the well-known laws.

Lemma 8.4.1 *Addition and multiplication are associative and commutative, and $\cdot$ distributes over $+$.*

(i) $\vdash (x + y) + z = x + (y + z)$
(ii) $\vdash x + y = y + x$
(iii) $\vdash x(yz) = (xy)z$
(iv) $\vdash xy = yx$
(v) $\vdash x(y + z) = xy + xz$
(vi) $\vdash x^{y+z} = x^y x^z$
(vii) $\vdash (x^y)^z = x^{yz}$.

Proof Routine. $\qquad\qquad\qquad\qquad\qquad\qquad\qquad\qquad\qquad\qquad\qquad\square$

Lemma 8.4.2

(i) $\vdash x = 0 \vee \exists y(x = Sy)$
(ii) $\vdash x + z = y + z \rightarrow x = y$
(iii) $\vdash z \neq 0 \rightarrow (xz = yz \rightarrow x = y)$
(iv) $\vdash x \neq 0 \rightarrow (x^y = x^z \rightarrow y = z)$
(v) $\vdash y \neq 0 \rightarrow (x^y = z^y \rightarrow x = z)$

Proof Routine. $\qquad\qquad\qquad\qquad\qquad\qquad\qquad\qquad\qquad\qquad\qquad\square$

Some useful facts are listed in the exercises.

Although the language of **PA** is modest, many of the usual relations and functions can be defined. The ordering is an important example.

Definition 8.4.3 $x < y := \exists z(x + Sz = y)$

We will use the following abbreviations:

$$
\begin{aligned}
&x < y < z &&\text{stands for } x < y \wedge y < z \\
&\forall x < y \, \varphi(x) &&,, \qquad \forall x(x < y \rightarrow \varphi(x)) \\
&\exists x < y \, \varphi(x) &&,, \qquad \exists x(x < y \wedge \varphi(x)) \\
&x > y &&,, \qquad y < x \\
&x \leq y &&,, \qquad x < y \vee x = y.
\end{aligned}
$$

Theorem 8.4.4

  (i) $\vdash \neg x < x$

  (ii) $\vdash x < y \land y < z \rightarrow x < z$

  (iii) $\vdash x < y \lor x = y \lor y < x$

  (iv) $\vdash 0 = x \lor 0 < x$

  (v) $\vdash x < y \rightarrow Sx = y \lor Sx < y$

  (vi) $\vdash x < Sx$

  (vii) $\vdash \neg x < y \land y < Sx$

 (viii) $\vdash x < Sy \leftrightarrow (x = y \lor x < y)$

  (ix) $\vdash x < y \leftrightarrow x + z < y + z$

  (x) $\vdash z \neq 0 \rightarrow (x < y \leftrightarrow xz < yz)$

 (xi) $\vdash x \neq 0 \rightarrow (0 < y < z \rightarrow x^y < x^z)$

 (xii) $\vdash z \neq 0 \rightarrow (x < y \rightarrow x^z < y^z)$

(xiii) $\vdash x < y \leftrightarrow Sx < Sy$.

Proof Routine. $\square$

Quantification with an explicit bound can be replaced by a repeated disjunction or conjunction.

Theorem 8.4.5 $\vdash \forall x < \bar{n}\varphi(x) \leftrightarrow \varphi(0) \land \cdots \land \varphi(\overline{n-1})$, $(n > 0)$, $\vdash \exists x < \bar{n}\varphi(x) \leftrightarrow \varphi(0) \lor \cdots \lor \varphi(\overline{n-1})$, $(n > 0)$.

Proof Induction on n. $\square$

Theorem 8.4.6

 (i) *Well-founded induction*

$$\vdash \forall x \big(\forall y < x \; \varphi(y) \rightarrow \varphi(x)\big) \rightarrow \forall x \varphi(x)$$

 (ii) *Least number principle (LNP)*

$$\vdash \exists x \varphi(x) \rightarrow \exists x \big(\varphi(x) \land \forall y < x \neg \varphi(y)\big).$$

Proof (i) Let us put $\psi(x) := \forall y < x \varphi(y)$. We assume $\forall x(\psi(x) \rightarrow \varphi(x))$ and proceed to apply induction on $\psi(x)$.

   Clearly $\vdash \psi(0)$.

   So let by the induction hypothesis $\psi(x)$.

   Now $\psi(Sx) \leftrightarrow [\forall y < Sx\varphi(y)] \leftrightarrow [\forall y((y = x \lor y < x) \rightarrow \varphi(y))] \leftrightarrow [\forall y((y = x \rightarrow \varphi(y)) \land (y < x \rightarrow \varphi(y)))] \leftrightarrow [\forall y(\varphi(x) \land (y < x \rightarrow \varphi(y)))] \leftrightarrow [\varphi(x) \land \forall y < x\varphi(y)] \leftrightarrow [\varphi(x) \land \psi(x)]$.

   Now $\psi(x)$ was given and $\psi(x) \rightarrow \varphi(x)$. Hence we get $\psi(Sx)$. This shows $\forall x \psi(x)$, and thus we derive $\forall x \varphi(x)$.

   (ii) Consider the contraposition and reduce it to (i). $\square$

In our further considerations the following notions play a role.

Definition 8.4.7

(i) *Divisibility*

$$x|y := \exists z (xz = y)$$

(ii) *Cut-off subtraction*

$$z = y \dot{-} x := (x < y \wedge x + z = y) \vee (y \leq x \wedge z = 0)$$

(iii) *Remainder after division*

$$z = rem(x, y) := (x \neq 0 \wedge \exists u (y = ux + z) \wedge z < x) \vee (x = 0 \wedge z = y)$$

(iv) x is *prime*

$$Prime(x) := x > 1 \wedge \forall yz (x = yz \to y = x \vee y = 1).$$

The right-hand sides of (ii) and (iii) indeed determine functions, as shown in the following.

Lemma 8.4.8

(i) $\vdash \forall xy \exists! z ((x < y \wedge z + x = y) \vee (y \leq x \wedge z = 0))$
(ii) $\vdash \forall xy \exists! z ((x \neq 0 \wedge \exists u (y = ux + z) \wedge z < y) \vee (x = 0 \wedge z = 0)).$

Proof In both cases induction on y. □

There is another characterization of the prime numbers.

Lemma 8.4.9

(i) $\vdash Prime(x) \leftrightarrow x > 1 \wedge \forall y (y|x \to y = 1 \vee y = x)$
(ii) $\vdash Prime(x) \leftrightarrow x > 1 \wedge \forall yz (x|yz \to x|y \vee x|z)$

Proof (i) is a mere reformulation of the definition.

(ii) $\to$ is a bit tricky. We introduce a bound on the product yz, and do wf-induction on the bound. Put $\varphi(w) = \forall yz \leq w (x|yz \to x|y \vee x|z)$. We now show $\forall w (\forall v < w \varphi(v) \to \varphi(w))$.

Let $\forall v < w \varphi(v)$ and assume $\neg \varphi(w)$, i.e. there are $y, z \leq w$ such that $x|yz, \neg x|y, \neg x|z$. We will "lower" the y such that the w is also lowered. Since $\neg x|y, \neg x|z$, we have $z \neq 0$. Should $y \geq x$, then we may replace it by $y = rem(x, y)$ and carry on the argument. So let $y < x$. Now we once more get the remainder, $x = ay + b$ with $b < y$. We consider $b = 0$ and $b > 0$.

If $b = 0$, then $x = ay$; hence $y = 1 \vee y = x$. If $y = 1$, then $x|z$. Contradiction.

If $y = x$ then $x|y$. Contradiction.

Now if $b > 0$, then $bz = (x - ay)z = xz - ayz$. Since $x|yz$, we get $x|bz$. Observe that $bz < yz < w$, so we have a contradiction with $\forall v < w \varphi(v)$. Hence by *RAA* we have established the required statement.

For $\leftarrow$ we only have to apply the established facts about divisibility. □

Can we prove in Peano arithmetic that there are any primes? Yes, for example $PA \vdash \forall x(x > 1 \to \exists y(Prime(y) \land y|x))$.

Proof Observe that $\exists y(y > 1 \land y|x)$. By the LNP there is a least such y:

$$\exists y\big(y > 1 \land y|x \land \forall z < y(z > 1 \to \neg z|y)\big).$$

Now it is easy to show that this minimal y is a prime.

Primes and exponents are most useful for coding finite sequences of natural numbers, and hence for coding in general. There are many more codings, and some of them are more realistic in the sense that they have a lower complexity. For our purpose, however, primes and exponents will do.

As we have seen, we can code a finite sequence $(n_0, \dots, n_{k-1})$ as the number $2^{n_0+1} \cdot 3^{n_1+1} \dots p_{k-1}^{n_{k-1}+1}$.

We will introduce some auxiliary predicates.

Definition 8.4.10 (Successive Primes) $Succprimes(x, y) := x < y \ \land Prime(x) \ \land Prime(y) \ \land \forall z(x < z < y \to \neg Prime(z))$.

The next step is to define the sequence of prime numbers $2, 3, 5, \dots, p_n, \dots$. The basic trick here is that we consider all successive primes with ascending exponents: $2^0, 3^1, 5^2, 7^3, \dots, p_x^x$. We form the product and then pick the last factor.

Definition 8.4.11 (The xth Prime Number, p_x) $p_x = y := \exists z(\neg 2|z \land \forall v < y \forall u \le y(Succprime(v, u) \to \forall w < z(v^w|z \to u^{w+1}|z)) \land y^x|z \land \neg y^{x+1}|z)$.

Observe that, as the definition yields a function, we have to show the following.

Lemma 8.4.12 $\vdash \exists z(\neg 2|z \land \forall v < y_0 \forall u \le y_0(Succprime(v, u) \to \forall w < z(v^w|z \to u^{w+1}|z)) \land y_0^x|z \land \neg y_0^{x+1}|z) \land \exists z(\neg 2|z \land \forall v < y_1 \forall u \le y_1(Succprime(v, u) \to \forall w < z(v^w|z \to u^{w+1}|z)) \land y_1^x|z \land \neg y_1^{x+1}|z) \to y_0 = y_1$.

The above definition just mimics the informal description. Note that we can bound the existential quantifier as $\exists z < y^{x^2}$. We have now justified the notation of sequence numbers as products of the form

$$p_0{}^{n_0+1} \cdot p_1{}^{n_1+1} \cdot \dots \cdot p_{k-1}{}^{n_{k-1}+1}.$$

The reader should check that according to the definition $p_0 = 2$. The decoding can also be defined. In general one can define the power of a prime factor.

Definition 8.4.13 (Decoding) $(z)_k = v := p_k^{v+1}|z \land \neg p_k^{v+2}|z$.

The length of a coded sequence can also be extracted from the code.

Definition 8.4.14 (Length) $lth(z) = x := p_x|z \land \forall y < z(p_y|z \to y < x)$.

Lemma 8.4.15 $\vdash Seq(z) \to (lth(z) = x \leftrightarrow (p_x | z \land \neg p_{x+1} | z))$.

We define separately the coding of the empty sequence: $\langle\rangle := 1$.

The coding of the sequence $(x_0, \ldots, x_{n-1})$ is denoted by $\langle x_0, \ldots, x_{n-1} \rangle$.

Operations like concatenation and restriction of coded sequences can be defined such that

$$\langle x_0 \ldots x_{n-1} \rangle * \langle y_0 \ldots y_{m-1} \rangle = \langle x_0, \ldots, x_{n-1}, y_0, \ldots, y_{m-1} \rangle$$
$$\langle x_0 \ldots x_n - 1 \rangle | m = \langle x_0 \ldots x_m - 1 \rangle,$$

where $m \leq n$. (Warning: here | is used for the restriction relation, do not confuse with divisibility.)

The tail of a sequence is defined as follows:

$$tail(y) = z \leftrightarrow \left(\exists x \big(y = \langle x \rangle * z \big) \lor \big(lth(y) = 0 \land z = 0 \big) \right).$$

Closed terms of **PA** can be evaluated in **PA**.

Lemma 8.4.16 *For any closed term t there is a number n such that* $\vdash t = \bar{n}$.

Proof External induction on t, cf. Lemma 3.3.3. Observe that n is uniquely determined. □

Corollary 8.4.17 $\mathbb{N} \models t_1 = t_2 \Rightarrow \vdash t_1 = t_2$ *for closed* t_1, t_2.

Gödel's theorem will show that in general "*true in the standard model*" (we will from now on just say "true") and provable in **PA** are not the same. However for a class of simple sentences this is correct.

Definition 8.4.18

(i) The class Δ_0 of formulas is inductively defined by:

$\varphi \in \Delta_0$ for atomic φ

$\varphi, \varphi \in \Delta_0 \Rightarrow \neg\varphi, \varphi \land \psi, \varphi \lor \psi, \varphi \to \psi \in \Delta_0$

$\varphi \in \Delta_0 \Rightarrow \forall x < y \varphi, \exists x < y \varphi \in \Delta_0$.

(ii) The class Σ_1 is given by:

$\varphi, \neg\varphi \in \Sigma_1$ for atomic φ

$\varphi, \psi \in \Sigma_1 \Rightarrow \varphi \lor \psi, \varphi \land \psi \in \Sigma_1$

$\varphi \in \Sigma_1 \Rightarrow \forall x < y \varphi, \exists x < y \varphi, \exists x \varphi \in \Sigma_1$.

A formula is called strict Σ_1 if it is of the form $\exists \vec{x} \varphi(\vec{x})$, where φ is Δ_0.

We will call formulas in the classes Δ_0 and Σ_1, Δ_0, Σ_1-formulas respectively. Formulas, provably equivalent to Σ_1-formulas, will also be called Σ_1-formulas.

For Σ_1-formulas we have that "true = provable".

Lemma 8.4.19 $\vdash \varphi$ *or* $\vdash \neg\varphi$, *for* Δ_0-*sentences* φ.

Proof Induction on φ.

(i) φ atomic. If $\varphi \equiv t_1 = t_2$ and $t_1 = t_2$ is true, see Corollary 8.4.17.

If $t_1 = t_2$ is false, then $t_1 = n$ and $t_2 = m$, where, say, $n = m + k$ with $k > 0$. Now assume (in **PA**) $t_1 = t_2$, then $\bar{m} = \bar{m} + \bar{k}$. By Lemma 8.4.2 we get $0 = \bar{k}$. But since $k = S(l)$ for some l, we obtain a contradiction. Hence $\vdash \neg t_1 = t_2$.

(ii) The induction cases are obvious. For $\forall x < t\varphi(x)$, where t is a closed term, use the identity $\forall x < \bar{n}\varphi(x) \leftrightarrow \varphi(0) \wedge \cdots \wedge \varphi(\overline{n-1})$. Similarly for $\exists x < t\varphi(x)$. $\square$

Theorem 8.4.20 (Σ_1-Completeness) $\models \varphi \Leftrightarrow \textbf{PA} \vdash \varphi$, *for Σ_1 sentences φ.*

Proof Since the truth of $\exists x\varphi(x)$ comes to the truth of $\varphi(\bar{n})$ for some n, we may apply the above lemma. $\square$

8.5 Representability

In this section we will give the formalization of all this in **PA**, i.e. we will show that definable predicates exist corresponding with the predicates introduced above (in the standard model)—and that their properties are provable.

Definition 8.5.1 (Representability)

• A formula $\varphi(x_0, \ldots, x_{n-1}, y)$ represents an n-ary function f if for all $k_0, \ldots, k_{n-1}$

$$f(k_0, \ldots, k_{n-1}) = p \quad \Rightarrow \quad \vdash \forall y\big(\varphi(\bar{k}_0, \ldots, \bar{k}_{n-1}, y) \leftrightarrow y = \bar{p}\big)$$

• A formula $\varphi(x_0, \ldots, x_{n-1})$ represents a predicate P if for all $k_0, \ldots, k_{n-1}$

$$P(k_0, \ldots, k_{n-1}) \quad \Rightarrow \quad \vdash \varphi(\bar{k}_0, \ldots, \bar{k}_{n-1})$$

and

$$\neg P(k_1, \ldots, k_n) \quad \Rightarrow \quad \vdash \neg\varphi(\bar{k}_0, \ldots, \bar{k}_{n-1})$$

• A term $t(x_0, \ldots, x_{n-1})$ represents f if for all $k_0, \ldots, k_{n-1}$

$$f(k_0, \ldots, k_{n-1}) = p \quad \Rightarrow \quad \vdash t(\bar{k}_0, \ldots, \bar{k}_{n-1}) = \bar{p}.$$

Lemma 8.5.2 *If f is representable by a term, then f is representable by a formula.*

Proof Let f be represented by t. Let $f(\mathbf{k}) = p$. Then $\vdash t(\bar{\mathbf{k}}) = \bar{p}$. Now define the formula $\varphi(\vec{x}, y) := t(\vec{x}) = y$. Then we have $\vdash \varphi(\bar{\mathbf{k}}, \bar{p})$, and hence $\bar{p} = y \rightarrow \varphi(\bar{\mathbf{k}}, y)$. This proves $\vdash \varphi(\bar{\mathbf{k}}, y) \leftrightarrow \bar{p} = y$. $\square$

Sometimes it is convenient to split the representability of functions into two clauses.

Lemma 8.5.3 *A k-ary function is representable by φ iff*

$$f(n_0 - n_{k-1}) = m \quad \Rightarrow \quad \vdash \varphi(\bar{n}_0, \ldots, \bar{n}_{k-1}, \bar{m}) \quad and \quad \vdash \exists! z \varphi(\bar{n}_0 \ldots \bar{n}_{k-1}, z).$$

Proof Immediate. Note that the last clause can be replaced by $\vdash \varphi(\overline{n_0} \ldots \overline{n_{k-1}}, z) \rightarrow z = \bar{m}$. $\square$

The basic functions of arithmetic have their obvious representing terms. However, quite simple functions cannot be represented by terms. For example, the sigma function is represented by $\varphi(x, y) := (x = 0 \wedge y = 0) \vee (\neg x = 0 \wedge y = 1)$, but not by a term. However we can easily show $\vdash \forall x \exists! y \varphi(x, y)$, and therefore we could conservatively add the sg to **PA** (cf. Theorem 4.4.6). Note that quite a number of useful predicates and functions have Δ_0-formulas as a representation.

Lemma 8.5.4 *P is representable* $\Leftrightarrow$ *K_P is representable.*

Proof Let $\varphi(\vec{x})$ represent P. Define $\psi(\vec{x}, y) = (\varphi(\vec{x}) \wedge (y = 1)) \vee (\neg \varphi(\vec{x}) \wedge (y = 0))$. Then ψ represents K_P, because if $K_P(\mathbf{k}) = 1$, then $P(\mathbf{k})$, so $\vdash \varphi(\bar{\mathbf{k}})$ and $\vdash \psi(\bar{\mathbf{k}}, y) \leftrightarrow (y = 1)$, and if $K_P(\mathbf{k}) = 0$, then $\neg P(\mathbf{k})$, so $\vdash \neg \varphi(\bar{\mathbf{k}})$ and $\vdash \psi(\bar{\mathbf{k}}, y) \leftrightarrow (y = 0)$. Conversely, let $\psi(\vec{x}, y)$ represent K_P. Define $\varphi(\vec{x}) := \psi(\vec{x}, 1)$. Then φ represents P. $\square$

There is a large class of representable functions; it includes the primitive recursive functions.

Theorem 8.5.5 *The primitive recursive functions are representable.*

Proof Induction on the definition of primitive recursive function. It is simple to show that the initial functions are representable. The constant function C_m^k is represented by the term $\bar{m}$, the successor function S is represented by $x + 1$ and the projection function P_i^k is represented by x_i.

The representable functions are closed under substitution and primitive recursion. We will indicate the proof for the closure under primitive recursion.

Consider

$$\begin{cases} f(\vec{x}, 0) = g(\vec{x}) \\ f(\vec{x}, y + 1) = h(f(\vec{x}, y), \vec{x}, y) \end{cases}$$

g is represented by φ, h is represented by ψ:

$$g(\vec{n}) = m \quad \Rightarrow \quad \begin{cases} \vdash \varphi(\vec{n}, m) \text{ and} \\ \vdash \varphi(\vec{n}, y) \rightarrow y = m \end{cases}$$

$$h(p, \vec{n}, q) = m \quad \Rightarrow \quad \begin{cases} \vdash \psi(p, \vec{n}, q, m) \text{ and} \\ \vdash \psi(p, \vec{n}, q, y) \rightarrow y = m. \end{cases}$$

Claim f is represented by $\sigma(\vec{x}, y, z)$, which is mimicking $\exists w \in Seq(lth(w) = y + 1 \wedge ((w)_0 = g(\vec{x}) \wedge \forall i \leq y((w)_{i+1} = h((w)_i, \vec{x}, i) \wedge z = (w)_y)))$.

$$\sigma(\vec{x}, y, z) := \exists w \in Seq(lth(w) = y + 1 \wedge \varphi(\vec{x}, (w)_0)$$
$$\wedge \, \forall i \leq y\big(\psi\big((w)_i, \vec{x}, i, (w)_{i+1}\big) \wedge z = (w)_y\big).$$

Now let $f(\vec{n}, p) = m$, then

$$\begin{cases} f(\vec{n}, 0) = g(\vec{n}) = a_0 \\ f(\vec{n}, 1) = h(f(\vec{n}, 0), \vec{n}, 0) = a_1 \\ f(\vec{n}, 2) = h(f(\vec{n}, 1), \vec{n}, 1) = a_2 \\ \vdots \\ f(\vec{n}, p) = h(f(\vec{n}, p - 1), \vec{n}, p - 1) = a_p = m. \end{cases}$$

Put $w = \langle a_0, \ldots, a_p \rangle$; note that $lth(w) = p + 1$.

$$g(\vec{n}) = f(\vec{n}, 0) = a_0 \quad \Rightarrow \quad \vdash \varphi(\vec{n}, a_0)$$
$$f(\vec{n}, 1) = a_1 \quad \Rightarrow \quad \vdash \psi(a_0, \vec{n}, 0, a_1)$$
$$\vdots$$
$$f(\vec{n}, p) = a_p \quad \Rightarrow \quad \vdash \psi(a_{p-1}, \vec{n}, p - 1, a_p).$$

Therefore we have $\vdash lth(w) = p + 1 \wedge \varphi(\vec{n}, a_0) \wedge \psi(a_0, \vec{n}, 0, a_1) \wedge \cdots \wedge \psi(a_{p-1}, \vec{n}, p - 1, a_p) \wedge (w)_p = m$ and hence $\vdash \sigma(\vec{n}, p, m)$.

Now we have to prove the second part: $\vdash \sigma(\vec{n}, p, z) \rightarrow z = m$. We prove this by induction on p.

(1) $p = 0$. Observe that $\vdash \sigma(\vec{n}, 0, z) \leftrightarrow \varphi(\vec{n}, z)$, and since φ represents g, we get $\vdash \varphi(\vec{n}, z) \rightarrow z = \overline{m}$.

(2) $p = q + 1$. Induction hypothesis: $\vdash \sigma(\vec{n}, q, z) \rightarrow z = \overline{f(\vec{n}, q)}(= \overline{m})$ $\sigma(\vec{n}, q + 1, z) \leftrightarrow \exists w \in Seq(lth(w) = q + 2 \wedge \varphi(\vec{n}, (w)_0) \wedge \forall i \leq y(\psi((w)_i, \vec{n}, i, (w)_{i+1}) \wedge z = (w)_{q+1}))$.

We now see that $\vdash \sigma(\vec{n}, q + 1, z) \rightarrow \exists u(\sigma(\vec{n}, q, u) \wedge \psi(u, \vec{n}, q, z)$.

Using the induction hypothesis we get $\vdash \sigma(\vec{n}, q + 1, z) \rightarrow \exists u(u = f(\vec{n}, q) \wedge \psi(u, \vec{n}, q, z))$ and hence $\vdash \sigma(\vec{n}, q + 1, z) \rightarrow \psi(f(\vec{n}, q), \vec{n}, q, z)$.

Thus by the property of ψ: $\vdash \sigma(\vec{n}, q + 1, z) \rightarrow z = f(\vec{n}, q + 1)$.

There is now one more step to show that all recursive functions are representable, for we have seen that all recursive functions can be obtained by a single minimization from a primitive recursive predicate.   □

Theorem 8.5.6 *All recursive functions are representable.*

Proof We show that the representable functions are closed under minimalization. Since representability for predicates is equivalent to representability for functions, we consider the case $f(\vec{x}) = \mu y P(\vec{x}, y)$ for a predicate P represented by φ, where $\forall \vec{x} \exists y P(\vec{x}, y)$.

Claim $\psi(\vec{x}, y) := \varphi(\vec{x}, y) \wedge \forall z < y \neg \varphi(\vec{x}, z)$ *represents* $\mu y P(\vec{x}, y)$.

$$m = \mu y P(\vec{n}, y) \quad \Rightarrow \quad P(\vec{n}, m) \wedge \neg P(\vec{n}, 0) \wedge \cdots \wedge \neg P(\vec{n}, m-1)$$
$$\Rightarrow \quad \vdash \varphi(\vec{n}, m) \wedge \neg \varphi(\vec{n}, 0) \wedge \cdots \wedge \neg \varphi(\vec{n}, m-1)$$
$$\Rightarrow \quad \vdash \varphi(\vec{n}, m) \wedge \forall z < m \neg \varphi(\vec{n}, z)$$
$$\Rightarrow \quad \vdash \psi(\vec{n}, m).$$

Now let $\varphi(\vec{n}, y)$ be given, then we have $\varphi(\vec{n}, y) \wedge \forall z < y \neg \varphi(\vec{n}, z)$. This immediately yields $m \geq y$. Conversely, since $\varphi(\vec{n}, m)$, we see that $m \leq y$. Hence $y = m$. This informal argument is straightforwardly formalized as $\vdash \varphi(\vec{n}, y) \to y = m$. $\square$

We have established that recursive sets are representable. One might perhaps hope that this can be extended to recursively enumerable sets. This happens not to be the case. We will consider the RE sets now.

Definition 8.5.7 $R(\vec{x})$ is *semi-representable* in T if $R(\vec{n}) \Leftrightarrow T \vdash \varphi(\vec{n})$ for some $\varphi(\vec{x})$.

Theorem 8.5.8 R *is semi-representable* $\Leftrightarrow$ R *is recursively enumerable.*

For the proof see p. 250.

Corollary 8.5.9 R *is representable* $\Leftrightarrow$ R *is recursive.*

Exercises

1. Show
$$\vdash x + y = 0 \to x = 0 \wedge y = 0$$
$$\vdash xy = 0 \to x = 0 \vee y = 0$$
$$\vdash xy = 1 \to x = 1 \wedge y = 1$$
$$\vdash x^y = 1 \to y = 0 \vee x = 1$$
$$\vdash x^y = 0 \to x = 0 \wedge y \neq 0$$
$$\vdash x + y = 1 \to (x = 0 \wedge y = 1) \vee (x = 1 \wedge y = 0).$$

2. Show that all Σ_1-formulas are equivalent to prenex formulas with the existential quantifiers preceding the bounded universal ones (hint: consider the combination $\forall x < t \exists y \varphi(x, y)$, this yields a coded sequence z such that $\forall x < t \varphi(x, (z)_x)$). That is in **PA** Σ_1-formulas are equivalent to strict Σ_1-formulas.

3. Show that one can contract similar quantifiers, e.g. $\forall x \forall y \varphi(x, y) \leftrightarrow \forall z \varphi((z)_0, (z)_1)$.

8.6 Derivability

In this section we define a coding for a recursively enumerable predicate $Thm(x)$, that says "x is a theorem". Because of the minimization and upper bounds on quantifiers, all predicates and functions defined along the way are primitive recursive. Observe that we are back in recursion theory, that is in informal arithmetic.

Coding of the Syntax   The function $\ulcorner - \urcorner$ codes the syntax. For the alphabet, it is given by

$\wedge$	$\rightarrow$	$\forall$	0	S	$+$	$\cdot$	exp	$=$	$($	$)$	x_i
2	3	5	7	11	13	17	19	23	29	31	p_{11+i}

Next we code the terms:

$$\ulcorner f(t_1, \ldots, t_n) \urcorner := \langle \ulcorner f \urcorner, \ulcorner (\urcorner, \ulcorner t_1 \urcorner, \ldots, \ulcorner t_n \urcorner, \ulcorner) \urcorner \rangle$$

Finally we code the formulas. Note that $\{\wedge, \rightarrow, \forall\}$ is a functionally complete set, so the remaining connectives can be defined.

$$\ulcorner (t = s) \urcorner := \langle \ulcorner (\urcorner, \ulcorner t \urcorner, \ulcorner = \urcorner, \ulcorner s \urcorner, \ulcorner) \urcorner \rangle$$

$$\ulcorner (\varphi \wedge \psi) \urcorner := \langle \ulcorner (\urcorner, \ulcorner \varphi \urcorner, \ulcorner \wedge \urcorner, \ulcorner \psi \urcorner, \ulcorner) \urcorner \rangle$$

$$\ulcorner (\varphi \rightarrow \psi) \urcorner := \langle \ulcorner (\urcorner, \ulcorner \varphi \urcorner, \ulcorner \rightarrow \urcorner, \ulcorner \psi \urcorner, \ulcorner) \urcorner \rangle$$

$$\ulcorner (\forall x_i \varphi) \urcorner := \langle \ulcorner (\urcorner, \ulcorner \forall \urcorner, \ulcorner x_i \urcorner, \ulcorner \varphi \urcorner, \ulcorner) \urcorner \rangle$$

$Const(x)$ and $Var(x)$ characterize the codes of constants and variables, respectively.

$$Const(x) := x = \ulcorner 0 \urcorner$$

$$Var(x) := \exists i \leq x (p_{11+i} = x)$$

$$Fnc1(x) := x = \ulcorner S \urcorner$$

$$Fnc2(x) := x = \ulcorner + \urcorner \vee x = \ulcorner \cdot \urcorner \vee x = \ulcorner exp \urcorner$$

$Term(x)$—x is a term—and $Form(x)$—x is a formula—are primitive recursive predicates according to the primitive recursive version of the recursion theorem. Note that we will code according to the standard function notation, e.g. $+(x, y)$ instead of $x + y$.

$$
\begin{aligned}
Term(x) := {} & Const(x) \vee Var(x) \\
& \vee \big(Seq(x) \wedge lth(x) = 4 \wedge Fnc1((x)_0) \\
& \quad \wedge (x)_1 = \ulcorner (\urcorner \wedge Term((x)_2) \wedge (x)_3 = \ulcorner) \urcorner \big) \\
& \vee \big(Seq(x) \wedge lth(x) = 5 \wedge Fnc2((x)_0) \\
& \quad \wedge (x)_1 = \ulcorner (\urcorner \wedge Term((x)_2) \wedge Term((x)_3) \wedge (x)_4 = \ulcorner) \urcorner \big)
\end{aligned}
$$

$$
Form(x) := \left\{
\begin{aligned}
& Seq(x) \wedge lth(x) = 5 \wedge (x)_0 = \ulcorner (\urcorner \wedge (x)_4 = \ulcorner) \urcorner \\
& \quad \wedge \big[(Term((x)_1) \wedge (x)_2 = \ulcorner = \urcorner \wedge Term((x)_3)) \\
& \qquad \vee (Form((x)_1) \wedge (x)_2 = \ulcorner \wedge \urcorner \wedge Form((x)_3)) \\
& \qquad \vee (Form((x)_1) \wedge (x)_2 = \ulcorner \rightarrow \urcorner \wedge Form((x)_3)) \\
& \qquad \vee ((x)_1 = \ulcorner \forall \urcorner \wedge Var((x)_2) \wedge Form((x)_3)) \big]
\end{aligned}
\right.
$$

All kinds of syntactical notions can be coded in primitive recursive predicates, for example $Free(x, y)$—x is a free variable in y, and $FreeFor(x, y, z)$—x is free for y in z.

$$Free(x, y) := \begin{cases} (Var(x) \land Term(y) \neg Const(y) \\ \quad \land \big(Var(y) \to x = y\big) \\ \quad \land \big(Fnc1((y)_0) \to Free(x, (y)_2)\big) \\ \quad \land \big(Fnc2((y)_0) \to \big(Free(x, (y)_2) \lor Free(x, (y)_3)\big)\big)\big) \\ \text{or} \\ (Var(x) \land Form(y) \\ \quad \land \big((y)_1 \neq \ulcorner \forall \urcorner \to \big(Free(x, (y)_1) \lor Free(x, (y)_3)\big)\big) \\ \quad \land \big((y)_1 = \ulcorner \forall \urcorner \to \big(x \neq (y)_2 \land Free(x, (y)_4)\big)\big)\big) \end{cases}$$

$$FreeFor(x, y, z) := \begin{cases} Term(x) \land Var(y) \land Form(z) \\ \quad \land \big[\big((z)_2 = \ulcorner = \urcorner\big) \\ \quad \lor \big((z)_1 \neq \ulcorner \forall \urcorner \land FreeFor(x, y, (z)_1) \land FreeFor(x, y, (z)_3)\big) \\ \quad \lor \big((z)_1 = \ulcorner \forall \urcorner \land \neg Free((z)_2, x) \\ \quad \land \big(Free(y, z) \to \big(Free((z)_2, x) \land Free(x, y, (z)_3)\big)\big)\big)\big] \end{cases}$$

) Having coded these predicates, we can define a substitution operator Sub such that $Sub(\ulcorner \varphi \urcorner, \ulcorner x \urcorner, \ulcorner t \urcorner) = \ulcorner \varphi[t/x] \urcorner$.

$$Sub(x, y, z) := \begin{cases} x & \text{if } Const(x) \\ x & \text{if } Var(x) \land x \neq y \\ z & \text{if } Var(x) \land x = y \\ \langle (x)_0, \ulcorner (\urcorner, Sub((x)_2, y, z), \ulcorner) \urcorner \rangle & \text{if } Term(x) \land Fnc1((x)_0) \\ \langle (x)_0, \ulcorner (\urcorner, Sub((x)_2, y, z), Sub((x)_3, y, z), \ulcorner) \urcorner \rangle \\ \quad \text{if } Term(x) \land Fnc2((x)_0) \\ \langle \ulcorner (\urcorner, Sub((x)_1, y, z), (x)_2, Sub((x)_3, y, z), \ulcorner) \urcorner \rangle \\ \quad \text{if } Form(x) \land FreeFor(x, y, z) \land (x)_0 \neq \ulcorner \forall \urcorner \\ \langle \ulcorner (\urcorner, (x)_1, (x)_2, Sub((x)_3, y, z), \ulcorner) \urcorner \rangle \\ \quad \text{if } Form(x) \land FreeFor(z, y, x) \land (x)_0 = \ulcorner \forall \urcorner \\ 0 & \text{else.} \end{cases}$$

Clearly Sub is primitive recursive (course-of-value recursion).

Coding of Derivability Our next step is to obtain a primitive recursive predicate Der that says that x is a derivation with hypotheses $y_0, \ldots, y_{lth(y)-1}$ and conclusion z. Before that we give a coding of derivations.

Initial derivation

$$[\varphi] = \langle 0, \varphi \rangle$$

$\land$I

$$\left[\begin{array}{cc} D_1 & D_2 \\ \varphi & \psi \\ \hline \multicolumn{2}{c}{(\varphi \land \psi)} \end{array} \right] = \left\langle \langle 0, \ulcorner \land \urcorner \rangle, \left[\begin{array}{c} D_1 \\ \varphi \end{array} \right], \left[\begin{array}{c} D_2 \\ \psi \end{array} \right], \ulcorner (\varphi \land \psi) \urcorner \right\rangle$$

∧E

$$\left[\begin{array}{c} D \\ \hline (\varphi \wedge \psi) \\ \hline \varphi \end{array}\right] = \left\langle \langle 1, \ulcorner \wedge \urcorner \rangle, \left[\begin{array}{c} D \\ (\varphi \wedge \psi) \end{array}\right], \ulcorner \varphi \urcorner \right\rangle$$

→ I

$$\left[\begin{array}{c} \varphi \\ D \\ \psi \\ \hline (\varphi \to \psi) \end{array}\right] = \left\langle \langle 0, \ulcorner \to \urcorner \rangle, \left[\begin{array}{c} D \\ \psi \end{array}\right], \ulcorner (\varphi \to \psi) \urcorner \right\rangle$$

→ E

$$\left[\begin{array}{cc} D_1 & D_2 \\ \varphi & (\varphi \to \psi) \\ \hline & \psi \end{array}\right] = \left\langle \langle 1, \ulcorner \to \urcorner \rangle, \left[\begin{array}{c} D_1 \\ \varphi \end{array}\right], \left[\begin{array}{c} D_2 \\ (\varphi \to \psi) \end{array}\right], \ulcorner \psi \urcorner \right\rangle$$

RAA

$$\left[\begin{array}{c} (\varphi \to \bot) \\ D \\ \bot \\ \hline \varphi \end{array}\right] = \left\langle \langle 1, \ulcorner \bot \urcorner \rangle, \left[\begin{array}{c} D \\ \bot \end{array}\right], \ulcorner \varphi \urcorner \right\rangle$$

∀I

$$\left[\begin{array}{c} D \\ \varphi \\ \hline (\forall x \varphi) \end{array}\right] = \left\langle \langle 0, \ulcorner \forall \urcorner \rangle, \left[\begin{array}{c} D \\ \varphi \end{array}\right], \ulcorner (\forall x \varphi) \urcorner \right\rangle$$

∀E

$$\left[\begin{array}{c} D \\ (\forall x \varphi) \\ \hline \varphi[t/x] \end{array}\right] = \left\langle \langle 1, \ulcorner \forall \urcorner \rangle, \left[\begin{array}{c} D \\ (\forall x \varphi) \end{array}\right], \ulcorner \varphi[t/x] \urcorner \right\rangle$$

For *Der* we need a device to cancel hypotheses from a derivation. We consider a sequence y of (codes of) hypotheses and successively delete items u.

$$Cancel(u, y) := \begin{cases} y & \text{if } lth(y) = 0 \\ Cancel(u, tail(y)) & \text{if } (y)_0 = u \\ \langle (y)_0, Cancel(u, tail(y)) \rangle & \text{if } (y)_0 \neq u. \end{cases}$$

Here $tail(y) = z \Leftrightarrow lth(y) > 0 \wedge \exists x (y = \langle x \rangle * z) \vee (lth(y) = 0 \wedge z = 0)$.

Now we can code *Der*, where $Der(x, y, z)$ stands for "x is the code of a derivation of a formula with code z from a coded sequence of hypotheses y". In the definition of *Der* $\bot$ is defined as $(0 = 1)$.

$$Der(x, y, z) := Form(z) \wedge \bigwedge_{i=0}^{lth(y)-1} Form((i)_v)$$

$$\wedge \left\| \left(\exists i < lth(y)(z = (y)_i \wedge x = \langle 0, z \rangle) \right) \right.$$

or

$$(\exists x_1 x_2 \leq x \exists y_1 y_2 \leq x \exists z_1 z_2 \leq x (y = y_1 * y_2$$
$$\wedge Der(x_1, y_1, z_1) \wedge Der(x_2, y_2, z_2)$$
$$\wedge z = \langle \ulcorner (\urcorner, z_1, \ulcorner \wedge \urcorner, z_2, \ulcorner) \urcorner \rangle \wedge x = \langle \langle 0, \ulcorner \wedge \urcorner \rangle, x_1, x_2, z \rangle))$$

or

$$(\exists u \leq x \exists x_1 \leq x \exists z_1 \leq x (Der(x_1, y, z_1)$$
$$\wedge (z_1 = \langle \ulcorner (\urcorner, z, \ulcorner \wedge \urcorner, u, \ulcorner) \urcorner \rangle) \vee (z_1 = \langle \ulcorner (\urcorner, u, \ulcorner \wedge \urcorner, z, \ulcorner) \urcorner \rangle))$$
$$\wedge x = \langle \langle 1, \ulcorner \wedge \urcorner \rangle, x_1, z \rangle))$$

or

$$(\exists x_1 \leq x \exists y_1 \leq x \exists u \leq x \exists z_1 \leq x (y = Cancel(u, y_1)$$
$$\vee y = y_1) \wedge Der(x_1, y_1, z_1) \wedge z = \langle \ulcorner (\urcorner, u, \ulcorner \to \urcorner, z_1, \ulcorner) \urcorner \rangle$$
$$\wedge x = \langle \langle 0, \ulcorner \to \urcorner \rangle, x_1, z_1 \rangle)$$

or

$$(\exists x_1 x_2 \leq x \exists y_1 y_2 \leq x \exists z_1 z_2 \leq x (y = y_1 * y_2$$
$$\wedge Der(x_1, y_1, z_1) \wedge Der(x_2, y_2, z_2)$$
$$\wedge z_2 = \langle \ulcorner (\urcorner, z_1, \ulcorner \to \urcorner, z, \ulcorner) \urcorner \rangle \wedge x = \langle \langle 1, \ulcorner \to \urcorner \rangle, x_1, x_2, z \rangle))$$

or

$$\left(\exists x_1 \leq x \exists z_1 \leq x \exists v \leq x (Der(x_1, y, z_1) \wedge Var(v) \right.$$
$$\wedge \bigwedge_{i=0}^{lth(y)-1} \neg Free(v, (y)_i) \wedge z = \langle \ulcorner \forall \urcorner, v, \ulcorner (\urcorner, z_1, \ulcorner) \urcorner \rangle$$
$$\left. \wedge x = \langle \langle 0, \ulcorner \forall \urcorner \rangle, x_1, z_1 \rangle) \right)$$

or

$$(\exists t \leq x \exists v \leq x \exists x_1 \leq x \exists z_1 \leq x (Var(v) \wedge Term(t)$$
$$\wedge Freefor(t, v, z_1) \wedge z = Sub(z_1, v, t)$$
$$\wedge Der(x_1, y, \langle \ulcorner \forall \urcorner, v, \ulcorner (\urcorner, z_1, \ulcorner) \urcorner \rangle) \wedge x = \langle \langle 1, \ulcorner \forall \urcorner \rangle, y, z \rangle))$$

or

$$(\exists x_1 \leq x \exists y_1 \leq x \exists z_1 \leq x (Der(x_1, y_1, \langle \ulcorner \perp \urcorner \rangle)$$
$$\wedge y = Cancel(\langle z, \ulcorner \to \urcorner, \ulcorner \perp \urcorner \rangle, y_1) \wedge x = \langle \langle 1, \ulcorner \perp \urcorner \rangle, x_1, z_1 \rangle)) \Big].$$

Coding of Provability The axioms of Peano arithmetic are listed on p. 236. However, for the purpose of coding derivability we have to be precise; we must include the axioms for identity. They are the usual ones (see Sect. 3.6 and Lemma 3.10.2), including the "congruence axioms" for the operations:

$$(x_1 = y_1 \wedge x_2 = y_2) \rightarrow \big(S(x_1) = S(y_1) \wedge x_1 + x_2 = y_1 + y_2$$
$$\wedge\, x_1 \cdot x_2 = y_1 \cdot y_2 \wedge x_1^{x_2} = y_1^{y_2}\big).$$

These axioms can easily be coded and put together in a primitive recursive predicate $Ax(x)$—*x is an axiom*. The provability predicate $Prov(x, z)$—*x is a derivation of z from the axioms of* $\mathcal{PA}$—follows immediately.

$$Prov(x, z) := \exists y \leq x \left(Der(x, y, z) \wedge \bigwedge_{i=0}^{lth(y)-1} Ax\big((y)_i\big) \right).$$

Finally we can define $Thm(x)$—*x is a theorem*. Thm is recursively enumerable.

$$Thm(z) := \exists x Prov(x, z).$$

Having at our disposition the provability predicate, which is Σ_1^0, we can finish the proof of "semi-representable = RE" (Theorem 8.5.8).

Proof For convenience let R be a unary recursively enumerable set.

$\Rightarrow$: R is semi-representable by φ. $R(n) \Leftrightarrow\, \vdash \varphi(\overline{n}) \Leftrightarrow \exists y Prov\,(\ulcorner\varphi(\overline{n})\urcorner, y)$. Note that $\ulcorner\varphi(\overline{n})\urcorner$ is a recursive function of n. $Prov$ is primitive recursive, so R is recursively enumerable.

$\Leftarrow$: R is recursively enumerable $\Rightarrow R(n) = \exists x P(n, x)$ for a primitive recursive P. $P(n, m) \Leftrightarrow\, \vdash \varphi(\overline{n}, \overline{m})$ for some φ. $R(n) \Leftrightarrow P(n, m)$ for some $m \Leftrightarrow\, \vdash \varphi(\overline{n}, \overline{m})$ for some $m \Rightarrow\, \vdash \exists y \varphi(\overline{n}, y)$. Therefore we also have $\vdash \exists y \varphi(\overline{n}, y) \Rightarrow R(n)$. So $\exists y \varphi(\overline{n}, y)$ semi-represents R. $\square$

8.7 Incompleteness

Theorem 8.7.1 (Fixpoint Theorem) *For each formula* $\varphi(x)$ *(with* $FV(\varphi) = \{x\}$*) there exists a sentence* ψ *such that* $\vdash \varphi(\ulcorner\psi\urcorner) \leftrightarrow \psi$.

Proof Popular version: consider a simplified substitution function $s(x, y)$ which is the old substitution function for a fixed variable: $s(x, y) = Sub(x, \ulcorner x_0 \urcorner, y)$. Then define $\theta(x) := \varphi(s(x, x))$. Let $m := \ulcorner\theta(x)\urcorner$, then put $\psi := \theta(\overline{m})$. Note that $\psi \leftrightarrow \theta(\overline{m}) \leftrightarrow \varphi(\overline{s(m, m)}) \leftrightarrow \varphi(s(\ulcorner\theta(x)\urcorner, \overline{m})) \leftrightarrow \varphi(\ulcorner\theta(\overline{m})\urcorner) \leftrightarrow \varphi(\ulcorner\psi\urcorner)$.

This argument would work if there were a function (or term) for s in the language. This could be done by extending the language with sufficiently many functions ("all primitive recursive functions" surely will do). Now we have to use representing formulas.

Formal version: let $\sigma(x, y, z)$ represent the primitive recursive function $s(x, y)$. Now suppose $\theta(x) := \exists y(\varphi(y) \wedge \sigma(x, x, y))$, $m = \ulcorner\theta(x)\urcorner$ and $\psi = \theta(\overline{m})$. Then

$$\psi \leftrightarrow \theta(\overline{m}) \leftrightarrow \exists y\big(\varphi(y) \wedge \sigma(\overline{m}, \overline{m}, y)\big) \tag{8.1}$$

$$\vdash \forall y\big(\sigma(\overline{m}, \overline{m}, y) \leftrightarrow y = \overline{s(m, m)}\big)$$

$$\vdash \forall y\big(\sigma(\overline{m}, \overline{m}, y) \leftrightarrow y = \overline{\ulcorner\theta(\overline{m})\urcorner}\big) \tag{8.2}$$

By logic (1) and (2) give $\psi \leftrightarrow \exists y(\varphi(y) \wedge y = \overline{\ulcorner \theta(\overline{m}) \urcorner})$ so $\psi \leftrightarrow \varphi(\overline{\ulcorner \theta(\overline{m}) \urcorner}) \leftrightarrow \varphi(\overline{\ulcorner \psi \urcorner})$. $\qquad \square$

Definition 8.7.2

(i) **PA** (or any other theory T of arithmetic) is called *ω-complete* if $\vdash \exists x \varphi(x) \Rightarrow \vdash \varphi(\overline{n})$ for some $n \in \mathbb{N}$.

(ii) T is *ω-consistent* if there is no φ such that $(\vdash \exists x \varphi(x)$ and $\vdash \neg \varphi(\overline{n})$ for all $n)$ for all φ.

Theorem 8.7.3 (*Gödel's First Incompleteness Theorem*) *If* **PA** *is ω-consistent then* **PA** *is incomplete.*

Proof Consider the predicate $Prov(x, y)$ represented by the formula $\overline{Prov}(x, y)$. Let $Thm(x) := \exists y \overline{Prov}(x, y)$. Apply the Fixpoint Theorem to $\neg \overline{Thm}(x)$: there exists a φ such that $\vdash \varphi \leftrightarrow \neg \overline{Thm}(\ulcorner \varphi \urcorner)$. φ, the so-called *Gödel sentence*, says in **PA**: "I am not provable."

Claim 1 *If $\vdash \varphi$ then* **PA** *is inconsistent.*

Proof $\vdash \varphi \Rightarrow$ there is an n such that $Prov(\ulcorner \varphi \urcorner, n)$, hence $\vdash \overline{Prov}(\ulcorner \varphi \urcorner, \overline{n}) \Rightarrow \vdash \exists y \overline{Prov}(\ulcorner \varphi \urcorner, y) \Rightarrow \vdash \overline{Thm}(\ulcorner \varphi \urcorner) \Rightarrow \vdash \neg \varphi$. Thus **PA** is inconsistent. $\qquad \square$

Claim 2 *If $\vdash \neg \varphi$ then* **PA** *is ω-inconsistent.*

Proof $\vdash \neg \varphi \Rightarrow \vdash \overline{Thm}(\ulcorner \varphi \urcorner) \Rightarrow \vdash \exists x \overline{Prov}(\ulcorner \varphi \urcorner, x)$. Suppose **PA** is ω-consistent; since ω-consistency implies consistency, we have $\nvdash \varphi$ and therefore $\neg Prov(\ulcorner \varphi \urcorner, n)$ for all n. Hence $\vdash \neg \overline{Prov}(\ulcorner \varphi \urcorner, \overline{n})$ for all n. Contradiction. $\qquad \square$

Remarks In the foregoing we made use of the representability of the provability predicate, which in turn depended on the representability of all recursive functions and predicates.

For the representability of $Prov(x, y)$, the set of axioms has to be recursively enumerable. Thus Gödel's First Incompleteness Theorem holds for all recursively enumerable theories in which the recursive functions are representable. Therefore one cannot make **PA** complete by adding the Gödel sentence, the result would again be incomplete.

In the standard model $\mathbb{N}$ either φ, or $\neg \varphi$ is true. The definition enables us to determine which one. Notice that the axioms of **PA** are true in $\mathbb{N}$, so $\models \varphi \leftrightarrow \neg \overline{Thm}(\ulcorner \varphi \urcorner)$. Suppose $\mathbb{N} \models \overline{Thm}(\ulcorner \varphi \urcorner)$, then $\mathbb{N} \models \exists x \overline{Prov}(\ulcorner \varphi \urcorner, x) \Leftrightarrow \mathbb{N} \models \overline{Prov}(\ulcorner \varphi \urcorner, \overline{n})$ for some $n \Leftrightarrow \vdash \overline{Prov}(\ulcorner \varphi \urcorner, \overline{n})$ for some $n \Leftrightarrow \vdash \varphi \Rightarrow \vdash \neg \overline{Thm}(\ulcorner \varphi \urcorner) \Rightarrow \mathbb{N} \models \neg \overline{Thm}(\ulcorner \varphi \urcorner)$. Contradiction. Thus φ is true in $\mathbb{N}$. This is usually expressed as "there is a true statement of arithmetic which is not provable".

Remarks It is generally accepted that **PA** is a *true* theory, that is $\mathbb{N}$ is a model of **PA**, and so the conditions on Gödel's theorem seem to be superfluous. However, the fact that **PA** is a true theory is based on a semantical argument. The refinement consists in considering arbitrary theories, without the use of semantics.

The incompleteness theorem can be freed of the ω-consistency condition. For that purpose we introduce Rosser's predicate:

$$Ros\,(x) := \exists y \big(Prov\,(neg\,(x),y) \wedge \forall z < y \neg Prov\,(x,z)\big),$$

with $neg(\ulcorner \varphi \urcorner) = \ulcorner \neg \varphi \urcorner$. The predicate following the quantifier is represented by $\overline{Prov}\,(\overline{neg}\,(x),y) \wedge \forall z < y \neg \overline{Prov}\,(x,z)$. An application of the Fixpoint Theorem yields a ψ such that

$$\vdash \psi \leftrightarrow \exists y \big[\,\overline{Prov}\,(\ulcorner \neg \psi \urcorner, y) \wedge \forall z < y \neg \overline{Prov}\,(\ulcorner \psi \urcorner, z)\big]. \tag{8.3}$$

Claim **PA** *is consistent* $\Rightarrow \nvdash \psi$ *and* $\nvdash \neg \psi$.

Proof (i) Suppose $\vdash \psi$, then there exists an n such that $Prov\,(\ulcorner \psi \urcorner, n)$, so

$$\vdash \overline{Prov}\,(\ulcorner \psi \urcorner, \overline{n}). \tag{8.4}$$

From (8.3) and (8.4) it follows that $\vdash \exists y < \overline{n}\,\overline{Prov}\,(\ulcorner \neg \psi \urcorner, y)$, i.e. $\vdash \overline{Prov}\,(\ulcorner \neg \psi \urcorner, \overline{0})$ $\vee \cdots \vee \overline{Prov}\,(\ulcorner \neg \psi \urcorner, \overline{n-1})$. Note that $\overline{Prov}$ is Δ_0, thus the following holds: $\vdash \sigma \vee \tau \Leftrightarrow \vdash \sigma$ or $\vdash \tau$, so $\vdash \overline{Prov}\,(\ulcorner \neg \psi \urcorner, \overline{0})$ or ... or $\vdash \overline{Prov}\,(\ulcorner \neg \psi \urcorner, \overline{n-1})$ hence $Prov\,(\ulcorner \neg \psi \urcorner, i)$ for some $i < n \Rightarrow \vdash \neg \psi \Rightarrow$ **PA** is inconsistent.

(ii) Suppose $\vdash \neg \psi$, then $\vdash \forall y [\overline{Prov}\,(\ulcorner \neg \psi \urcorner, y) \rightarrow \exists z < y \overline{Prov}\,(\ulcorner \psi \urcorner, z)]$ also $\vdash \neg \psi \Rightarrow Prov\,(\ulcorner \neg \psi \urcorner, n)$ for some $n \Rightarrow \vdash \overline{Prov}\,(\ulcorner \neg \psi \urcorner, \overline{n})$ for some $n \Rightarrow \vdash \exists z < \overline{n}\,\overline{Prov}\,(\ulcorner \psi \urcorner, z) \Rightarrow$ (as in the above) $Prov\,(\ulcorner \psi \urcorner, k)$ for some $k < n$, so $\vdash \psi \Rightarrow$ **PA** is inconsistent. $\qquad\qquad\square$

We have seen that truth in $\mathbb{N}$ does not necessarily imply provability in **PA** (or any other (recursively enumerable) axiomatizable extension). However, we have seen that **PA** is Σ_1^0-complete, so truth and provability still coincide for simple statements.

Definition 8.7.4 A theory T (in the language of **PA**) is called Σ_1^0-*sound* if $T \vdash \varphi \Rightarrow \mathbb{N} \vDash \varphi$ for Σ_1^0-sentences φ.

We will not go into the intriguing questions of the foundations or philosophy of mathematics and logic. Accepting the fact that the standard model $\mathbb{N}$ is a model of **PA**, we get consistency, soundness and Σ_1^0-soundness for free. It is an old tradition in proof theory to weaken assumptions as much as possible, so it makes sense to see what one can do without any semantic notions. The interested reader is referred to the literature.

We now present an alternative proof of the incompleteness theorem. Here we use the fact that **PA** is Σ_1^0-sound.

Theorem 8.7.5 *PA is incomplete.*

Consider an RE set X which is not recursive. It is semi-represented by a Σ_1^0-formula $\varphi(x)$. Let $Y = \{n | \mathbf{PA} \vdash \varphi(\bar{n})\}$.

By Σ_1^0-completeness we get $n \in X \Rightarrow \mathbf{PA} \vdash \varphi(\bar{n})$. Since Σ_1^0-soundness implies consistence, we also get $\mathbf{PA} \vdash \neg\varphi(\bar{n}) \Rightarrow n \notin X$, hence $Y \subseteq X^c$. The provability predicate tells us that Y is RE. Now X^c is not RE, so there is a number k with $k \in (X \cup Y)^c$. For this number k we know that $\mathbf{PA} \nvdash \neg\varphi(\bar{k})$ and also $\mathbf{PA} \nvdash \varphi(\bar{k})$, as $\mathbf{PA} \vdash \varphi(\bar{k})$ would imply by Σ_1^0-soundness that $k \in X$.

As a result we have established that $\neg\varphi(\bar{k})$ is true but not provable in $\mathbf{PA}$, i.e. $\mathbf{PA}$ is incomplete.

We almost immediately get the undecidability of $\mathbf{PA}$.

Theorem 8.7.6 *PA is undecidable.*

Proof Consider the same set $X = \{n | PA \vdash \varphi(\bar{n})\}$ as above. If $\mathbf{PA}$ were decidable, the set XS would be recursive. Hence $\mathbf{PA}$ is undecidable. □

Note that we get the same result for any axiomatizable Σ_1^0-sound extension of $\mathbf{PA}$. For stronger results see Smoryński's *logical number theory* that f with $f(n) = \ulcorner \varphi(\bar{n}) \urcorner$ is primitive recursive.

Remarks The Gödel sentence γ "I am not provable" is the negation of a strict Σ_1^0-sentence (a so-called Π_1^0-sentence). Its negation cannot be true (why?). So $\mathbf{PA} + \neg\gamma$ is not Σ_1^0-sound.

We will now present another approach to the undecidability of arithmetic, based on effectively inseparable sets.

Definition 8.7.7 Let φ and ψ be existential formulas: $\varphi = \exists x \varphi'$ and $\psi = \exists x \psi'$. The *witness comparison* formulas for φ and ψ are given by:

$$\varphi \leq \psi := \exists x \big(\varphi'(x) \wedge \forall y < x \neg \psi'(y)\big)$$
$$\varphi < \psi := \exists x \big(\varphi'(x) \wedge \forall y \leq x \neg \psi'(y)\big).$$

Lemma 8.7.8 (Informal Reduction Lemma) *Let φ and ψ be strict Σ_1^0, $\varphi_1 := \varphi \leq \psi$ and $\psi_1 := \psi < \varphi$. Then*

(i) $\mathbb{N} \vDash \varphi_1 \rightarrow \varphi$
(ii) $\mathbb{N} \vDash \psi_1 \rightarrow \psi$
(iii) $\mathbb{N} \vDash \varphi \vee \psi \leftrightarrow \varphi_1 \vee \psi_1$
(iv) $\mathbb{N} \vDash \neg (\varphi_1 \wedge \psi_1)$.

Proof Immediate from the definition. □

Lemma 8.7.9 (Formal Reduction Lemma)   *Let* φ, ψ, φ_1 *and* ψ_1 *be as above.*

(i) $\vdash \varphi_1 \to \varphi$

(ii) $\vdash \psi_1 \to \psi$

(iii) $\mathbb{N} \vDash \varphi_1 \Rightarrow \vdash \varphi_1$

(iv) $\mathbb{N} \vDash \psi_1 \Rightarrow \vdash \psi_1$

(v) $\mathbb{N} \vDash \varphi_1 \Rightarrow \vdash \neg \psi_1$

(vi) $\mathbb{N} \vDash \psi_1 \Rightarrow \vdash \neg \varphi_1$

(vii) $\vdash \neg(\varphi_1 \wedge \psi_1)$.

Proof (i)–(iv) are direct consequences of the definition and Σ_1^0-completeness.

(v) and (vi) are exercises in natural deduction (use $\forall uv(u < v \vee v \leq u)$) and (vii) follows from (v) (or (vi)). $\square$

Theorem 8.7.10 (Undecidability of **PA**)   *The relation* $\exists y Prov(x, y)$ *is not recursive. Popular version:* $\vdash$ *is not decidable for* **PA**.

Proof Consider two effectively inseparable recursively enumerable sets A and B with strict Σ_1^0-defined formulas $\varphi(x)$ and $\psi(x)$. Define $\varphi_1(x) := \varphi(x) \leq \psi(x)$ and $\psi_1(x) := \psi(x) < \varphi(x)$.

$$\text{Then} \quad n \in A \Rightarrow \mathbb{N} \vDash \varphi(\overline{n}) \wedge \neg\psi(\overline{n})$$
$$\Rightarrow \mathbb{N} \vDash \varphi_1(\overline{n})$$
$$\Rightarrow \vdash \varphi_1(\overline{n})$$
$$\text{and} \quad n \in B \Rightarrow \mathbb{N} \vDash \psi(\overline{n}) \wedge \neg\varphi(\overline{n})$$
$$\Rightarrow \mathbb{N} \vDash \psi_1(\overline{n})$$
$$\Rightarrow \vdash \neg\varphi_1(\overline{n}).$$

Let $\widehat{A} = \{n \mid \vdash \varphi_1(\overline{n})\}$, $\widehat{B} = \{n \mid \vdash \neg\varphi_1(\overline{n})\}$, then $A \subseteq \widehat{A}$ and $B \subseteq \widehat{B}$. **PA** is consistent, so $\widehat{A} \cap \widehat{B} = \emptyset$. $\widehat{A}$ is recursively enumerable, but because of the effective inseparability of A and B not recursive. Suppose that $\{\ulcorner\sigma\urcorner \mid \vdash \sigma\}$ is recursive, i.e. $X = \{k \mid Form(k) \wedge \exists z Prov(k, z)\}$ is recursive. Consider f with $f(n) = \ulcorner\varphi_1(\overline{n})\urcorner$, then $\{n \mid \exists z Prov(\ulcorner\varphi_1(\overline{n})\urcorner, z)\}$ is also recursive, i.e. $\widehat{A}$ is a recursive separator of A and B. Contradiction. Thus X is not recursive. $\square$

From the undecidability of *PA* we immediately get once more the incompleteness theorem.

Corollary 8.7.11 *PA is incomplete.*

Proof (a) If **PA** were complete, then from the general theorem "complete axiomatizable theories are decidable" it would follow that **PA** was decidable.

(b) Because $\widehat{A}$ and $\widehat{B}$ are both recursively enumerable, there exists an n with $n \notin \widehat{A} \cup \widehat{B}$, i.e. $\nvdash \varphi(\overline{n})$ and $\nvdash \neg\varphi(\overline{n})$. $\square$

Remark The above results are by no means optimal; one can represent the recursive functions in considerably weaker systems, and hence prove their incompleteness. There are a number of subsystems of *PA* which are finitely axiomatizable, for example the *Q* of Raphael Robinson (cf. Smoryński, *Logical number theory*, p. 368 ff.), which is incomplete and undecidable. Using this fact one easily

gets the following.

Corollary 8.7.12 (Church's Theorem) *Predicate logic is undecidable.*

Proof Let $\{\sigma_1, \ldots, \sigma_n\}$ be the axioms of Q, then $\sigma_1, \ldots, \sigma_n \vdash \varphi \Leftrightarrow \vdash (\sigma_1 \wedge \cdots \wedge \sigma_n) \rightarrow \varphi$. A decision method for predicate logic would thus provide one for Q. □

Remark

(1) Since **HA** is a subsystem of **PA** the Gödel sentence γ is certainly independent of **HA**. Therefore $\gamma \vee \neg\gamma$ is not a theorem of **HA**. For if **HA** $\vdash \gamma \vee \neg\gamma$, then by the disjunction property for **HA** we would have **HA** $\vdash \gamma$ or **HA** $\vdash \neg\gamma$, which is impossible for the Gödel sentence. Hence we have a specific theorem of **PA** which is not provable in **HA**.
(2) Since **HA** has the existence property, one can go through the first version of the proof of the incompleteness theorem, while avoiding the use of ω-consistency.

Exercises

1. Show that f with $f(n) = \ulcorner t(\overline{n}) \urcorner$ is primitive recursive.
2. Show that f with $f(n) = \ulcorner \varphi(\overline{n}) \urcorner$ is primitive recursive.
3. Find out what $\varphi \rightarrow \varphi \leq \varphi$ means for a φ as above.

References

J. Barwise (ed.). Handbook of Mathematical Logic. North-Holland, Amsterdam, 1977

G. Boolos. The Logic of Provability. Cambridge University Press, Cambridge, 1993

E. Börger. Computability, Complexity, Logic. North-Holland, Amsterdam, 1989

C.C. Chang, J.J. Keisler. Model Theory. North-Holland, Amsterdam, 1990

D. van Dalen, Intuitionistic Logic. In Gabbay, D. and F. Guenthner (eds.) Handbook of Philosoph-
    ical Logic, vol. 5 (2nd edn.). Kluwer Academic, Dordrecht, pp. 1–114 (2002)

M. Davis. Computability and Unsolvability. McGraw Hill, New York, 1958

M. Dummett. Elements of Intuitionism. Oxford University Press, Oxford, 1977 (2nd edition 2000)

J.Y. Girard. Proof Theory and Logical Complexity. I. Bibliopolis, Napoli, 1987

J.Y. Girard, Y. Lafont, P. Taylor. Proofs and Types. Cambridge University Press, Cambridge, 1989

J.Y. Girard. Le point aveugle, cours de logique, tome 1 : vers la perfection. Hermann, Paris, 2006

J.Y. Girard. Le point aveugle, cours de logique, tome 2 : vers l'imperfection. Hermann, Paris, 2006

P. Hinman. Recursion-Theoretic Hierarchies. Springer, Berlin, 1978

S.C. Kleene. Introduction to Meta-mathematics. North-Holland, Amsterdam, 1952

G. Mints. A Short Introduction to Intuitionistic Logic. Kluwer Academic, Dordrecht 2000

S. Negri, J. von Plato. Structural Proof Theory. Cambridge University Press, Cambridge, 2001

P. Odifreddi. Classical Recursion Theory. North-Holland, Amsterdam, 1989

D. Prawitz. Natural Deduction. A Proof Theoretical Study. Almqvist & Wiksell, Stockholm, 1965

A. Robinson. Non-standard Analysis. North-Holland, Amsterdam, 1965

H. Schwichtenberg, S.S. Wainer. Proofs and Computations. Perspectives in Logic. Association for
    Symbolic Logic and Cambridge University Press, Cambridge, 2012,

J.R. Shoenfield. Mathematical Logic. Addison-Wesley, Reading, MA, 1967

J.R. Shoenfield. Recursion Theory. Lecture Notes in Logic, vol. 1. Springer, Berlin, 1993

C. Smoryński. Self-reference and Modal Logic. Springer, Berlin, 1985

C. Smoryński. Logical Number Theory I. Springer, Berlin, 1991 (volume 2 is forthcoming)

M.H. Sørensen, P. Urzyczyn. Lectures on the Curry–Howard Isomorphism. Elsevier, Amsterdam,
    2006

K.D. Stroyan, W.A.J. Luxemburg. Introduction to the Theory of Infinitesimals. Academic Press,
    New York, 1976

A.S. Troelstra, D. van Dalen. Constructivism in Mathematics I, II. North-Holland, Amsterdam,
    1988

A.S. Troelstra, H. Schwichtenberg. Basic Proof Theory. Cambridge University Press, Cambridge,
    1996

D. van Dalen, *Logic and Structure*, Universitext, DOI 10.1007/978-1-4471-4558-5,
© Springer-Verlag London 2013

Index